Computer Simulation of Porous Materials
Current Approaches and Future Opportunities

Inorganic Materials Series

Series editors:
Duncan W. Bruce, *University of York, UK*
Dermot O'Hare, *University of Oxford, UK*
Richard I. Walton, *University of Warwick, UK*

Titles in the Series:
1: Pre-combustion Carbon Dioxide Capture Materials
2: Post-combustion Carbon Dioxide Capture Materials
3: Solar Energy Capture Materials
4: Bioinspired Inorganic Materials: Structure and Function
5: Energy Storage and Conversion Materials
6: Two-dimensional Inorganic Nanomaterials for Conductive Polymer Nanocomposites
7: The Chemistry of Inorganic Biomaterials
8: Computer Simulation of Porous Materials: Current Approaches and Future Opportunities

How to obtain future titles on publication:
A standing order plan is available for this series. A standing order will bring delivery of each new volume immediately on publication.

For further information please contact:
Book Sales Department, Royal Society of Chemistry, Thomas Graham House, Science Park, Milton Road, Cambridge, CB4 0WF, UK
Telephone: +44 (0)1223 420066, Fax: +44 (0)1223 420247,
Email: booksales@rsc.org
Visit our website at www.rsc.org/books

Computer Simulation of Porous Materials

Current Approaches and Future Opportunities

Edited by

Kim Jelfs
Imperial College London, UK
k.jelfs@imperial.ac.uk

ROYAL SOCIETY
OF **CHEMISTRY**

Inorganic Materials Series No. 8

Print ISBN: 978-1-78801-900-2
PDF ISBN: 978-1-83916-331-9
EPUB ISBN: 978-1-83916-332-6
Print ISSN: 2472-3819
Electronic ISSN: 2472-3827

A catalogue record for this book is available from the British Library

The Royal Society of Chemistry is a charity, registered in England and Wales, Number 207890, and a company incorporated in England by Royal Charter (Registered No. RC000524), registered office: Burlington House, Piccadilly, London W1J 0BA, UK, Telephone: +44 (0) 20 7437 8656.

For further information see our website at www.rsc.org

Printed in the United Kingdom by CPI Group (UK) Ltd, Croydon, CR0 4YY, UK

Series Preface

Materials chemistry has grown hugely in the past few decades and continues to evolve rapidly. This is especially the case since the subject now finds itself central to many of the major challenges that face global society, such as those concerning energy and the environment; but applications of materials are also found in many everyday situations, with ongoing developments in technology around electronics and the miniaturisation of devices. Although the study of materials is a highly interdisciplinary topic, the role of innovative, fundamental chemistry continues to lie at the heart of the discovery of new materials, the understanding of their atomic-scale structure and the relationship of structure to a useful property.

With the *Inorganic Materials Series*, our aim is to provide a series of themed volumes, each reflecting the diversity of the topic as it continues to develop over the coming years, with many chosen to relate closely to emerging practical applications. The use of the title *Inorganic Materials* is by no means limiting, but instead emphasises the study of the chemistry of all elements in the Periodic Table and how they impact the development of new materials. The chapters in each book will have a pedagogical flavour with a target audience from first-year postgraduate student upwards, but with wide subject coverage ranging from continuous inorganic solids, to molecular and soft matter systems. We will use expert, active researchers as editors and contributors to provide an up-to-date perspective of their field. At the same time our aim is to provide an international perspective, so to reflect the diversity and interdisciplinarity of the now very broad area of the subject.

Duncan W. Bruce, York, UK
Dermot O'Hare, Oxford, UK
Richard I. Walton, Warwick, UK

Inorganic Materials Series No. 8
Computer Simulation of Porous Materials: Current Approaches and Future Opportunities
Edited by Kim Jelfs

Published by the Royal Society of Chemistry, www.rsc.org

Preface

This book brings together two rapidly developing fields, those of porous materials and computational modelling. Porous materials have been used by mankind for millennia, with their use on an industrial scale arising in the last hundred years, as humans have increasingly been able to tailor material structure and properties to suit the application demands. This book focuses on microporous materials, which have pore dimensions similar to the size of small molecules, leading to applications as sorbents, in molecular separations, (photo)catalysis, optoelectronics and more, with increasing examples of multifunctionality. Zeolites have a long and industrious record, particularly in the area of ion exchange agents and catalysis, and there is intense research effort in the development of a wide range of functional porous materials, including metal–organic frameworks (MOFs), polymers and porous molecules. Each materials class can have unique advantages and disadvantages, from ease and cost of synthesis and processing, their chemical and physical stability, to the ability to finely tailor and control structure, and to tune properties.

In the last few decades, computational modelling has played an increasing role in the field of porous materials, providing rationalisation of properties that could not be garnered from experimental studies alone, as well as moving towards the prediction of structure, synthesis method and properties *a priori*, before any experiment begins in the laboratory. There is currently a significant opportunity in the field of porous materials, as an increasing number of researchers, including those focused on experimental work, have a very high degree of computer literacy, with software coding and data visualisation as components of undergraduate degrees. More than ever before, software and data are open-source, opening new doors to accelerating discovery. There is also considerable excitement and interest in the application of artificial intelligence techniques, in particular machine

Inorganic Materials Series No. 8
Computer Simulation of Porous Materials: Current Approaches and Future Opportunities
Edited by Kim Jelfs

Published by the Royal Society of Chemistry, www.rsc.org

learning, in all areas of science, and researchers in the porous materials community are no exception to this and are looking to accelerate high-throughput experimental screening, computational property screening or experimental design through data-driven methods.

This book is timely in providing a foundation for understanding the key areas where computational modelling can and is going to impact the broad field of porous materials. This includes structure prediction of the full range of (micro)porous material classes, the calculation of sorption and diffusion in porous materials, the challenging task of predicting catalytic and optical performance, and the simulation of mechanical properties as well as more exotic behaviour such as negative thermal expansion. Finally, the book closes with a focus on the future opportunities for porous materials modelling, including in particular the application of machine learning for the rapid prediction of properties and data-driven targeting of optimal synthesis procedures.

Readers of this book will be placed at the cutting edge of the field of porous materials, able to apply transformative methods in their own area of interest, and to identify opportunities where computational modelling can transform how they research and discover new porous materials.

Kim E. Jelfs
London, UK

Contents

Inorganic Materials Series No. 8
Computer Simulation of Porous Materials: Current Approaches and Future Opportunities
Edited by Kim Jelfs

Published by the Royal Society of Chemistry, www.rsc.org

CHAPTER 1

Introduction to Computational Modelling of Microporous Materials

KIM E. JELFS

Department of Chemistry, Molecular Sciences Research Hub, Imperial College London, White City Campus, Wood Lane, W12 0BZ, United Kingdom
Email: k.jelfs@imperial.ac.uk

1.1 Introducing Porous Materials Modelling

Porous materials are materials that have void spaces within them that can be occupied by guests to produce a range of different properties and functionalities. Depending upon the size of their pores, porous materials are classified as microporous (pores smaller than 2 nm), mesoporous (pores between 2 and 50 nm) or macroporous (pores larger than 50 nm). Here, we will focus mostly upon microporous materials, as these pores can host small molecules and atoms and therefore provide functionality across a range of applications from encapsulation, separation, catalysis, sensing and more. As a note on terminology, the term *sorption* is generally used in the discussion of porous materials as these materials will both *adsorb* guests onto surfaces as well as *absorb* guests into the material's bulk.

The computational modelling of materials is increasingly widespread, and porous materials modelling is no exception to this. Computational modelling in porous materials has already had decades of application, especially in the field of zeolites, with significant impact in rationalising the structure

Inorganic Materials Series No. 8
Computer Simulation of Porous Materials: Current Approaches and Future Opportunities
Edited by Kim Jelfs

Published by the Royal Society of Chemistry, www.rsc.org

and property of the materials, frequently providing additional atomic-level insight into the materials that could not be uncovered through experimental characterisation alone. This *a posteriori* rationalisation of experimental observations built a foundation upon which many predictive studies have been built, for example, screening materials in advance for useful properties prior to experimental realisation. A commonly used term is computational materials design, but we urge caution in not overusing the term 'design' when a more appropriate term is 'screening' or 'discovery'. We should leave the term 'design' for those occasions when our insight, knowledge and software truly have enabled design. Computational screening, with consideration of both whether a hypothetical material can be experimentally realised and have target properties, is powerful in its own right.

Ever increasing computing power, increasing computational literacy and programming ability, combined with open-source data and algorithms hold great potential for the further impact of modelling in the field of porous materials. As in all computational fields, there is significant emphasis on the potential for advances through the use of artificial intelligence, notably machine learning. There is enormous value in all porous material scientists understanding, even if not directly applying, the potential of computational modelling for the advancement of their own research. Beyond technological advancements, we can also expect an increasing focus on understanding and predicting more complex behaviour, for example, consideration of multi-functionality in a material, such as the interplay of porosity with optoelectronic properties, and the consideration of the importance and utilisation of defects as an opportunity for property tuning.

In this book, the key foundations for porous materials modelling are laid out, along with the major areas where modelling is applied, successes so far, as well as the future prospects for the field. In this chapter (Chapter 1), the key classes of microporous materials are introduced, along with the applications of porous materials. This is followed by the introduction of the key types of modelling approaches that are applied in the area of porous materials. This chapter should therefore provide the foundation for accessing the following chapters that focus on specific application areas of porous materials modelling. Chapter 2 (Addicoat) outlines the approaches and successes in the structure prediction of porous materials, covering both bottom-up approaches from building blocks and top-down approaches from libraries of topological nets. Examples are discussed from zeolites, metal–organic frameworks, polymers and porous molecules. With a known structure of a material, either from a simulation or an experimental crystal structure, many further simulations analysing properties and function can be performed. Chapter 3 (Evans) discusses the analysis of the mechanical properties of porous materials. How to analyse the strength and flexibility of porous materials is covered, along with the potential to uncover unusual properties, such as negative thermal expansion. The modelling of both sorption and diffusion behaviour in porous materials is covered next (Glover and Besley, Chapter 4). This spans sorption simulations of guest uptake, pore analysis, and molecular dynamics and enhanced sampling techniques for guest diffusion. Modelling the spectroscopic and catalytic properties of porous materials is then covered (Morales-Vidal and Ortuño, Chapter 5), including a discussion of how to choose appropriate methods, followed by examples of mechanistic investigations, focused upon metal–organic frameworks and covalent organic frameworks. Finally, the future for the field, in particular through the application of artificial intelligence techniques, is covered (Jensen and Olivetti, Chapter 6).

1.2 An Overview of Microporous Material Classes

There are a large range of microporous materials and these can have a variety of chemical compositions, whether inorganic or organic in nature, as well as hybrid materials containing both inorganic and organic components. Microporous materials also vary by their degree of crystallinity, with some having both short- and long-range order, and amorphous porous materials lacking any long-range order at all. In this section, the key classes of porous material that are discussed in later chapters will be introduced.

1.2.1 Zeolites

Zeolites are naturally occurring crystalline aluminosilicate minerals that are typically found in regions where there has been volcanic activity. The name *zeolite* originates from how the rock appeared to 'boil' upon heating, with the ancient Greek *zein* for 'to boil' and *lithus* for 'rock', so literally 'zeolite' means 'boiling rock'. It was not until the 1940s that Barrer found a method to synthesise zeolites in a laboratory. Barrer was able to synthesise zeolites *via* mimicking the conditions where they form naturally, in particular the high temperature and pressure, along with organic template molecules that would direct the formation to a targeted zeolite structure. There are now more than two hundred zeolite structures that are either naturally occurring or have been synthesised in the laboratory.[1]

Zeolites are composed of $[SiO_4]^{4-}$ and $[AlO_4]^{5-}$ tetrahedra, and these primary building units, known as T-sites, are linked by the bridging oxygen atoms to produce corner-sharing tetrahedra. Through the multitude of ways that these tetrahedra can be linked and arranged to form crystalline arrays, there are millions of hypothetical zeolite topologies. There are a series of what are known as 'secondary building units' or SBUs that can be used to characterise a zeolite topology. Examples of SBUs include n-membered rings (where $n = 3, 4, 5, 6\ldots$), and then the equivalent doubled rings, double 4-ring (D4R), double 5-ring (D5R), double 6-ring (D6R) and so on that have n bridging oxygens linking the two rings, as well as a small number of other SBUs linking different sized rings. Each zeolite has a three letter framework type code assigned by the International Zeolite Association,[1] which will be unique to a zeolite topology. An example of the zeolite with the framework type code MFI is shown in Figure 1.1.

The microporosity of zeolites originates either from the channel formed in an SBU or from the channels formed between SBUs when they are connected

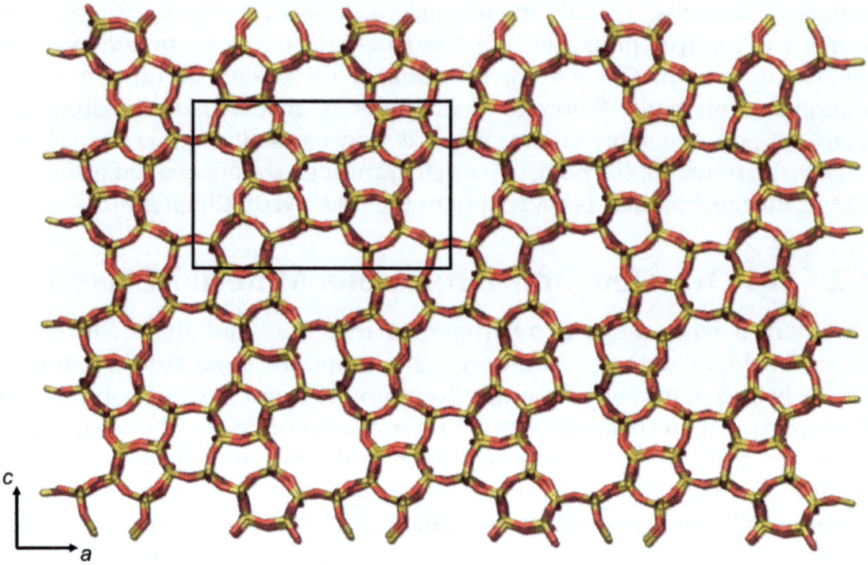

Figure 1.1 An example of a zeolite structure. This is the zeolite MFI viewed down the *b*-axis, with the *a*- and *c*-axes of the unit cell (shown in black) labelled. Ten membered-ring channels run down the *a*- and *b*-axes.

into a zeolite topology. The pore dimensions of zeolites vary across the full range of the definition of microporous materials. The first SBU where a small guest can feasibly pass through is the 6MR, with a diameter of ~2 Å. Not only will the pore dimensions of different zeolite structures vary, but obviously also the shape of the channels, such as straight or curved channels, and some with larger cavities at points in a channel or side pockets connected to a main channel. Zeolites vary further by their pore dimensionality. There are zeolites with only 1-dimensional channels, those with 2-dimensional channels and those with 3-dimensional interconnected channels.

In terms of chemical composition, while an all-siliceous zeolite would only contain $[SiO_4]^{4-}$ tetrahedra, typically zeolites are aluminosilicates, with some $[SiO_4]^{4-}$ substituted for $[AlO_4]^{5-}$ tetrahedra. While the chemical composition may vary, all zeolites with the same framework structure (topology) will still have the same framework type code. For example, the code MFI refers to the structure that is common to both the all-siliceous silicialite-1 material (see Figure 1.1) and for ZSM-5 that contains both silicon and aluminium. In every instance where an Al^{3+} ion substitutes a Si^{4+} ion, there needs to be a 'charge-compensating' cation to maintain the zeolite structure's overall charge neutrality. These cations are generally not part of the framework itself but are more loosely bound in the zeolite's pores. The cations can be the organic templates used in the synthesis, or metal cations such as sodium or calcium, with the latter ions generally being solvated by water unless the zeolite has been dehydrated. Alternatively, the compensating cation can be a proton, and this then imparts a high acidity to the zeolite structure, that can be used in catalytic applications. Overall, the majority of zeolite applications stem from the presence of compensating cations and thus as a result of the aluminium substitution in the framework. The Si/Al ratio for a zeolite can range from 1 to infinity, with the lower limit of 1 a result of Lowenstein's rule that states that $[AlO_4]^{5-}$ tetrahedra will not form direct Al–O–Al bridges.

Beyond zeolites, there are a range of related materials known as 'zeotypes'. Zeotypes can adopt the same framework structures as zeolites, but have different chemical compositions. For example, there is the family of aluminophosphates (ALPOs) consisting of Al^{3+} and P^{5+} ions (without the need for any compensating cation), SAPOs consisting of silicon, aluminium and phosphorus, zinc phosphates, and germanium sulfides. Through altering the chemical composition and in particular through the possibility to incorporate transition metal ions, this brings access to a further range of properties, particularly in terms of catalysis.

The applications for zeolites stem in particular from their chemical and physical stability, the presence of charge compensating cations, and the fact that zeolite pore dimensions are commensurate with small guest molecules and zeolite structures are available with a wide range of pore dimensions, shapes and topologies, allowing one to 'pick' an appropriate zeolite for a specific target application. Naturally occurring zeolites are typically very cheap and available on a large scale, and some synthetic zeolites are also available at a reasonably low cost. Zeolites have three major applications, several of which involve multimillion tonne productions. These applications are in ion exchange, catalysis and as molecular sieves (see Figure 1.2).

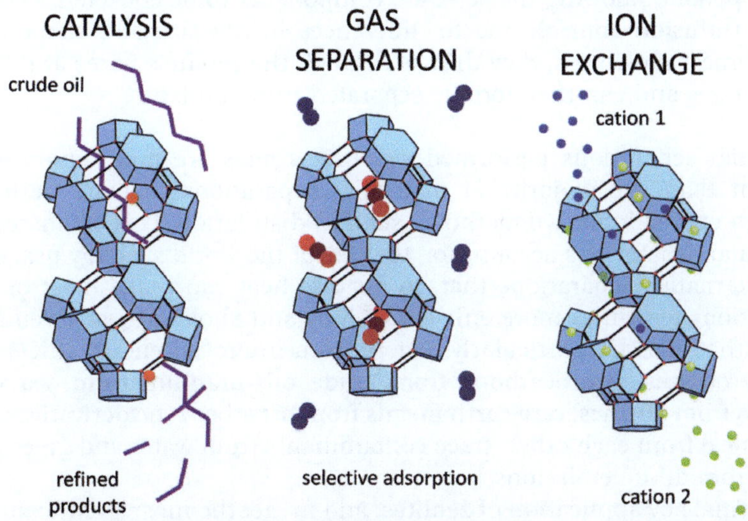

Figure 1.2 The main applications of zeolites.
Reproduced from ref. 2 with permission from the Royal Society of Chemistry.

In ion exchange, the charge compensating cations in the zeolite are able to be exchanged with other cations, as the cations remain loosely bound. The selectivity of a zeolite towards a cation is determined by the topology and chemical composition, as this will affect the ability of an ion to diffuse through the pores and the strength of the binding at the sorption site. Zeolite A has been used in ion exchange on a large scale in washing detergents, where it acts as a water softener by selectively removing Ca^{2+} ions from hard water. Zeolites have also been used to remove radioactive ions from nuclear wastes or after nuclear disasters, or toxic ions from the waste water of heavy industry.

Zeolites are highly hydrophilic materials and will absorb significant quantities of water even at raised temperatures. This has led to their use as drying agents, for example, in hospitals for drying gases used for medical treatments or on a large scale in chemical industry, such as drying liquid propane. Zeolites can perform molecular separations, thus acting as 'molecular sieves', by several different mechanisms:

(i) **Size exclusion**: excluding molecules with dimensions larger than the zeolite's pores, and thus separating the larger molecules from a mixture.
(ii) **Sorption differences**: where two different components of a mixture have different enthalpies of sorption within a zeolite. For example, in pressure swing adsorption, a mixture would be forced into a zeolite under pressure, then the pressure will be lowered slightly and only the more weakly bound guest component will be released from the zeolite. A further pressure reduction will release the other component, allowing the separate components to be collected.
(iii) **Diffusion control**: due to differences in the size and shape of the guest molecules, they diffuse through the zeolite's pores at different rates and can therefore be separated from a mixture.

Molecular separations performed *via* these routes are much more energy efficient than the majority of molecular separations that are performed through energy intensive methods, such as distillation. It is estimated that molecular separations account for 10–15% of the world's energy usage, and that alternative separations that do not use heat can make most of these separations ten times more efficient.[3] Lively and Sholl suggest seven separations that can still particularly benefit from improved energy efficiency in their processes: hydrocarbons from crude oil, uranium from sea water, alkenes from alkanes, rare-earth metals from ores, benzene derivatives (such as xylenes) from each other, trace contaminants from water and greenhouse gases from dilute emissions.[3]

The final key application of zeolites, and in fact the most significant, is in catalysis, specifically acid catalysis being used for hydrocarbon cracking. When a proton is the charge compensating cation in a zeolite, then the resulting Brønsted acid sites are extremely acidic, multiple times the strength of concentrated sulfuric acid. Furthermore, the selectivity of the zeolites, as discussed above, can be used to have selective catalytic results. This selectivity can be based on reactant size, product size or the size of the transition state for the reaction. Particularly important zeolite catalysis includes methanol to light olefin conversion, benzene alkylation and xylene isomerisation.

The computational modelling of zeolites has a long history over many decades. Many of the approaches for modelling both zeolite structure and properties have inspired the approaches later used in other types of porous materials, such as metal–organic frameworks. The scope of zeolite modelling is broad, including modelling structures, synthesis, nucleation, growth and dissolution of surfaces, catalysis, sorption and diffusion.[2] Much of the early work on modelling zeolites was facilitated by their comparably simple range of interactions that needed to be described in structures containing only two or three elements (silicon, oxygen and aluminium), for which there were molecular mechanics based forcefield descriptions that required only a few parameters to reliably reproduce their behaviour. Early simulations of zeolite structures often treated the zeolite structure itself as fixed and rigid, only considering guest binding and guest flexibility, a useful tool to reduce the computational cost of the screening. This 'rigid body approximation' approach is still often used for other porous material characterisations, although host flexibility and pore breathing motions have been shown to be critical for fully understanding or predicting many guest behaviours, including in zeolites. As computing power has increased, the calculation possibilities upon zeolite systems have expanded, to include large-scale electronic structure calculations as well as high-throughput simulations on hundreds of thousands of materials at the forcefield level.

1.2.2 Metal–Organic Frameworks (MOFs)

Metal–organic frameworks (MOFs) are a 'hybrid' form of porous material, that consist of both an inorganic and organic component, linked by strong coordination bonds. The IUPAC definition of MOFs is that they are a 'coordination network with organic ligands containing potential voids', where a coordination network itself is a 'coordination compound extending through repeating coordination entities in two or three dimensions'. MOFs were originally synthesised and reported as coordination compounds in the 1990s by Robson,[4] with Yaghi later giving them the name of metal–organic frameworks. MOFs are a field of intense research interest, with tens of thousands of structures reported, massively outweighing the ~200 zeolite structures that have been synthesised and covering a much broader range of the hypothetical topologies that could be adopted. Indeed, X-ray diffraction crystal structures now make up a sizeable proportion of all crystal structures deposited in the Cambridge Structure Database. The considerable attention on MOFs stems from the great chemical diversity of the structures that can be synthesised by combining different inorganic and organic components, resulting in a wide range of interesting, and potentially tunable, performance across sorption, catalysis, optical and electronic properties.

The inorganic component of MOFs is typically a metal ion or cluster as a node, connected by organic ligands to form open frameworks that are both porous and crystalline. An example MOF structure is shown in Figure 1.3 of MOF-5, which consists of a zinc-based metal cluster (Zn_4O) as the node and 1,4-benzodicarboxylic acid as the struts. This forms a cubic topology that is known by a three-letter code, *pcu*, and a resultant 3-dimensional pore network. MOF-5 has an alternative name of IRMOF-1, with the 'IR' standing for 'isoreticular'. The term reticular chemistry for a series of MOFs is used where they have the same topology, but different pore sizes – for example, by changing the length of the organic ligand, but keeping the same metal cluster as the node.[5] Another example of a reticular series of MOFs is for the MIL series from the Materiaux de l'Institut Lavoisier, which consists of terephthalate ligands, but with a variety of different metal centres (*e.g.* V^{3+}, Cr^{3+}, Al^{3+}, Fe^{3+} and many other transition metals).[6] Isoreticular chemistry provides a route to tuning of properties with ligand choice, such as pore size with ligand size, or sorption properties with the functionalisation of ligands to add functional groups that strongly interact with particular guest molecules. Increasing ligand length to increase pore size does not allow one to easily reach very large pore sizes, as the frameworks will

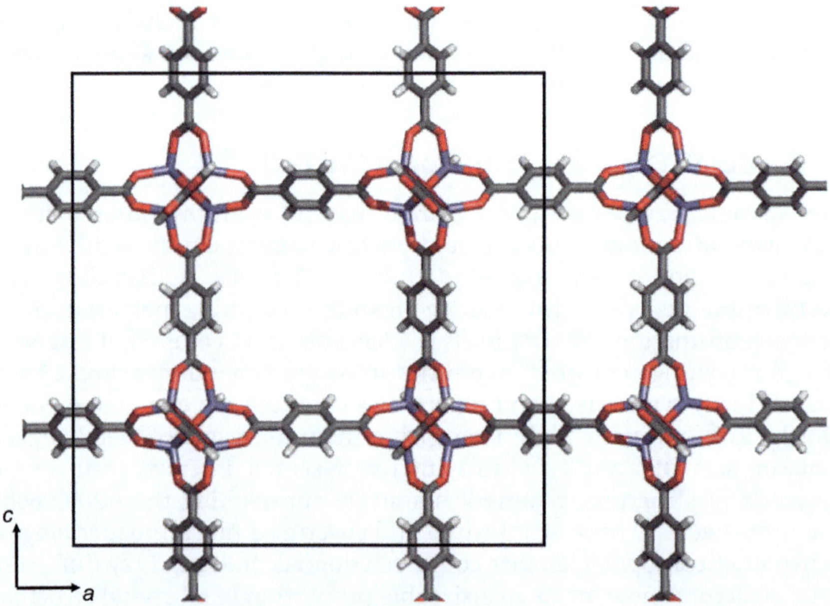

Figure 1.3 An example of a metal–organic framework structure. This is MOF-5 viewed down the *b*-axis, with the *a*- and *c*-axes of the unit cell (shown in black) labelled.

often interpenetrate, removing the potential void space, once the voids become sufficiently large to accommodate a second interpenetrated repeat of the framework.

Compared to zeolite structures, which typically have surface areas up to ~1000 $m^2 g^{-1}$, MOFs can reach significantly larger porosity, with the current record for a Brunauer–Emmett–Teller (BET) surface area of over 7000 $m^2 g^{-1}$.[7] Another distinction between MOFs and zeolites is that, broadly speaking, MOFs have a much greater degree of structural flexibility and the ability therefore to have a structural response to external stimuli such as temperature, pressure and guest sorption. For this reason, MOFs are frequently referred to as 'soft' porous crystals.[8] Often the structural response is significant, as will be discussed further in Chapter 3, with expansion and contractions that double or halve the volume, or, as has been a recent focus, with the comparatively unusual feature of negative thermal expansion.[9]

While MOFs have already exhibited large numbers of interesting properties, one of the hurdles to commercial applications has been issues with both chemical and physical stability, with many MOFs being kinetically unstable in the presence of air or water. A notable exception to instability comes from zeolitic-imidazolate frameworks (ZIFs), which are topologically equivalent to zeolites, but with zinc as the node (in place of silicon in a zeolite) and an imidazole as the ligand (in place of oxygen), where the O–Si–O angle is closely reproduced by the Zn–(centre of imidazole)–Zn angle. ZIFs generally have excellent chemical stability, and can even be refluxed in many solvents, water and alkaline solutions.[10]

MOFs hold great potential for commercial application, but thus far there are limited examples of these. One exception is the use of a MOF for the storage and controlled release of 1-methylcyclopropene (1-MCP), which can slow the ripening of fruit by inhibiting the adsorption of ethylene on the fruit's surface. This MOF is therefore used in the packaging of fruit to prolong the fruits' lifetime; for apples this can increase the lifetime by many months. A second commercial application of MOFs is in the storage of hazardous gases used in the electronics industry. Beyond these commercial applications, MOFs are being investigated for a wide range of applications, including applications in storage and separations, as discussed already for zeolites. For storage and separation, MOFs have potential advantages over zeolites in their diversity and hence tunability of pore sizes and shapes, but currently struggle to compete on factors such as cost, stability and ease of synthesis, though this can be expected to change over time. MOFs show significant promise for the storage of methane or hydrogen in cars powered by fuel cells and renewable technology, where the MOFs have potential advantages in their low weight compared to alternatives. MOFs could also become components of batteries and some MOFs can conduct protons and could therefore be proton conducting membranes within fuels cells. MOFs are also being investigated for use in drug delivery, due to the possibility of loading a large number of drug molecules within the MOFs' pores and then tuning the material for a desired drug release profile.

MOFs can be used in catalysis, either being directly involved in the reaction through the use of transition metal ions in their frameworks, or by the MOF acting as a host to support a homogeneous catalyst. MOFs are being investigated for photocatalysis, for example visible-light driven hydrogen evolution, or the reduction of carbon dioxide to hydrocarbons. While currently these are preliminary studies that typically use sacrificial donors, there are potential advantages in MOF tunability through changing the framework components, such as the fact that MOF frameworks are open, allowing easy diffusion of reactants and products into and out of the material, and that it is relatively easy to chemically functionalise a MOF so as to incorporate other molecular components, for example, through post-functionalisation methods. The potential for strong stimuli-response behaviour of MOFs (compared to zeolites) also opens the potential for application as chemical sensors. As further MOFs are synthesised and characterised, there is also the potential for finding structures with particularly promising high-temperature superconductivity or electrical conductivity or as supercapacitors. Finally, there is considerable interest in the use of defects in MOFs, *via* 'defect engineering' as another route to control and optimise their performance.

Many of the approaches earlier applied to zeolites have now been applied to MOFs, often on a large-scale, such as high-throughput studies of sorption or guest selectivity, using the increased computing power available in the last 10 years. In some cases, large-scale screening of MOFs for function outside of sorption or guest uptake is still hindered by computing power, given the inclusion of metals in their structure and a vast chemical diversity, which means that reliable forcefields (FFs) are only available for a handful of MOFs where they have been specifically parameterised. This often results in the need for the use of electronic structure methods, particularly density functional theory (DFT) calculations, which obviously come with an associated significant increase in computational cost compared to molecular mechanics-based methods. However, for many MOF properties, such as electronic, optical and catalytic properties, the simulation of these properties these will require electronic structure calculations.

1.2.3 Covalent Organic Frameworks (COFs)

Covalent organic frameworks (COFs) are crystalline and porous frameworks that only consist of organic elements and are much rarer than MOFs. The lack of strongly directional coordination bonds compared to MOFs makes the design and realisation of such materials with only covalent bonds challenging, even when trying to apply the principles of reticular chemistry. A notable challenge is in obtaining a crystalline COF solid compared to the amorphous organic polymers discussed in the next section. Common chemistries involved in the synthesis of COFs are boron condensation, triazine trimerisation and imine condensation.

Many of the COFs reported to date are 2-dimensional structures, for example, containing hexagonal layers stacked upon each other with varying

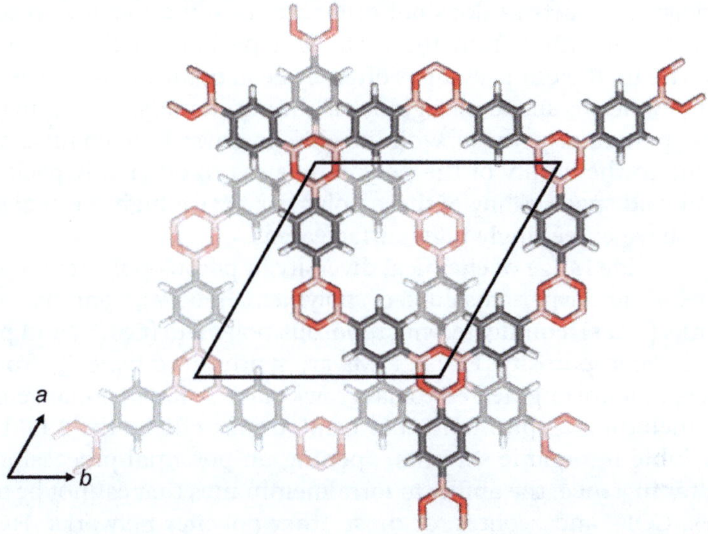

Figure 1.4 An example of a covalent organic framework (COF) structure. This is COF-1 viewed down the *b*-axis, with the *a*- and *c*-axes of the unit cell (shown in black) labelled.

degrees of order in the stacking layers. An example of a 2-dimensional COF is shown in Figure 1.4, COF-1, which is composed of boronate units connected by benzene rings and is synthesised from 1,4-benzenediboronic acid.[11] There are also some examples of 3-dimensional COFs, such as COF-108,[12] another COF material composed only of C–C, C–O, C–B and B–O bonds. At the time of synthesis, COF-108 had the lowest density of any reported crystalline material.[12] The strong covalent bonding in COFs can produce very stable materials, even at high temperatures and, again, there is an opportunity for tuning of properties through the massive possible space of potential organic components that could be used and tailored as precursors in COFs.

COFs have similar potential applications to that of MOFs, across gas storage, such as methane and hydrogen, gas separation, sensing due to the potential optical properties such as fluorescence, or electrical conduction being responsive to guests. Additionally, there is considerable interest in the use of COFs in photocatalysis, such as for water splitting or carbon dioxide reduction. The potential advantages of COFs include the opportunity to tune performance through the vast array of options for chemical components, and the fact that they are inherently lighter compared to zeolites and MOFs as they only contain light atoms and no metals.

1.2.4 Porous Polymer Networks

Porous polymer networks, as with COFs, are (micro)porous and contain only light atoms connected by covalent bonds, but unlike COFs, are not crystalline, but are instead amorphous, lacking long-range order. The porosity of porous polymer networks does not originate from the topology resulting in void spaces, but rather from the inefficient packing of the polymer components. The inefficient packing is often by design, for example, through the awkward geometric shape or rigidity of the monomers, as the majority of polymers, particularly those with flexible hydrocarbon chains, are non-porous due to the ability of the polymer chains to efficiently pack so as to remove all void space. Many of these polymers have a high thermal stability, and some have exceedingly high surface areas.

There is a wide range of chemical diversity in porous polymers reported to date, including hyper-crosslinked polymers (HCPs), porous aromatic frameworks (PAFs), conjugated microporous polymers (CMPs) and polymers of intrinsic microporosity (PIMs). PIMs are synthesised typically from linear monomers that are engineered to have 'awkward' 3-dimensional geometries, typically including a spiro-unit, that cannot pack efficiently.[13] PIMs are typically soluble in organic solvents, opening up potential processing opportunities, for instance, the ability to form membranes that cannot be achieved for MOFs, COFs and zeolites, or most other polymer networks. HCPs generate porosity through the high degree of cross-linking between chains to create a rigid framework with open pores.[14] CMPs are a type of HCP where a large number of aromatic rings and/or alkyne groups help infer both a relatively rigid open framework as well as a conjugated network that produces useful interactions with light.[15] Finally, PAFs typically have a component that has aromatic groups and a tetrahedral carbon, for example PAF-1, a model of which is shown in Figure 1.5, and which has a BET surface area above 5000 $m^2 g^{-1}$.[16]

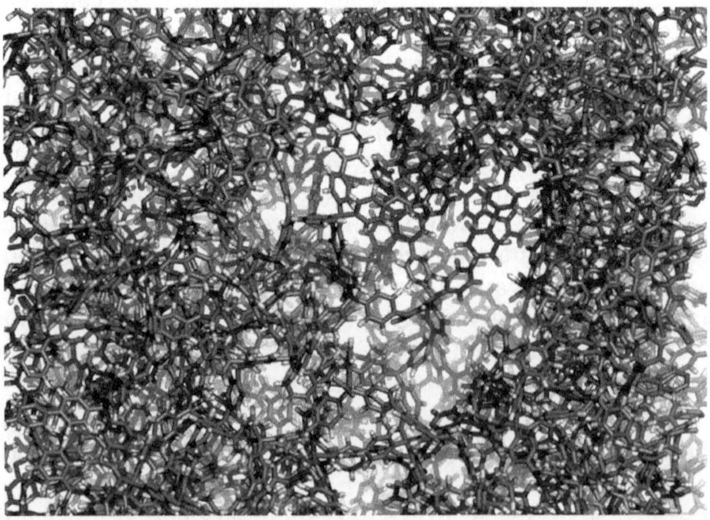

Figure 1.5 An example of a porous polymer network. This is a model porous aromatic framework, PAF-1.

The potential applications of porous polymers include those already mentioned for other porous materials. This includes in guest storage, as the polymers are inherently light as they consist only of light elements and in molecular separations, especially when processable into membranes or due to their high chemical and thermal stability. Porous polymers can perform catalysis, for example, through the inclusion of an active site in their precursors or *via* incorporation into a pore. CMPs in particular have been heavily investigated for potential use in water splitting for renewable energy.

A significant advantage of modelling porous polymers comes from their amorphous nature, which means that there will be no single crystal X-ray diffraction crystal structure that can be used to rationalise or understand the properties of the material. Therefore, structure prediction can play a big role here in producing realistic structural models, as discussed in Chapter 2. Large models with at least hundreds of atoms are required to rationalise or predict sorption and separation behaviour in systems which lack long range order, necessitating the use of molecular mechanics simulations. However, for understanding or predicting optoelectronic behaviour in these polymers, electronic structure calculations are required, therefore necessitating the use of smaller molecular or cluster models.

1.2.5 Porous Molecular Materials

Thus far, all the porous materials discussed are linked by a common feature, they are extended networks that have continuous chemical bonding in three dimensions. By contrast, porous molecular materials are built from discrete individual molecules that are not chemically bonded to each other but are held together by intermolecular forces that are much weaker than coordination or covalent bonding.[17] Porous molecular materials are much rarer than porous network materials, as in general molecules will form more stable, low-energy arrangements that minimise the void space between molecules and are non-porous as a result. In the solid state, there are two key ways that a porous molecular solid can be formed, either through intrinsic porosity, or through extrinsic porosity. Intrinsic porosity arises when a molecule has some internal void space, for example in a 'cage' molecule (as shown in Figure 1.6) or due to the geometric shape of the molecule, for instance the bowl-shaped calixarene family or belt-like cucurbiturils. Extrinsic porosity arises where there is inefficient packing between the molecules in the solid state. An example from the 1960s is Dianin's molecule, also known as an 'organic zeolite', which packs so as to leave hourglass-shaped 1-dimensional channels running along one axis that can host small guests.[18] Even intrinsically porous systems often make use of pockets of extrinsic porosity in the solid state to form interconnected networks of intrinsic pores.

In the following sections, we firstly discuss porous molecular materials that are made solely from organic components (Section 1.2.5.1) separately from hybrid porous molecular materials that contain metals as well as organic components (Section 1.2.5.2).

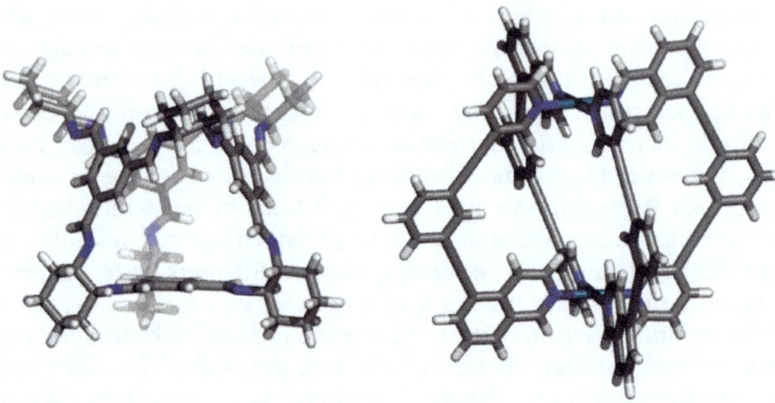

Figure 1.6 An example of an organic cage (left) and metal–organic cage (right). The organic cage is **CC3**[19] and the metal–organic cage is a Pd_2L_4 cage,[20] where L is the ditopic ligand.

1.2.5.1 Porous Organic Cages

Porous organic cages are porous molecular materials that are polycyclic molecules with the shape of a cage that have 3-dimensional structures with three or more molecular windows.[21] An example organic cage, **CC3**,[19] is shown on the left-hand side of Figure 1.6. Early research in this area by Cram and Warmuth reported nanocontainer cages formed through the condensation reaction of cavitands.[22,23] This has been followed by others, including Skowronek and Gwaronski,[24] Mastalerz[21] and Cooper,[25] with a particular focus on the use of imine chemistry in the synthesis of the molecular cages. While originally porous molecular systems reported very low porosity (with BET surface areas below ~400 $m^2 g^{-1}$), certainly compared with MOFs, more recent research interest in the field has resulted in significant increases in surface area, for instance a BET surface area of >3700 $m^2 g^{-1}$ from a large porous molecular cage from Mastalerz and co-workers.[26]

Porous organic cages are typically synthesised *via* dynamic covalent chemistry, and most often imine condensation reactions, although boronic acid condensation, sulfide exchange and alkyne metathesis reactions have also been used. Dynamic covalent chemistry has the advantage of reversibility, so that these highly symmetric structures can form rather than a statistical mixture of oligomers, or an amorphous polymer structure. Organic cages have been formed in a variety of different shapes and underlying topologies (*e.g.* related to cubes, tetrahedra and so on).[27] The discrete nature of cages means they can, in general, be more easily processed than network materials due to their solubility in common solvents. Membranes can be formed directly from cages, or cages can be included in a membrane, thin-film, or coating as an additive. Considering the imine chemistry typically involved in the synthesis of organic cages, several systems have been shown to be remarkably chemically and physically stable, for example, being resistant to many hours of boiling in water[28] or to exposure to highly alkaline or acidic environments through chemical modification of the systems.[29]

Organic cages have interesting potential uses in the liquid state, such as sensing, for example, with the liquid fluorescing until a guest explosive molecule enters the cage cavity and quenches the cage's fluorescence, allowing detection of the explosive.[30] Cages are an example of a system that can be designed to be a 'porous liquid', where a porous liquid is a material that exhibits the properties of a microporous material while also having the transport properties of a fluid.[31] Porous organic cages were adapted to form the first example of a type II porous liquid in 2015, where a type II porous liquid is a solution of molecular pores.[32] This was achieved *via* two different routes, firstly through functionalisation of a cage with crown ether groups that increased the solubility of the system, but were too large to penetrate the molecular pores, or, alternatively and more successfully, by scrambling a mixture of closely related, but differently functionalised molecules that could not effectively crystallise and thus were liquid at room temperatures. Such porous liquids have potential in the separation of gas mixtures, where their unique liquid state compared to other microporous solids has the potential to be exploited from an engineering perspective.

Both crystalline and amorphous solid-state cage materials have been researched for potential applications, particularly in separations. Porous organic cages have already exhibited potential for the separation of noble gases, xylenes, hydrocarbons such as alkane/alkene mixtures, air pollutants and hydrogen isotopes.[33–36] Cages are also being explored for other applications, such as for water desalination by reverse osmosis[37] and as proton conduction membranes.[38] To compete with other porous (network) materials for commercial applications, particularly on cost, porous cages will need to take advantage of their unique aspects, in particular their modular nature for processability and tunability.

The same modelling approaches that have been used for zeolites first, and more recently MOFs, have also been applied to porous organic cages. In particular, given the small size of the pores in cage systems, the need for consideration of the cages' flexibility for accurate guest–host modelling is to be expected.[39] The limited number of reported organic cages, particularly in terms of those with X-ray diffraction crystal structures, means that large-scale screening for cages has not been carried out in the way that it has for MOFs. While for MOFs, hypothetical structures can be constructed (see Chapter 2), this is much less trivial for cage systems; a small change in the cage could completely change the way in which the system would pack in the solid state, so it is not as easy to construct reasonable solid-state structures for hypothetical cage molecules. Crystal structure prediction techniques have been used to predict the solid-state structure with great success,[40] but this is not currently a computationally cheap or quick technique. However, the modular nature can also be viewed more optimistically – it opens the door to screening the systems in a different way, by analysing the individual molecules in a system and finding descriptors for their solid-state properties through analysis of the single molecules alone, which is much computationally cheaper.[41,42] Finally, the lack of metals in organic cage systems means that forcefield-based approaches are typically used in modelling them in the solid-state for behaviour such as guest sorption and diffusion. For optoelectronic properties, electronic structure based methods obviously need to be applied, but again there is the modular advantage; the analysis can be carried out on a single molecule and thus has significantly lower computational cost than a solid-state structure. However, many cage molecules themselves contain a very large number of atoms, meaning that even molecular-level excited state calculations are computationally demanding.

1.2.5.2 Metal–Organic Polyhedra

Metal–organic polyhedra (MOPs) are self-assembled from metal ions and ligands to form discrete molecular architectures, in contrast to MOFs, which are built from the same components, but have extended networks.[43–46] MOPs have been reported in a wide range of polygon topologies, including tetrahedra, octahedra and cubes, as well as larger polyhedra. An example of a MOP combined from ditopic ligands and palladium is shown on the right-hand side of Figure 1.6. MOPs are also termed 'molecular flasks' and can be used to encapsulate guests and to influence the reactivity of guests; for example, stabilising species, changing reaction selectivity or accelerating reactions. Compared to porous organic cages, the metals within the polyhedra provide additional catalytic opportunities. As with porous organic cages, MOPs are also being investigated as porous liquids.[47]

There has been very little modelling of MOPs in the sense of what would typically be thought of as 'porous materials modelling', rather, most effort has focused on electronic structure calculations of the systems, in particular to rationalise the self-assembly outcomes. Computational investigation of MOPs based on geometric analysis to predict assembly,[48] guest recognition[49] and catalytic reactivity[50] has been conducted, but these studies are relatively few in number.

1.3 An Overview of Modelling Approaches

In the following chapters, the theory and approaches for modelling specific aspects of porous materials are discussed in detail. In this chapter, we provide a light touch overview of the main techniques used in porous materials modelling, which aims at being little more than an introduction of key methods. The key to all computational chemistry techniques is a mathematical description of the system that allows the calculation of the system's energy and how that depends on the position of the atoms in the system. From such a description, we can derive insight into a system's properties. For porous materials, this involves the calculation of the energetics driving the assembly of the material, host/guest interactions, optoelectronic, mechanical, and catalytic properties and so on. In terms of the level of theory, there are statistical mechanics simulations that use simple models to reproduce the macroscopic properties of a porous system, for instance, the flow of guest media through a material's pores. If modelling at a classical level of theory, forcefield potentials are used, and for a quantum mechanical description, there are then multiple options of electronic structure calculation.

1.3.1 Structural Characterisation

Structural characterisation of the pore system of a porous material is key to understanding the material's properties, and there are many characterisation tools available, and an increasing number of these are now open-source. This topic will be covered in more detail in Chapter 4. Structural characterisation includes determining the dimension and shapes of the pores, and their interconnectivity. The interconnectivity of the channels must always be considered in the context of the size and shape of the guest whose ability to sorb and diffuse through a pore is being considered. Structural descriptors can include the largest cavity diameter in the pore system, as well as the pore limiting diameter, the diameter of the largest sphere that can diffuse across the model. While structural characterisation is generally performed on a rigid model taken from either an X-ray diffraction crystal structure or a simulation, it is also relevant to consider whether thermal motions should at least be kept in mind.

1.3.2 Role of Flexibility in Porous Materials

In computational studies of porous materials, it has been common to apply the 'rigid host approximation', particularly for simulations of guest sorption (Section 1.3.7). This approximation means that the porous host itself is kept completely rigid, with fixed atomic positions and no consideration of the flexibility. This approximation can be convenient for many reasons: (i) it substantially reduces the computational cost as only guest/host and guest/guest interactions need to be considered; (ii) it simplifies simulations considerably, making them possible on a reasonable timescale, for example, making it much easier to reach convergence than if thousands or more host configurations needed consideration; and (iii) an accurate mathematical description of the host structure's flexibility may not be available (for example, for a MOF structure). However, it has long been recognised, even in early zeolite studies – and arguably zeolites are the least 'flexible' of the porous materials discussed above – that consideration of host flexibility is often key to an accurate description of the system. Consideration of host flexibility is particularly important when the guest size is commensurate with that of the host's pores.

The term pore limiting envelope (PLE) has been introduced as a useful descriptor of a pore's diameter beyond the static pore diameter that would be considered from inspection of a crystal structure of a porous material alone, particularly considering an X-ray diffraction structure would be measured at a low temperature. The PLE is a histogram of pore diameters typically measured by sampling a dynamical simulation where the material's pores have been able to 'breathe'. This consideration can be key to understanding experimental observations of guest diffusion that would appear to be implausible from inspection of the pores in the static crystal structure alone.[39]

1.3.3 Molecular Mechanics

Molecular mechanics is an overarching term for the use of classical simulations that treat atoms and bonds as balls and springs, respectively, and do not explicitly consider the motion of electrons. The potentials that are used to describe the interaction of the atoms are known by several terms, including FF potentials, classical potentials or interatomic potentials (IP). The classical approximation allows for much larger systems to be considered than would otherwise be the case with the same computational resource if electronic motion was being considered.

Molecular mechanics works by dividing the potential energy of the system into a series of different contributions including self-energy, two-body terms, three-body terms, five-body terms and so on (although the series is normally truncated at the four-body terms). The potentials are divided into intermolecular and intramolecular contributions and the exact form of the potentials chosen is dependent on the system and properties being modelled. The typical form of a FF potential is:

$$V = E_{\text{bond stretching}} + E_{\text{bond bending}} + E_{\text{torsional}} + E_{\text{electrostatic}} + E_{\text{van der waals}} \tag{1.1}$$

where the first three terms are intramolecular interactions, and the remaining terms are intermolecular. The intramolecular terms consist of the two-body bond stretching term, normally related to Hooke's law, and where the equilibrium bond length and the 'spring' strength of the bond are the input parameters in the FF for a given bond pairing. An alternative bond stretching term that is frequently used is the Morse potential. The three-body term for bond bending generally has a similar form, and the final intramolecular term, the four-body term to describe torsional arrangements can take many forms to describe the potential energy surface related to the rotation about the dihedral angle of a central bond. The non-bonding intermolecular terms consist of the Coulombic electrostatic interactions and van der Waal forces, the latter including both long-range dispersion and short-range repulsion. For van der Waals forces there are a range of simple mathematical descriptions that can be used, included the commonly applied 12-6 Lennard-Jones potential or Buckingham potential. The parameters used to describe the potential must then be provided for each atom pair in the system to be modelled.

There are an enormous number of available FFs, both open-source and commercial, that have been applied to the modelling of porous materials.

The exact choice of FF will depend on the porous material system, for example, for zeolites there are FFs that can describe the Si–O–Si angles and the polarisability of the oxygen atoms well (the latter through additional terms to those in eqn (1.1) above). Furthermore, the FF choice will depend on which property or properties one is trying to consider; for example, if calculating sorption properties in a material, and one is assuming the host porous material is rigid, then only host/guest and guest/guest potentials are required. Often there are multiple literature potentials for a specific guest, for example, even a diatomic gas guest can be described as a single sphere or with additional parameters to describe, for example, the molecule's quadrupole. For some systems, in particular MOFs, the development of FFs for these systems is an ongoing area of research, given the difficulty of accurately describing the metal coordination environments in particular, as well as the incredible breadth of coordination chemistry involved in the thousands of MOFs reported. As is often the case with FF selection, a trade-off must be made between accuracy for a single system and transferability across a broader range of systems. Again, the exact choice will depend upon the specific research aim. Finally, we note that while FFs by definition do not implicitly include the consideration of electronic motion, and thus cannot describe bond breaking and bond formation, there are a subset of FFs known as 'reactive forcefields' where the bond breaking and bond formation have been parameterised with additional terms, opening the possibility for specific studies of reactions with a FF-based approach.

1.3.4 Electronic Structure Methods

Electronic structure calculations apply a quantum mechanical description of a chemical system, rather than the classical description in FF calculations. Electronic structure simulations are used particularly for the analysis of optoelectronic and catalytic properties of porous materials (as discussed in Chapter 5), but also for more reliable structural and energetic features across porous materials modelling, where computational resource allows. The most commonly used electronic structure method on porous materials is DFT. We can expect to see the continuing trend of an increasing number of porous materials studies using DFT calculations in the future with continuing access to greater computational power. It is worth noting that while molecular mechanics calculations themselves are computationally cheaper by orders of magnitude, it is not trivial to parameterise a new system if an accurate FF is not already available. Furthermore, in large-scale screening studies of thousands or hundreds of thousands of systems, the chemical diversity is likely to challenge the transferability of a FF. Therefore, the potential ability to 'brute force' DFT simulations does have its advantages.

DFT is a way of solving the Schrödinger equation using the electron density instead of the wavefunction. Within DFT, the energy and by extension other properties can be expressed as a functional, a function of a function, of the electron density. DFT is in principle exact; however, the exact form of the functional, generally known as the density or exchange–correlation functional, is unknown and in practice we have to choose one of a number of possible approximations for the density functional. The simplest approximation is the local density approximation (LDA), where the functional only depends on the value of the electron density for all points in space. The generalised gradient approximation, or GGA, goes beyond LDA by not only including the density at all points in space but also the gradient of the electron density. GGA is probably the most commonly used density functional when describing porous solids. Beyond GGA, there are among others, hybrid functionals, which include a contribution of exact exchange from wavefunction theory, and LDA+U/GGA+U, where LDA or GGA is combined with a Hubbard-like term acting on the d- or f-electrons. Use of hybrid functionals or LDA+U/GGA+U especially improves the description of the electronic, optical and magnetic properties of materials. Choosing the right functional is one of the biggest challenges when using DFT and might involve comparing the predictions of DFT to experimental data, where available.

In many porous materials, dispersion is a significant contribution to the materials' cohesive energy and hence one would ideally describe the effect of dispersion accurately. However, many of the standard classes of density functionals – LDA, GGA, hybrid functionals and LDA+U/GGA+U – do not correctly describe the attractive long-range dispersion. The simplest and most commonly used approach to correct this issue is the DFT+D method by Grimme and co-workers, which adds pair-wise C_6R^{-6} terms to provide a description of dispersion. There are different parametrisations of this method, the latest of which is DFT+D4[51] and the most commonly used one currently is DFT+D3,[52] which have been parameterised for a range of density functionals, *e.g.* PBE+D3. Beyond DFT+D, there are methods like Tkatchenko and Scheffler's TS-vdW-DFT.[53]

The time-dependent extension of DFT (TD-DFT) allows for the prediction of optical absorption and, after relaxation of the excited state, luminescence spectra of porous materials. Within the linear response version of TD-DFT, which is the most routinely used and available in many DFT codes, the equivalent of the density functional in ground-state DFT is the exchange–correlation kernel, which in principle depends not only on the density but also time/frequency. However, in practice this exchange–correlation kernel is generally approximated by a ground-state density functional, ignoring the time/frequency dependence, in the so-called adiabatic approximation. Hybrid density functionals or range-separated hybrid density functionals, where the amount of exact exchange varies with the interelectron separation, are usually preferred for TD-DFT calculations as they minimise problems in the description of charge-transfer excited states, such as metal–ligand excitations.

The cost of DFT calculations can be reduced by switching to density functional tight-binding (DFTB)[54] or extended tight-binding (xTB)[55] methods in which the (exchange–correlation) energy is expanded in a Taylor series. Different DFTB and xTB methods exist, differing in the order at which the Taylor series is terminated, the basis-set used and how different terms are approximated, amongst other ways. Extensions of DFTB methods to excited states exist.

The density in DFT calculations, as well as the Kohn–Sham orbitals used to solve the density, are expanded in terms of a basis-set. The most commonly used basis-sets for DFT calculations are atom-centred Gaussian or Slater basis-sets and delocalised plane-wave basis-sets. The former traditionally find application in molecular calculations and the latter in periodic calculations on crystalline materials. The difference between Gaussian and Slater basis-sets is the exact form of the basis-functions used. The Slater basis-functions are essentially the atomic orbitals of a free atom while the Gaussian basis-functions are approximation to those atomic orbitals in terms of Gaussian functions. Even though Gaussian basis-sets require more basis-functions than Slater basis-sets for a similar description of the density and Kohn–Sham orbitals, and struggle to reproduce the cusp of the electron density at the nuclear positions, most molecular codes use Gaussian basis-sets because multi-centre integrals of basis-functions, which pop up frequently in DFT, are much easier to calculate for Gaussian basis-functions. Independent of the class of basis-set used, increasing the quality of the basis-set, *i.e.* using more basis-functions, gives more accurate results but increases the computational cost of the calculation. As such there is typically a trade-off between the quality of the results and the cost of the calculations, which specifically becomes an issue for calculations on large molecules (or large cluster models of a solid) and solids with large unit cells.

1.3.5 Molecular Dynamics

Molecular dynamics (MD) is a deterministic method that explores the phase space of a system over time through Newton's equations of motion. The output of the simulation is a trajectory of the system's coordinates (and forces) over time. This allows us to understand in porous materials how either the host, the guest, or the host and guest in concert evolve in a system over time. The equations of motion are not directly integrated, but rather a finite differences method is used to determine the particles' trajectory. In porous materials, these simulations will normally be carried out using molecular mechanics approaches. However, *ab initio* molecular dynamics (AIMD) simulations of porous systems with modest numbers of atoms have been carried out, in particular where bond breaking and formation were of interest.

Depending upon which features in a system are of interest, MD simulations can be carried out in a variety of ensembles. An NVT ensemble, with a constant number of particles, volume of the simulation cell and temperature could be used to equilibrate a model at a particular temperature. The *microcanonical* or NVE ensemble conserves the energy of the system, with the volume and the number of particles also kept constant but allowing the temperature and pressure to fluctuate. An MD simulation using an NVE ensemble could be used to study the diffusion of a guest through a material's pores. A variety of thermostats can be used to keep the temperature of the system constant, and equivalently there are barostats for pressure allowing for simulations using the NVT and NPT ensembles, and so on.

1.3.6 Enhanced Sampling

While molecular mechanics based MD simulations can observe events routinely happening on a picosecond to nanosecond timescale in porous materials, this may not be sufficient to observe 'rare events' that may occur on timescales longer than tens of nanoseconds. As an example, in a system where the pores of the material are commensurate with a guest's dimensions, a 10 ns MD simulation may not once observe the guest diffusing through the pore neck of the material. However, the guest would diffuse on a longer timescale, and would therefore be observed to diffuse through the material on an experimental timescale – thus this could be an event that we would like to be able to rationalise, determine the energetic barriers towards it occurring, and be able to predict in advance of experimental observation. This is an example of a situation in which an enhanced sampling technique could be applied to address the timescale limitation of a regular MD simulation. Enhanced sampling methods to explore the response of porous materials to external pressure and temperature are discussed in Chapter 3. Examples of enhanced sampling techniques include metadynamics simulations, umbrella sampling and variational enhanced sampling. Broadly speaking, these approaches look to accelerate the dynamics of simulations, for instance, by applying an external bias, so that the configurations of the system related to rare events are observed more frequently and the free energy differences between the configurations can be calculated.

1.3.7 Grand Canonical Monte Carlo Simulations

Grand Canonical Monte Carlo (GCMC) simulations are used in porous materials to simulate the sorption of guests within a host. GCMC simulations can be used to produce representative guest-loaded configurations and levels, for example, to predict guest uptake or selectivity, and can also reproduce adsorption isotherms. Monte Carlo (MC) simulations in general are not deterministic simulations (unlike MD), but rather make use of random numbers to perform random moves and orientation changes to generate successive configurations of a system. The algorithms involved in MC make use of importance sampling, normally *via* a Boltzmann factor, to ensure the majority of the simulation is spent sampling low-energy configurations. The grand canonical ensemble keeps the chemical potential, system volume and temperature fixed, with the MC moves being guest particle addition, particle deletion and particle displacement (*e.g.* translation or rotation). If guest mixtures are being simulated, then an additional move can be an identity swap of two particles. For larger guests, for example long chain hydrocarbons, additional move types may need to be considered to increase the likelihood of a successful guest insertion into the host. GCMC simulations almost always use the rigid host approximation due to computational cost considerations when trying to equilibrate a guest/host mixture or multiple mixtures at different loading levels, although there are hybrid-GCMC simulation approaches that allow the system to dynamically evolve during MD simulations upon guest loading. These simulations will be discussed in detail in Chapter 4.

1.3.8 Machine Learning (ML)

Artificial intelligence techniques, in particular machine learning (ML), will be discussed in their application to porous materials in Chapter 6. ML is highly topical at the moment and is an approach that seeks to create data-driven predictive models. Supervised ML is the most commonly applied of ML techniques to porous materials so far and is in essence about fitting a function to data through pattern recognition. Examples of supervised ML algorithms include decision tree algorithms and random forest models. For supervised ML, one requires a sufficient quantity of high-quality data, based on either prior experimental or simulated results. An example for porous materials would be a dataset of sorption properties across a material class. Given input of a new material in this class, you would want to have trained the model to accurately predict that material's sorption behaviour, without resorting to either experiments or computational calculations of the properties. A key part of the process of training a supervised machine learning model is to find adequate structural descriptors for the material. These could, for example, be the chemical composition of the material, a 2D representation of the connectivity of the atoms in the porous material host, or a 3D representation of the structure of the material, or at least of a key environment within the material. During the training of the ML model, a function is fitted to give the best performance for that data based on an appropriate metric with which to judge the performance. Generally, to assess the performance of the model, the data will have been divided into a training set, upon which the model is trained, and a test set which can be used for validation of the model on unknown data.

Another area of ML relevant to porous materials is the unsupervised ML used in natural language processing (NLP). The use of NLP specifically applied to the extraction of useful data from the chemical literature through text mining has the potential to unlock further data-driven predictions. The chemical literature is a vast array of information that it is no longer possible to manually extract on a large scale. The data is also reported in a variety of formats – text, tables, figures in chemical journals as well as within patents *etc.* NLP is an approach to train algorithms to extract targeted features of the chemical literature. One example in porous materials of the application of NLP is to extract information on porous material synthesis procedures and then to use that data to predict laboratory synthesis procedure outcomes. In Chapter 6, the future potential and directions of artificial intelligence techniques in porous material modelling are discussed.

1.4 Summary

This chapter has given a general overview to the field of computational porous materials modelling. The variety of different microporous material types has been discussed and contrasted, and the key areas where porous materials modelling is of value have been highlighted. The introduction to key terms and computational methods provides a foundation for reading the remaining chapters that focus on specific computational modelling areas within the field.

References

1. *Atlas of Zeolite Framework Types*, ed. L. McCusker and D. Olson, Elsevier, 6th edn, 2007.
2. V. Van Speybroeck, K. Hemelsoet, L. Joos, M. Waroquier, R. G. Bell and C. R. A. Catlow, *Chem. Soc. Rev.*, 2015, **44**(20), 7044.
3. D. S. Sholl and R. P. Lively, *Nature*, 2016, **532**, 435.
4. B. F. Hoskins and R. Robson, *J. Am. Chem. Soc.*, 1990, **112**, 1546.
5. O. M. Yaghi, M. O'Keefe, N. W. Ockwig, H. K. Chae, M. Eddaoudi and J. Kim, *Nature*, 2003, **423**, 705.
6. F. Millange, C. Serre and G. Ferey, *Chem. Commun.*, 2002, **8**, 822.
7. O. K. Farha, I. Eryazici, N. C. Jeong, B. G. Hauser, C. E. Wilmer, A. A. Sarjeant, R. Q. Snurr, S. T. Nguyen, A. O. Yazaydın and J. T. Hupp, *J. Am. Chem. Soc.*, 2012, **134**, 15016.
8. S. Horike, S. Horike, S. Shimomura, S. Shimomura and S. Kitagawa, *Nat. Chem.*, 2009, **1**, 695.
9. J. D. Evans, J. P. Dürholt, S. Kaskel and R. Schmid, *J. Mater. Chem. A*, 2019, **272**, 90.
10. K. S. Park, Z. Ni, A. P. Côté, J. Y. Choi, R. Huang, F. J. Uribe-Romo, H. K. Chae, M. O'Keefe and O. M. Yaghi, *Proc. Natl. Acad. Sci. U. S. A.*, 2006, **103**, 10186.
11. A. P. Côté, A. I. Benin, N. W. Ockwig, M. O'Keefe, A. J. Matzger and O. M. Yaghi, *Science*, 2005, **310**, 1166.
12. M. G. Rabbani and H. M. El-Kaderi, *Chem. Mater.*, 2011, **23**, 1650.
13. P. M. Budd, B. S. Ghanem, S. Makhseed, N. B. McKeown, N. B. McKeown, K. J. Msayib and C. E. Tattershall, *Chem. Commun.*, 2004, **2**, 230.
14. R. Dawson, A. I. Cooper and D. J. Adams, *Prog. Polym. Sci.*, 2012, **37**, 530.
15. R. Dawson, A. Laybourn, A. Laybourn, R. Clowes, Y. Z. Khimyak, Y. Z. Khimyak, D. J. Adams and A. I. Cooper, *Macromolecules*, 2009, **42**, 8809.
16. Y. Li, T. Ben, B. Zhang, Y. Fu and S. Qiu, *Sci. Rep.*, 2013, **3**, 2420.
17. J. Liang, S. Xing, P. Brandt, A. Nuhnen, C. Schlüsener, Y. Sun and C. Janiak, *J. Mater. Chem. A*, 2020, **8**(38), 19799.
18. R. Barrer and V. Shanson, *J. Chem. Soc., Chem. Commun.*, 1976, **9**, 333.
19. T. Tozawa, J. T. A. Jones, S. I. Swamy, S. Jiang, D. J. Adams, S. Shakespeare, R. Clowes, D. Bradshaw, T. Hasell, S. Y. Chong, C. Tang, S. Thompson, J. Parker, A. Trewin, J. Bacsa, A. M. Z. Slawin, A. Steiner and A. I. Cooper, *Nat. Mater.*, 2009, **8**, 973.

20. J. E. M. Lewis, A. Tarzia, A. J. P. White and K. E. Jelfs, *Chem. Sci.*, 2020, **11**, 677.
21. M. Mastalerz, *Angew. Chem. Int. Ed.*, 2010, **49**, 5042.
22. B. S. Park, C. B. Knobler and D. J. Cram, *Chem. Commun.*, 1998, 55.
23. X. Liu and R. Warmuth, *J. Am. Chem. Soc.*, 2006, **128**, 14120.
24. P. Skowronek and J. Gawronski, *Org. Lett.*, 2008, **10**, 4755.
25. T. Hasell and A. I. Cooper, *Nat. Rev. Mater.*, 2016, **1**, 10686.
26. G. Zhang, O. Presly, F. White, I. M. Oppel and M. Mastalerz, *Angew. Chem. Int. Ed.*, 2014, **53**, 1516.
27. V. Santolini, M. Miklitz, E. Berardo and K. E. Jelfs, *Nanoscale*, 2017, **9**, 5280.
28. T. Hasell, M. Schmidtmann, C. A. Stone, M. W. Smith and A. I. Cooper, *Chem. Commun.*, 2012, **48**, 4689.
29. M. Liu, M. A. Little, K. E. Jelfs, J. T. A. Jones, M. Schmidtmann, S. Y. Chong, T. Hasell and A. I. Cooper, *J. Am. Chem. Soc.*, 2014, **136**, 7583.
30. K. Acharyya and P. S. Mukherjee, *Chem. Commun.*, 2014, **50**, 15788.
31. N. O'Reilly, N. Giri and S. L. James, *Chem. – Eur. J.*, 2007, **13**, 3020.
32. N. Giri, M. G. Del Pópolo, G. Melaugh, R. L. Greenaway, K. Rätzke, T. Koschine, L. Pison, M. F. C. Gomes, A. I. Cooper and S. L. James, *Nature*, 2015, **527**, 216.
33. A. Kewley, A. Stephenson, L. Chen, M. E. Briggs, T. Hasell and A. I. Cooper, *Chem. Mater.*, 2015, **27**, 3207.
34. T. Mitra, K. E. Jelfs, M. Schmidtmann, A. Ahmed, S. Y. Chong, D. J. Adams and A. I. Cooper, *Nat. Chem.*, 2013, **5**, 276.
35. L. Chen, P. S. Reiss, S. Y. Chong, D. Holden, K. E. Jelfs, T. Hasell, M. A. Little, A. Kewley, M. E. Briggs, A. Stephenson, K. M. Thomas, J. A. Armstrong, J. Bell, J. Busto, R. Noel, J. Liu, D. M. Strachan, P. K. Thallapally and A. I. Cooper, *Nat. Mater.*, 2014, **13**, 954.
36. M. Liu, L. Zhang, M. A. Little, V. Kapil, M. Ceriotti, S. Yang, L. Ding, D. L. Holden, R. Balderas-Xicohténcatl, D. He, R. Clowes, S. Y. Chong, G. Schütz, L. Chen, M. Hirscher and A. I. Cooper, *Science*, 2019, **366**, 613.
37. X. Kong and J. Jiang, *J. Phys. Chem. C*, 2018, **122**, 1732.
38. M. Liu, L. Chen, S. Lewis, S. Y. Chong, M. A. Little, T. Hasell, I. M. Aldous, C. M. Brown, M. W. Smith, C. A. Morrison, L. J. Hardwick and A. I. Cooper, *Nat. Commun.*, 2016, **7**, 12750.
39. D. Holden, S. Y. Chong, L. Chen, K. E. Jelfs, T. Hasell and A. I. Cooper, *Chem. Sci.*, 2016, **7**, 4875.
40. J. T. A. Jones, T. Hasell, X. Wu, J. Bacsa, K. E. Jelfs, M. Schmidtmann, S. Y. Chong, D. J. Adams, A. Trewin, F. Schiffman, F. Cora, B. Slater, A. Steiner, G. M. Day and A. I. Cooper, *Nature*, 2011, **474**, 367.
41. J. D. Evans, K. E. Jelfs, G. M. Day and C. J. Doonan, *Chem. Soc. Rev.*, 2017, **46**, 3286.
42. K. E. Jelfs and A. I. Cooper, *Curr. Opin. Solid State Mater. Sci.*, 2013, **17**, 19.
43. D. J. Tranchemontagne, Z. Ni, M. O'Keefe and O. M. Yaghi, *Angew. Chem. Int. Ed.*, 2008, **47**, 5136.

44. Y. Inokuma, M. Kawano and M. Fujita, *Nat. Chem.*, 2011, **3**(5), 349.
45. D. Zhang, T. K. Ronson and J. R. Nitschke, *Acc. Chem. Res.*, 2018, **51**, 2423.
46. R. Chakrabarty, P. S. Mukherjee and P. J. Stang, *Chem. Rev.*, 2011, **111**, 6810.
47. L. Ma, C. J. E. Haynes, A. B. Grommet, A. Walczak, C. C. Parkins, C. M. Doherty, L. Longley, A. Tron, A. R. Stefankiewicz, T. D. Bennett and J. R. Nitschke, *Nat. Chem.*, 2020, **12**, 270.
48. M. Yoneya, T. Yamaguchi, S. Sato, M. Fujita and M. Fujita, *J. Am. Chem. Soc.*, 2012, **134**, 14401.
49. B. P. Hay, C. Jia and J. Nadas, *Comput. Theor. Chem.*, 2014, **1028**, 72.
50. T. A. Young, V. Martí-Centelles, J. Wang, P. J. Lusby and F. Duarte, *J. Am. Chem. Soc.*, 2020, **142**, 1300.
51. E. Caldeweyher, S. Ehlert, A. Hansen, H. Neugebauer, S. Spicher, C. Bannwarth and S. Grimme, *J. Chem. Phys.*, 2019, **150**, 154122.
52. S. Grimme, J. Antony, S. Ehrlich, S. Ehrlich, H. Krieg and H. Krieg, *J. Chem. Phys.*, 2010, **132**, 154104.
53. A. Tkatchenko and M. Scheffler, *Phys. Rev. Lett.*, 2009, **102**, 073005.
54. M. Elstner and G. Seifert, *Philos. Trans. R. Soc., A*, 2014, **372**, 20120483.
55. C. Bannwarth, E. Caldeweyher, S. Ehlert, A. Hansen, P. Pracht, J. Seibert, S. Spicher and S. Grimme, *Wiley Interdiscip. Rev.: Comput. Mol. Sci.*, 2020, **140**, 18A301.

CHAPTER 2

Structure Prediction of Porous Materials

MATTHEW A. ADDICOAT

School of Science and Technology, Nottingham Trent University, Clifton Lane, Nottingham NG11 8NS, UK
Email: matthew.addicoat@ntu.ac.uk

2.1 Why Predict Structures of Porous Materials?

In the last few decades materials science has seen an explosion of interest in porous materials for use in a variety of applications such as gas storage and separation, catalysis and sensing.[1,2]

Recent interest has been driven by the deliberate synthesis of porous materials, including several families of materials such as coordination polymers,[3,4] Metal–Organic Frameworks (MOFs),[5,6] Covalent Organic Frameworks (COFs),[7,8] Hydrogen-bonded Organic Frameworks (HOFs),[9–11] zeolites[12] and Zeolitic Imidazolate Frameworks (ZIFs).[13,14] These families of materials are all (a) crystalline and (b) extended solids; however, neither crystallinity nor structural extent/periodicity are a prerequisite for porosity. Metal–Organic Polyhedra (MOPs)[15] and Porous Organic Cages (POCs)[16,17] may be considered the non-periodic cousins of MOFs and COFs, respectively. Equally, amorphous porous materials have been developed for many of the same applications as the crystalline materials above.[18,19] Porosity in amorphous materials may be intrinsic, as in amorphous solids composed of shape-persistent POCs; extrinsic, arising due to inefficient packing, such as in Polymers of Intrinsic Microporosity (PIMs),[20–22] or both. In addition to the already dizzying array of structures within each family, hybrids of various

Inorganic Materials Series No. 8
Computer Simulation of Porous Materials: Current Approaches and Future Opportunities
Edited by Kim Jelfs

Published by the Royal Society of Chemistry, www.rsc.org

materials, such as MOPs used as components of MOFs,[23] cage-COFs[24] and zeolite-ZIFs[25] have been synthesised.

These comparatively new families of porous materials possess an important feature that distinguishes them from older porous materials such as porous carbons[26,27] and silica,[28,29] in that they are composed of discrete molecular components, bound together by coordination, covalent or hydrogen bonds. Building materials from modular molecular components allows for the deliberate incorporation of functionality, either at the time of synthesis or by post-synthetic modification.[30–33] The modular synthesis of these materials represents both an opportunity and a challenge for their application: by correct choice of components, highly specific pore geometries and functionalities can be realised—*e.g.* in crystalline molecular reaction vessels.[34–36] But the same modular synthesis leads to effectively infinite tunability and a 'needle-in-a-haystack' problem. While there have been high-throughput syntheses of MOFs[37,38] and porous cages,[39] synthesis is both time- and resource-intensive and cannot even begin to span the trillions of possible combinations of molecular 'modules'; consequently, the bulk of materials discovery has been serendipitous.[40]

Therefore, to address this challenge, namely, to answer, "Of all the possible porous materials, *which* one has the properties required for *this* application?", computational methods have come to the forefront. Several high-throughput studies[41–43] have screened up to trillions of possible materials in order to identify promising candidates for applications such as CH_4 and H_2 storage and CO_2 capture. Many databases of porous materials have been developed for screening purposes, including: several for MOFs; the MOF subset of the Cambridge Structural Database,[44] the Computation-Ready Experimental database,[45] and the hMOF set of hypothetical MOFs[46] (the latter two are accessible online at https://mof.tech.northwestern.edu/); for COFs[47] and zeolites.[48]

The combinatorial explosion of possible porous materials is one clear reason we would wish to computationally predict their structures. A second reason is that even in the case where the synthetic reagents are known, the result is not always clear. Different synthetic conditions (*e.g.* the use of modulators[49,50]) frequently result in significantly different material structures. Many syntheses, especially those of COFs or amorphous porous materials, do not produce crystals of appropriate quality to be able to determine their structures using single crystal X-ray diffraction, and in those cases computation is an invaluable tool to infer material structure in conjunction with experimental characterisation.

The task for the computational materials scientist, in all these cases, whether simulating one structure or many, is to begin with knowledge of only the starting materials and from this, predict the structure formed. This chapter starts from that basis and discusses the methods that one might use to predict the structure of porous materials, both crystalline or amorphous, in 0 to 3 dimensions.

We start with the *building blocks*.

2.1.1 Building Blocks

As molecules are composed of atoms, molecular materials are composed of molecular fragments, often termed Secondary Building Units (SBUs) or, more generally, building blocks. To computationally build the structures of porous materials, therefore, the first requirement is to define a library of building blocks that can be attached together in some way analogous to assembling atoms into a molecule. A small example of such a library is shown in Figure 2.1.

Even from a small library, with only tens or hundreds of building blocks, the number of framework structures that can be formed by combining two

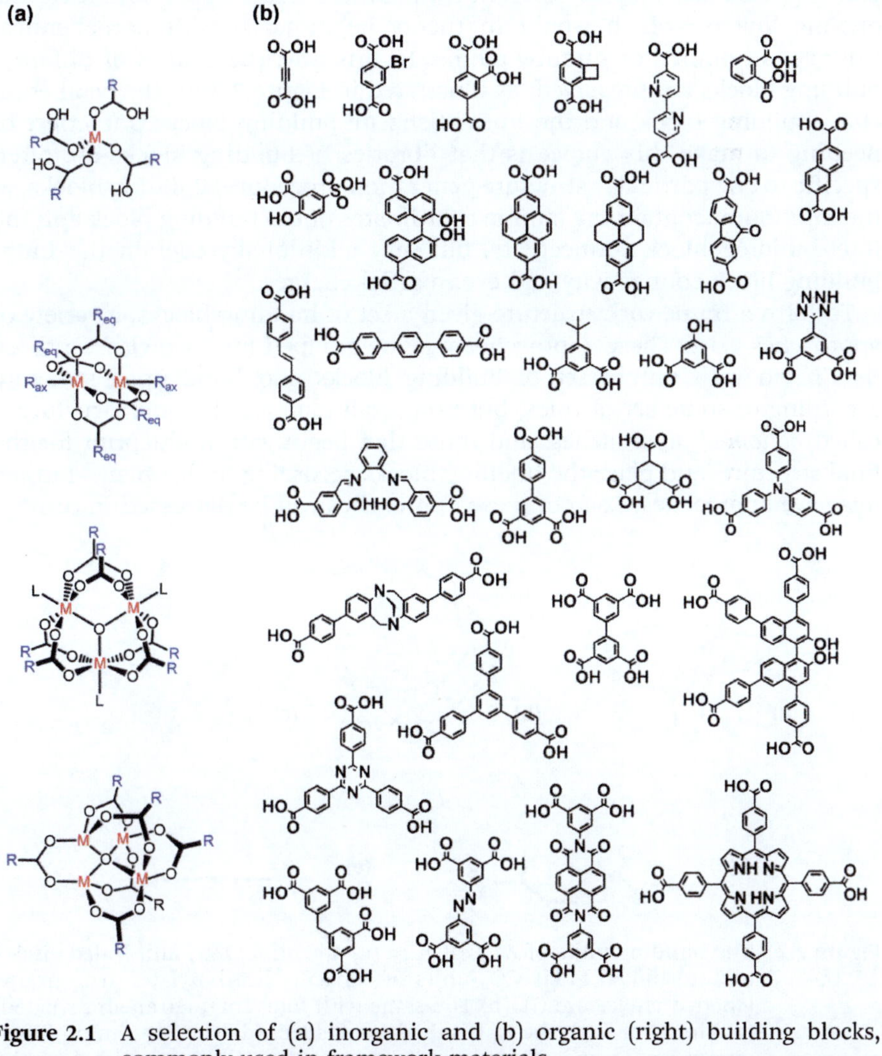

Figure 2.1 A selection of (a) inorganic and (b) organic (right) building blocks, commonly used in framework materials.
Adapted from ref. 51 with permission from American Chemical Society, Copyright 2012.

(or more) building blocks easily reaches into the millions. Given that the organic component(s) of framework structures may be derived from the entire universe of organic chemistry, the number of possible framework structures is truly vast. A recent study searched a space of 1.65 *trillion* hypothetical MOFs in order to identify MOFs that could be applied to CO_2 uptake.[42]

While in Figure 2.1 the organic building blocks are shown with carboxylate ends, as carboxylate species are commonly used in the synthesis of MOFs, it is not a requirement of any computational structure building program to exactly mirror the reagents used in synthesis. Especially in building MOFs and COFs, it is common to incorporate the carboxylate groups with the metal atoms (which are normally derived from a metal salt reagent) and leave the organic linkers (which would in theory be radicals) with non-chemical 'connection points' or dummy atoms. In this way, the shapes of different building blocks are preserved, as illustrated in Figure 2.2 for the octahedral OZn_4 building block and the linear benzene building block. One effect of needing to make this choice is that libraries of building blocks are often specific to one particular structure generation procedure. Building blocks, at the minimum, contain the atomic coordinates of the building block and the inter-building block connectivity, but may additionally contain the intra-building block connectivity and even partial charges.[52]

To form a framework structure given a set of building blocks, a variety of approaches exist. These approaches may be grouped into two classes: those that begin with only a set of building blocks and build up a structure according to some set of rules, but no overall plan for the final structure—called *bottom-up* approaches; and those that begin with a blueprint for the final structure, and place the building blocks according to that plan—termed *top-down* approaches. Each of these approaches will be discussed in turn.

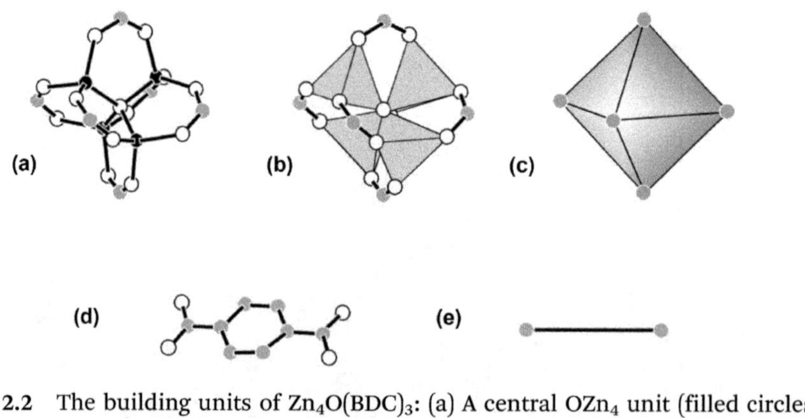

Figure 2.2 The building units of $Zn_4O(BDC)_3$: (a) A central OZn_4 unit (filled circles are Zn) linked to six CO_2 units of benzene dicarboxylate, BDC groups (shaded circles are C). (b) The same with four ZnO_4 tetrahedra shaded. (c) The same with the C_6 octahedron shaded. (d) The BDC anion (open circles are O and shaded circles are C). (e) The same with just the carboxylate C atoms joined by a link.
Reproduced from ref. 53 with permission from Elsevier, Copyright 2000.

2.1.2 Bottom-up Approaches

The first class of approaches, termed *bottom-up* approaches, require only 'local' information about the building blocks themselves and their connectivity. A building block employed in a bottom-up procedure contains information about the atomic geometry plus information about the points of connection, including both the connection site itself and the possible orientations of the building block, as shown in Figure 2.3. Optionally, building blocks may also contain information about which connection sites on the building blocks may combine with each other—*e.g.* The two open metal sites of a paddlewheel building block require different connecting atoms to the four carboxylate connection points.

Given a library of suitably encoded building blocks, a bottom-up algorithm follows, in some form, the following steps:

1. Choose an initial building block from library.
2. Choose a second building block from library.
   (a) Choose a connection point on both building blocks and, if appropriate, an orientation for the connection.
   (b) Check for overlap of atomic coordinates, and if none, connect the building blocks. If overlap is found (*i.e.* two building blocks partially coincide), choose a new orientation, connection point or building block.
   (c) If no building block can be placed, but the structure is not complete, then discard partial structure and begin again at step 1.
3. Continue choosing and connecting building blocks until no further building blocks can be placed.

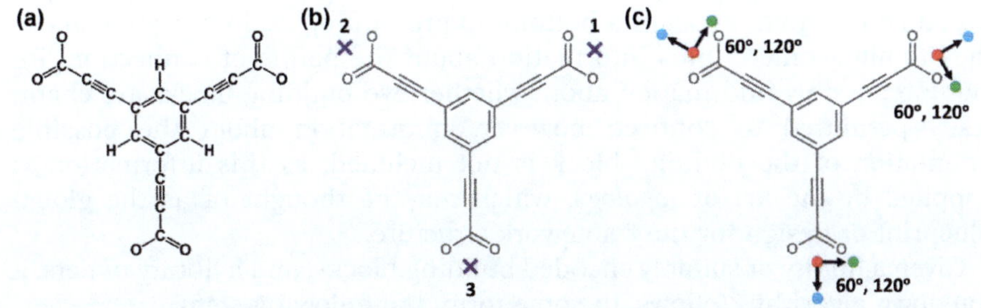

Figure 2.3 A building block encoded for use in a bottom-up structure generation procedure including (a) the atom composition and geometry, (b) topological information *via* numbered connection sites, and (c) geometrical information in the form of orientation vectors. To allow for rotational degrees of freedom (*i.e.* for building blocks that can rotate relative to one another), a list of angles for alternative orientations is also included. Reproduced from ref. 46 with permission from Springer Nature, Copyright 2012.

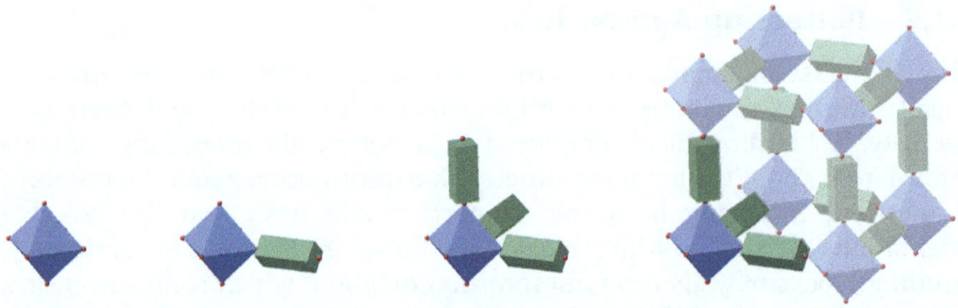

Figure 2.4 Schematic showing the bottom-up approach to building hypothetical framework structures. Individual building blocks are 'snapped' together at their connection points (red spheres) until they form a periodic connected structure. In the final picture, the pale building blocks are periodic images.

A bottom-up procedure is illustrated schematically in Figure 2.4. The termination of a bottom-up structure generation procedure depends on the desired structure. If a crystalline framework structure is desired, then all connection points of each building block should be occupied, and at some point in the procedure, three periodic boundaries need to be imposed by connecting two already placed building blocks. In the case of building an amorphous structure, the generation procedure may terminate when there is no further void space to place new building blocks or when a desired structure density is achieved.

2.1.3 Top-down Approaches

In contrast with bottom-up approaches are *top-down* approaches. Top-down approaches employ a library of building blocks as well as a library of *nets* or *topologies*. Building blocks encoded for a top-down procedure require largely the same information as for a bottom-up procedure, *viz.* information about the atomic geometry plus information about the points of connection, including possibly information about whether two building blocks are chemically permitted to connect; however, information about the possible orientation of the building block is not included, as this information is supplied by the *net* or *topology*, which may be thought of as the global blueprint or design for the framework structure.

Given a library of suitably encoded building blocks, and a library of nets, a top-down algorithm follows, in some form, the following steps:

1. Translate the net definition, and determine the 'shape' of each vertex if necessary—*e.g.* a 4-connected vertex may connect in a square planar geometry and therefore represent an appropriate slot for a paddlewheel building block, or it may connect in a tetrahedral geometry and therefore accommodate the tetraphenyl methane linker of COF-102.[54]
2. Scale the net to the appropriate size to fit the desired building blocks.

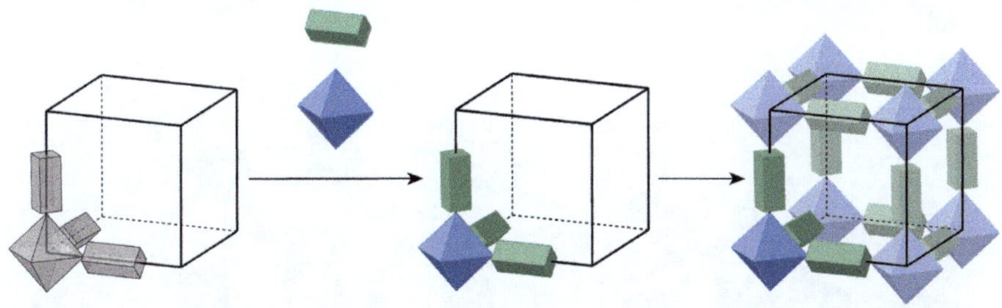

Figure 2.5 Schematic showing the top-down approach to building hypothetical framework structures. On the left a net with four empty (grey) slots, building blocks with matching connectivity are inserted in the slots. In the final picture, the pale building blocks are periodic images.

3. Insert the building blocks into the correct slots.
4. Create inter-building block bonds.
5. Refine the structure.

A top-down procedure is illustrated schematically in Figure 2.5.

Given the importance of nets to the top-down construction of porous framework materials, it is useful to spend some time considering them.

2.1.3.1 What is a Net?

The concept of a *net* is central and fundamental to reticular chemistry and has been well-known since the work of Wells[55] and O'Keeffe in the 1970s and 1980s.[56] Since then, several articles and reviews,[57–59] aimed at both mathematical and chemical readers, define the concepts and their realisation in crystal chemistry and the interested reader is referred to those resources. For the purpose at hand, understanding how a net may be used in order to predict structures of (in the vast majority of cases) crystalline compounds, and some of the caveats that arise, a brief introduction follows.

If one considers the two structures in Figure 2.6, diamond (C) and sphalerite (ZnS), it is clear that their atomic composition is different – diamond has only a single element but sphalerite has two elements, the size of each unit cell is different (*viz.* 3.56 Å and 5.41 Å, for diamond and ZnS, respectively), yet it is clear from a simple visual inspection that they somehow 'look the same', in that the components of both structures have the same connectivity. This underlying connectivity, which is a mathematical entity, independent of the chemical identities of the components, is called the *net*. The study of chemistry and material having the same net is *reticular* chemistry and any number of structures that have the same net are said to be *isoreticular*.[60]

A net is often referred to as a 'special kind of graph'. It ignores the atomic composition of the structure and focuses only on the overall connectivity of

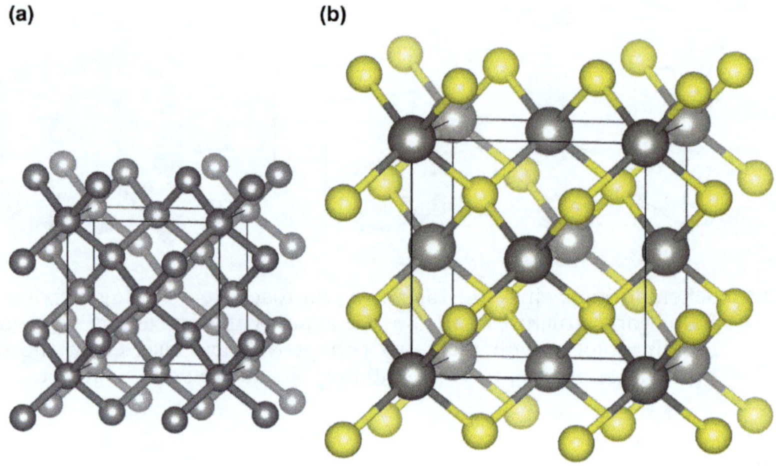

Figure 2.6 (a) The unit cell of diamond and (b) the unit cell of ZnS, sphalerite.

the structure.[61,62] In graph theoretical terms, a net is *simple*, meaning that any pair of vertices are linked by at most one *undirected* edge and there are no edges that link any vertex to itself. A net is also a *connected* graph, meaning that every vertex in the net can be reached by a continuous path of edges, without 'backtracking'. A net may be finite (*i.e.* zero-periodic, a polyhedron), one-, two- or three-periodic.[63]

An *embedding* of a net is a realisation in space of the net, where each vertex is assigned coordinates and the 'size' of the net is fixed by lattice parameters. Of most relevance to crystal chemistry, a *faithful embedding* of a net is where all vertices are distinct points and edges do not intersect or touch vertices. Two further specific embeddings are of particular importance to crystal chemistry: firstly, if a net has an embedding where all edges are of equal length and that length corresponds to the shortest distance between vertices, this is a termed *sphere packing*. Secondly, the most common embedding used in chemistry is the *equilibrium* or *barycentric* placement. In the barycentric placement, all vertices and edges of the net are placed to achieve the maximum possible symmetry and the coordinates of each vertex are the average of the coordinates of its neighbours. Most databases of nets (see Section 2.1.3.3) present barycentric embedding.

These definitions and features of nets have some direct effects in crystal chemistry. One example of this is the MFU-4 MOF,[64] which crystallises in a cubic fashion. The octahedral inorganic unit is the $[M^{oct}M_4^{tet}Cl_4]$, 'Kuratowski-type' building block and the linear linker is the benzobistriazolate building block. In this case, while the organic linker connects in a linear manner, two bonds are formed between each octahedral and linear building block. This violates the property of nets that multiple edges are not permitted between vertices, and so rather than a primitive cubic net (**pcu**), MFU-4 can best be described by the **snf** net.

2.1.3.2 Deriving Nets for Chemistry

From the definition of nets above (Section 2.1.3.1), it should be clear that the spatial distance between building blocks (vertices) is irrelevant when determining the net corresponding to a given crystal structure – *i.e.* many structures can and do exhibit a given net. Importantly, and particularly for the case of framework crystalline materials composed of molecular units (*e.g.* MOFs and COFs), the net is not uniquely defined for a given crystal (atomic) structure – *i.e.* one structure may be described by more than one net, depending on how the periodic structure is broken into fragments. In this case, where a different decomposition of the same crystal structure results in a different net, it is clear that the nets themselves must be related. A notable example is the PCN-61 MOF,[65] which may be decomposed into a 4-connected unit (the paddlewheel) and the 6-connected organic organic linker, in which case having the **nts** net. However, the organic linker is commonly split into two distinct 3-connected (triangular) building blocks and, in this case, the structure is described by the **ntt** net. This approach of combining smaller building blocks into one large building block can also be employed to generate very highly connected nets – *e.g.* the (3,24)-connected **rht** net.[66]

Two software packages are of note at this point. Firstly, Systre[67] is a program designed to analyse nets as they occur in crystal chemistry. The ToposPro[68] program is a very powerful package, and one of its key functions is that it can decompose a given crystal structure into its fragment components and then analyse the connectivity of those fragments to determine the net. Both of these software packages work with the sources of nets in Section 2.1.3.3.

2.1.3.3 Sources of Nets for Chemistry

The *Atlas of Zeolite Framework Types*[69] and its online counterpart, the Database of Zeolite Structures[48] was designed and implemented in 1996 by Baerlocher and McCusker on behalf of the Structure Commission of the International Zeolite Association (IZA-SC). Structures in the Atlas/Database are derived from real (natural or synthetic) zeolites by assuming a possibly hypothetical, SiO_2 composition and fixing bond lengths to standard values. Structures in the Atlas are assigned a three letter framework type code, printed in bold, uppercase letters, and normally related to the name of the material that was the first (proto) type of the structure (referred to as a 'type material').

To date, the Database of Zeolite Structures contains 241 ordered and 11 partially disordered zeolite framework types.

The Reticular Chemistry Structure Resource (RCSR),[70] available online at http://rcsr.net/ contains 0-, 1-, 2- and 3-periodic nets, separated in each case into 'default' (non-woven) nets and 'weaving' nets. The RSCR follows the *Atlas of Zeolite Framework Types*,[69] in assigning each net a three letter identifier code, with an optional extension. RCSR net identifiers are always printed in boldface lowercase letters (*i.e.* the diamond net of connected tetrahedra is indicated by **dia**). While many nets have been derived by systematic mathematical enumeration,[67] which is beyond the scope of this book, many nets have been derived from known chemical compounds, in which case their three letter identifier relates to the source compound, *e.g.* **dia**, diamond; **qtz**, quartz; **cds**, $CdSO_4$. It is important to note that the topologies of zeolites appear in both the *Atlas of Zeolite Framework Types* and the RCSR, but they may not necessarily have the same code in each. For example, the **ABW** net in the *Atlas of Zeolite Framework Types* is given the RCSR symbol **sra** as it is derived from $SrAl_2$. Four RCSR-encoded nets, related to the diamond (**dia**) net are shown in Figure 2.7 and it is useful to understand what each of them represent.

The **dia** net is the base net and corresponds to the structure of diamond, each vertex in the net is connected to four others in a tetrahedral geometry and all vertices and edges are equal. An *augmented* net is derived from the

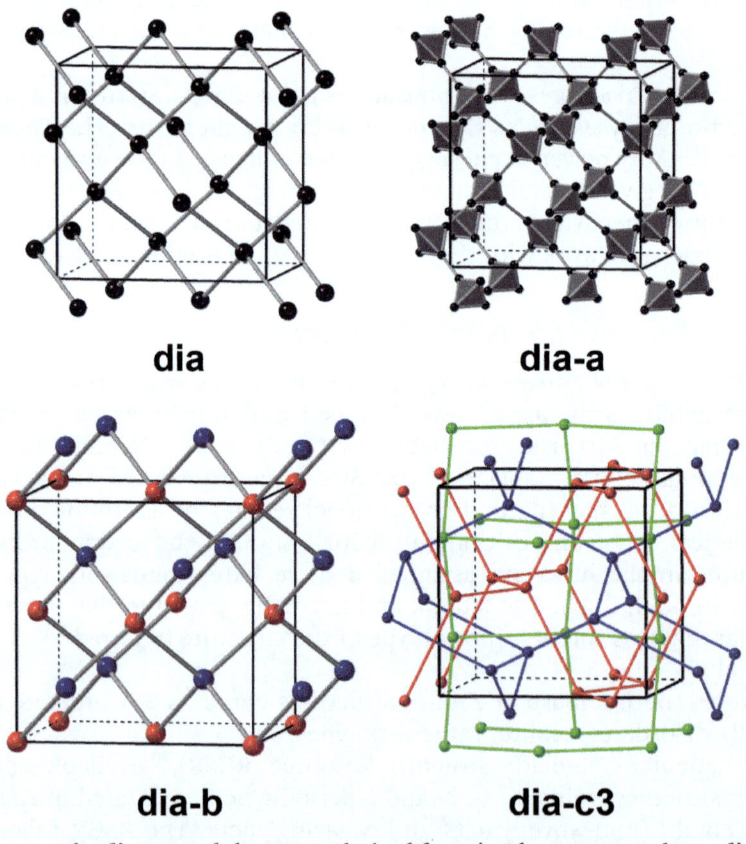

Figure 2.7 The **dia** net and three nets derived from it: The augmented net, **dia-a**, the binary net, **dia-b** and the three-fold catenated net **dia-c3**. In Figure 2.6, diamond has the **dia** net, while sphalerite is an example of the **dia-b** net.

base net, by taking each vertex and replacing it with a polygon or polyhedron corresponding to the geometry of the vertex – *i.e* replacing each tetrahedral vertex in the **dia** net and with a *tetrahedron*. Such augmented nets are indicated with the suffix **-a** applied to the original symbol, in this case **dia-a**. Augmented nets are particularly useful to visualise synthesised framework structures or indeed to target particular syntheses, where the molecular building blocks employed correspond to the polygons/polyhedra and the bonds between building blocks represent the non-polyhedral edges. A *binary* net is derived from the original net by colouring vertices (corresponding to different building blocks), such that no vertex is bonded to a vertex of the same colour. In Figure 2.7, the **dia-b** net is shown with each 'red' vertex being connected to four blue vertices and vice-versa. Binary nets may also extend to three more components, in which case an additional number is added to the symbol, *i.e.* the **dia-b3** net has the same geometry as the **dia** net, but three distinct 'colour' vertices. Where multiple possibilities for colouring exist, an asterisk is further added to the symbol, *i.e.* the **dia-b3*** is a second colouring of the **dia-b3** net. One further type of derived net, which occurs commonly is the *catenated* net, indicated by a **-c** suffix, *i.e.* **dia-c**. Catenated nets contain one or more distinct nets intergrown in the void space of the other(s), the vertices in each net connect only to others in the same net – *i.e.* in Figure 2.7, which shows the three-fold catenated **dia-c3** net, each 'red' vertex only connects to other red vertices and so on. As in the symbols for binary nets, a number is affixed to the symbol to show the number of catenated nets and where there are multiple different geometry catenations with the same number of nets, they are indicated by an asterisk.

To date, RCSR currently contains approximately 2900 3-periodic, 200 2-periodic, 8 1-periodic and 130 0-periodic non-weaving nets.

A third source of nets is the Euclidean Patterns in Non-Euclidean Tilings (EPINET) project[71,72] Unlike the methods largely employed to enumerate nets in the RCSR,[73,74] which determine 3-dimensional nets directly, EPINET uses a complementary approach, whereby nets are built in 2-dimensional non-Euclidean space and are then projected into 3-dimensional space. This approach produces many nets that direct approaches do not (and indeed, produces many nets that are unlikely to be relevant to crystal chemistry[62]), while reproducing many nets featured in the RCSR. EPINET nets are identified by 'sqc' followed by an ordinal number – *e.g.* sqc183, which corresponds to the **pts** net in the RCSR.

The EPINET project has currently enumerated 14532 nets.

2.2 Structure Generation of Crystalline Network Materials

Several codes exist to generate structures of crystalline network materials and have been published. In 2004, the Automated Assembly of Secondary Building Units (AASBU) approach that had previously been designed for inorganic structures[75] was adapted to hybrid organic–inorganic frameworks.[76,77] The AASBU approach uses building blocks that contain 'sticky atoms', which repel like building blocks and attract, unlike building blocks, *via* a Lennard-Jones potential. The building blocks are placed randomly in a simulation cell and then crystallise into an ordered structure *via* a simulated annealing Monte Carlo procedure.

The first published framework generator explicitly designed for molecular framework materials was published in 2012.[46] It employed a bottom-up, 'tinker-toy' algorithm to generate hypothetical MOF structures. Using 102 different building blocks, comprised of 5 inorganic connectors, 42 organic linkers (terminated by either a carboxylate group or nitrogen atom) and 13 functional groups, 137 953 hypothetical MOF structures, including 54 617 interpenetrated MOFs, were generated and screened for their methane storage capacity. Later analysis[78] of these structures revealed that only six nets – **pcu, sra, dia, tbo, nbo** and **fcu** – were represented, and that the **pcu** net accounted for 90 of the hypothetical frameworks. Two reasons were cited for this lack of topological diversity: firstly that the database of building blocks contained limited structural diversity – *e.g.* of the 5 inorganic connectors, only 3 were topologically distinct and 64% of the linkers were 2-connected (ditopic) linear linkers. Secondly, their algorithm restricted the selection of building blocks to those of the same connectivity. A third consideration, not highlighted in the analysis of Sikora *et al.*, relates to the complexity of the net: as the number of building blocks required to create the net increases, the probability of forming that net correspondingly decreases. The highly complex **mtn-e** net of MIL-101[79] contains 272 inorganic nodes and 816 ditopic linkers, and it seems exceedingly unlikely that all 1088 building blocks could be placed correctly without a top-down view of the desired design. Indeed, the hypothetical MOF structures generated by Wilmer *et al.* contain a maximum of 4 ZnO_4, 12 Zn/Cu paddlewheels, 8 V_2O_3 and only 2 Zr_6O_6 building blocks.[80]

2.2.1 Software

For this reason, subsequent published structure generation programs for crystalline network materials have all employed a top-down approach. Several groups have published codes, with varying availability. At the time of publication these include: Weaver by Schmid and co-workers,[81] Zeo++ by Martin and Haranczyk,[82] the Automated Topological Generator For Framework Structures (AuToGraFS)[83] and the Topologically Based Crystal Constructor (ToBaCCo) by Gómez-Gualdrón and co-workers.[84]

2.2.2 From a Net to a Crystal Structure

2.2.2.1 Choice of Net

With thousands of nets available in the RCSR alone, at first glance it appears daunting to search every possible net, for every combination of building blocks. However, the task of identifying the correct net is made simpler by the observation that 'simpler' nets are more likely to form. A typical synthesis generally employs a single type of linker, which is analogous to a net with a single type of edge. Such nets are termed *edge-transitive*[85] and it has been shown that edge-transitive nets are most likely to form given a particular combination of building blocks.[86] Likewise, using one high-coordinated building block for the node favours nets with only one type of vertex. Indeed, an analysis of nets synthesised up to 2009 showed that only three edge-transitive nets (**pcu**, **dia** and **bcu**) accounted for over 15% of published structures and that the 20 most common nets all possessed low (≤3) transitivity,[87] and an analysis five years later estimated less than 10% of MOFs did not follow the principle of minimal transitivity.[88]

2.2.2.2 Building Blocks and Embedding

The crystal structure is one of infinite *embeddings* of a net in Euclidean space. To generate a crystal structure from a net, one somehow needs to restore/recreate the chemical information with respect to the atoms (*e.g.* atomic positions, interatomic distances and bond and torsion angles) as well as the lattice (*e.g.* lattice parameters). The relationship between a net and the crystal structure is shown in Figure 2.8.

Considering first the building blocks: as the same building block can appear in many places in many different framework structures, most structure generators begin with a library or database of building blocks, leading to two distinct sources of error in the final structure; errors in the geometry of the building block itself and error in the bonding between building blocks. It is important to note at this point, computationally built structures are almost always generated in the $P1$ space group (*i.e.* with no symmetry).

As previously discussed, the building block data contains, at a minimum, the Cartesian coordinates of the building block, and often the internal bonding data as well. The structure generator published by Wilmer *et al.*[46] used building blocks cut from the crystal structures of real MOFs to create hypothetical MOFs. More commonly, building block databases contain symmetrised building blocks. The reason for this choice, and how the

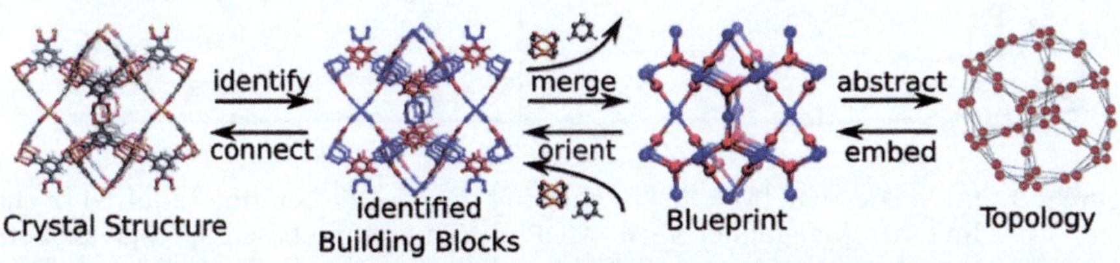

Figure 2.8 Relationship between a topology and a crystal structure *via* its embedding.
Adapted from ref. 89 with permission from the Royal Society of Chemistry.

building block geometry affects the final structure, may be seen in the example of the square-octahedral **soc**-MOF. MOFs with the **soc** topology are constructed from two components, a 6-connected (octahedral) μ_3-oxygen-centred trinuclear metal carboxylate cluster, $[M_3O(O_2C-)_6]$, and a 4-connected (square) linker; 3,3′,5,5′-azobenzenetetracarboxylate ($ABTC^{4-}$) linker. **soc**-MOFs have been synthesised for the trivalent metal ions, M = Al, Fe, Ga and In.[90] Using the same linker but different metal ions leads to lattice constants of 21.49, 21.95 and 22.46 Å for Al, Fe and Ga/In, respectively. Given these different lattice constants, it is clear that the linker is required to bend to a greater degree in Al-**soc**-MOF, as shown in Figure 2.9a–c. The strong curvature of this linker highlights (a) how a building block extracted from one structure may not fit well in a second, even related structure and (b) that a 'preformed' linker would require an extra orientation step in order to be oriented correctly (*i.e.* that all linkers bent either outwards or inwards with respect to a given pore). For these two reasons, the symmetrised linker

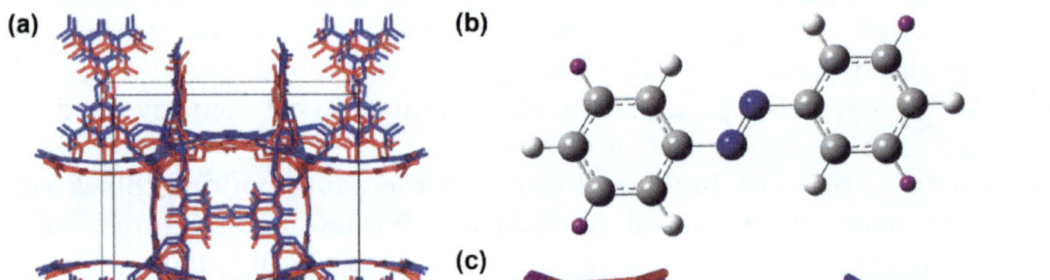

Figure 2.9 (a) M-**soc**-MOF (M = Al, In) showing the greater bending required of the linker in the smaller Al-**soc**-MOF. (b) The symmetrised 3,3′,5,5′-azobenzenetetracarboxylate ($ABTC^{4-}$) linker. (c) The linker extracted from Al-**soc**-MOF (red), In-**soc**-MOF (blue) overlaid on the symmetrised linker (grey), aligned at the azo functionality. (d) The **soc** net. (e) AuToGraFS-assembled **soc**-MOF, prior to optimisation.

is typically the 'neutral' choice, yielding the most acceptable error in the generated structure.

The second source of error is that of the bonding between building blocks. There are two underlying causes of errors in joining building blocks together. Firstly, as above, individual building blocks are supplied in a symmetrised geometry, in effect distorted from their native conformation. The second source of error arises from the embedding. As illustrated in Figure 2.8, the net (topology), which is simply a graph, is converted first to a blueprint – a set of slots for the building blocks. The most common method employs the maximum symmetry barycentric embedding of the net and simply scales all lattice vectors isotropically. This approach works remarkably well in simple cases, especially for so-called *edge transitive* nets, however, as shown in Figure 2.9d–e, for more complex nets, placing and aligning the building blocks in the blueprint slots results in a poor alignment of the connection points of the building blocks.

There are several approaches to remedying this error. In AuToGraFS,[83] the desired bonds between building blocks are defined, regardless of the degree of geometric fit, and these initially generated structures are then optimised using a purpose-built extension to the Universal Force Field, UFF4MOF.[91,92] Boyd and Woo developed a method using labelled quotient graphs to bypass the barycentric embedding of the net and generate a net where the geometry of the nodes fits exactly with the geometry of the building blocks. An example of the kind of net generated by this approach is shown in Figure 2.10.

A third approach, similar to that of Boyd and Woo to solving the geometric error introduced by embedding, is the TopoFF (Topological Force Field)

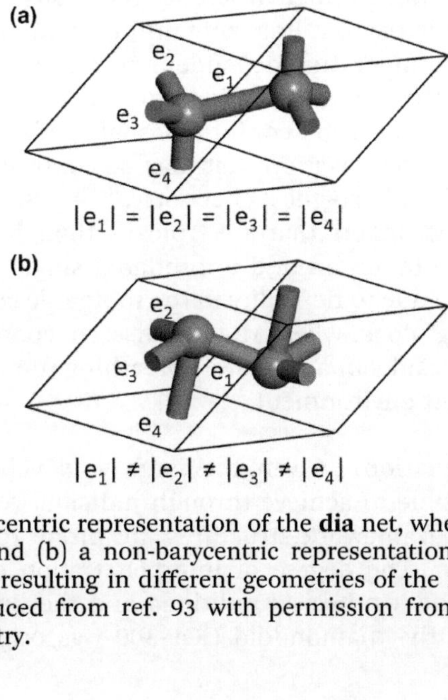

Figure 2.10 (a) Barycentric representation of the **dia** net, where all edge lengths are equal and (b) a non-barycentric representation with non-equal edge lengths resulting in different geometries of the nodes.
Reproduced from ref. 93 with permission from the Royal Society of Chemistry.

approach of Schmid and co-workers.[89] TopoFF begins with the default, isotropically-scaled barycentric embedding of the net, but then alters both the positions of the vertices, and the lattice parameters of the blueprint in order to minimise the RMSD of the coordination environment of each vertex (slot) in the net and the geometries of the building blocks.

2.2.2.3 Isomerisation

Even given the geometry and connectivity of the building blocks and a blueprint optimised to the dimensions of the building blocks, there are several forms of isomerisation possible in framework structures, and the structure of the framework may not be uniquely defined. Zaworotko defined *supramolecular isomerism* as the "existence of more than one type of network superstructures for the same molecular building blocks".[94] A consideration of the reverse problem – how to distinguish the structures of MOFs[95] developed a set of 'invariants' – structural descriptors for framework structures, which included atom types, ligand graph, ligand coordination mode, and several properties derived from the bond network and from several of its simplified versions, such as the dimensionality of the net, its topological indices and possible interpenetration. But even the sum of those descriptors does not yield a complete invariant – *i.e.* structures such as stereoisomers still could not be distinguished and thus would be incorrectly identified as identical.

Isomerisation due to building block flexibility. The molecular units used as building blocks in the construction of molecular framework materials may retain non-translational degrees of freedom that can influence the structure of the framework under different conditions. For instance, even after the connectivity is fixed, a linker retaining rotational degrees of freedom[96] can result in a temperature-dependent pore geometry due to rotation of the linker, or part thereof, and separate to any expected thermal expansion. Building blocks that exhibit conformational flexibility provide another means for framework isomerisation that may be controlled by external factors such as pressure[97] or host–guest chemistry,[98,99] as shown in Figure 2.11. While it is the organic linkers that are typically thought of as flexible, especially through the rotation around unhindered single bonds, framework isomers may also arise due to flexibility in the inorganic component, such as in the Bi_2O_{12} building block, where the nondiscrete coordination geometry around the Bi cation results in a reversibly breathing framework structure in response to the solvent environment.[100]

Interpenetration/Catenation. Given the large void volumes (even up to 90%[101]) that are possible to achieve through judicious 'choice' of building blocks and topology, framework structures are prone to interpenetration, also called catenation. The degree of interpenetration observed is highly dependent on both the synthesis conditions and the building blocks employed. For example, the diamondoid COF-300 was originally synthesised

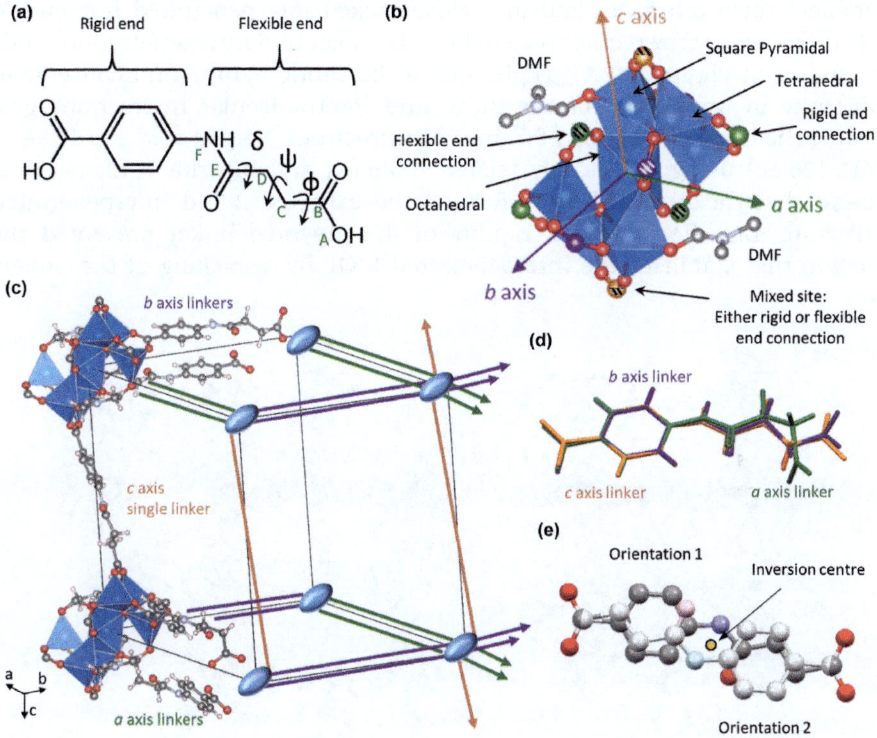

Figure 2.11 Impact of linker flexibility on the MOF unit cell. (a) Structure of linker with rotatable dihedral angles marked. (b) Structure and connectivity of an inorganic building block. (c) Unit cell, showing connection of the flexible linker along each crystallographic axis. (d) Conformers of flexible linker overlaid. (e) Two half-occupancy inversion-symmetry-related orientations of the *c*-axis linker.

in a 5-fold interpenetrated form,[102] but was later synthesised in a 7-fold interpenetrated structure.[103] These highly symmetric packings (catenations) are indicated by the RCSR symbols, **dia-c5** and **dia-c7**, respectively.

In addition to the degree of interpenetration, the packing of the interpenetrated frameworks can vary; indeed, relative motion of two or more interpenetrated frameworks is a well-known mechanism of framework flexibility and can play a key role in gas adsorption properties of the framework.[104] The packing of interpenetrated frameworks is often dictated by relatively subtle non-bonded interactions between linkers. For instance, Zhao and coworkers attached two crown ethers, benzo-12-crown-4 and benzo-15-crown-5, to the linkers of IRMOF-15, resulting in two doubly interpenetrated frameworks, MOF-BC-4 and MOF-BC-5, respectively. In both cases, the flexible crown ethers were too disordered to resolve by X-ray crystallography, the smaller crown ether resulted in a close-packed interpenetrated framework, while the larger crown ether resulted in a body-centred interpenetration mode as shown in Figure 2.12.[105] Highlighting the subtle nature (and consequent difficulty in prediction) of the intra- and intermolecular interactions governing the interpenetration of framework structures, Forgan *et al.* synthesised MIL-126(Sc) using the closely related biphenyl and bipyridyl linkers. They found that the biphenyl MOF formed the expected 2-fold interpenetrated structure, but the increased rigidity of the bipyridyl linker prevented the torsion that stabilises the interpenetrated MOF by π-stacking of the linkers

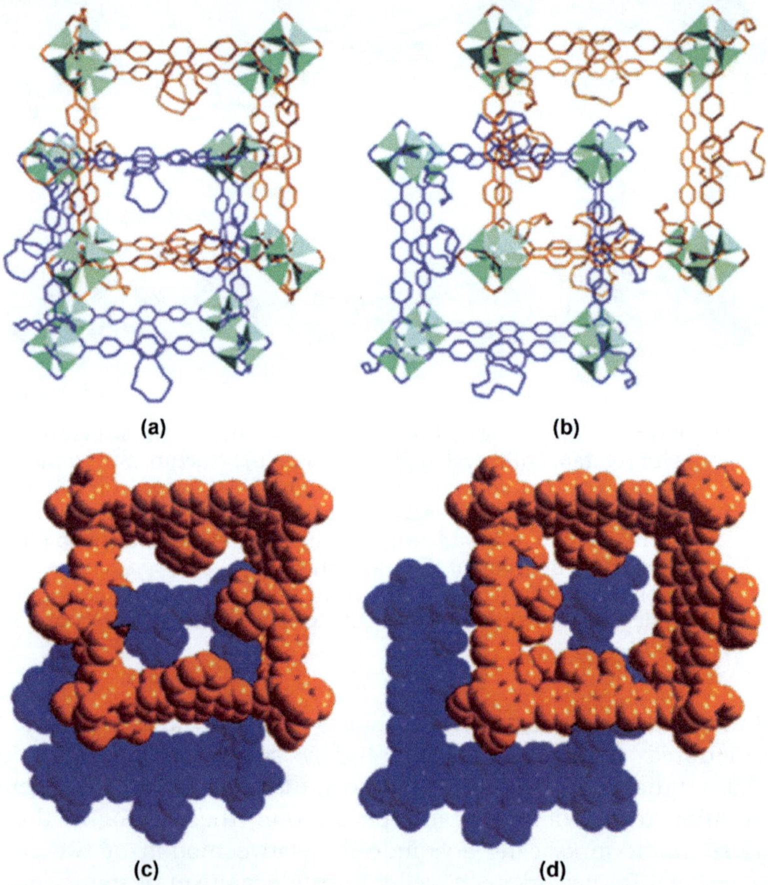

Figure 2.12 Crystal superstructures of (a) MOF-BC-4 and (b) MOF-BC-5. Hydrogen atoms and solvent molecules are omitted for the sake of clarity. Crown ether units with polyether chains and hydroquinone rings are coloured dark blue/brown, and the metal joints are coloured pale blue. The space filling representations for the crystal superstructures of (c) MOF-BC-4 and (d) MOF-BC-5.
Reproduced from ref. 105 with permission from the Royal Society of Chemistry.

and resulted in a non-interpenetrated framework.[106] While these effects were able to be rationalised after the fact, using data-mining and Density Functional Theory (DFT) calculations, they were surprising as several isostructural MOFs existed for the same pair of linkers.[107] Another synthetic strategy to obtain a non-interpenetrated MOF is to employ an organic linker with a bulky leaving group, such that interpenetration is sterically disallowed. In a subsequent step, the leaving group can be removed, resulting in a non-interpenetrated framework.

The examples above show the subtlety of even common types of interpenetration of 3D networks. In the case that single crystal X-ray structures are unable to be obtained, modelling will often proceed by building a non-interpenetrated net and then manually modelling various modes of interpenetration and comparing against, for example, a powder X-ray pattern.

1D and 2D networks may also interpenetrate and, unlike the case of 3D networks, which remain 3D when catenated, interpenetration of 1D and 2D networks can result in an increase in dimension of the overall structure – *e.g.* a 1D ladder may interpenetrate to form a 2D layer or a 3D network,[108,109] likewise, a 2D layer may interpenetrate to form a 3D structure.[110,111] 0D frameworks (as in Section 2.3), may also catenate and form higher dimensional structures. With reference to one of the most common nets for 2D frameworks, the square planar, **sql** net, Hyde and co-workers looked at three 2D-layer MOFs, MOF-14, MOF-388 and UTSA-28, and suggested an enumeration of possible weavings.[112] More generally, Proserpio and co-workers examined 1319 2-periodic interwoven (entangled) coordination polymers, characterising them by their Hopf and Extended Ring Nets (ERN)s.[113] The ERN is a powerful tool to characterise both any dimension increase and determine the degree and type of entanglement (interpenetration *vs.* polycatenation) as it contains both the periodicity of the overall ensemble and the periodicity of each component of the ensemble. They enumerated a total of 216 ERN topologically distinct modes of entanglement and coined the term *entanglement isomerism* to describe the phenomenon. In their analysis, they showed that 3/4 of the 1319 observed entangled structures were described by only 21 patterns – *i.e.* 10% of the extended ring nets they enumerated.

Isomerisation due to stacking. In addition to isomerisation within each layer, 2D layered networks also exhibit isomerism due to the non-covalent forces acting between layers.[114,115] Modelling 2D layered networks typically aims to distinguish between layers stacked directly on top of each other: eclipsed, AA stacking and layers stacked offset to each other by half the lattice vector; staggered or AB stacking. Hexagonal layers, in particular, may also exhibit ABC stacking. The stacking of layers is the result of a balance of London dispersion and Coulomb forces. The dispersion contribution, which dominates, is attractive between the layers, but causes a repulsive Coulomb force between the layers, especially in the region of highly polarised connectors between building blocks. The balance of these two forces results in a slightly offset stacking (slip-stacking) being most favourable, with layers typically offset by $\approx$1.4 Å with respect to the layer above/below. The majority of 2D layered frameworks stack in this straightforward manner and therefore the modelling of 2D layered networks typically only employs two or three layers. While computationally efficient, this approach does have limitations. Because the offset between layers typically has no preferred direction, 2D layered frameworks are likely to stack with offsets in different directions of each layer, so-called turbostratic disorder. Such disorder is not possible to model with only two layers, which, when stacked along the stacking direction, yield either a regular zigzag pattern or a continuous slanting pore.

In the case of modelling only the structure, this limitation can often be ignored, as the poor crystallinity of most layered frameworks means that structural determination typically relies on comparatively low-quality powder X-ray diffraction data, where the random offset of layers would only be observable as small effects on the relative intensity and width of some diffraction peaks. However, the chemical identity, and molecular structure of the building blocks employed in 2D layered frameworks can play a crucial role in directing the long-range structure of layered framework materials: stacking disorder can be reduced by careful engineering of interlayer interactions—*e.g.* employing sterically crowded, propeller-like building blocks as nodes or linkers can dictate the position of the adjacent layers[116,117] and similarly, a judicious choice of building blocks can favour interlayer hydrogen bonding, thus providing some degree of interlayer order.[118] Choosing a rigid planar building block over a structurally similar but non-planar building block has been shown to force layers to undulate, forming highly complex polycatenated stuctures.[119] On the other hand, highly flexible 2D layers, especially those that theoretically contain very large pores, may 'crumple' in a highly disordered manner as a way of achieving efficient space filling.[120]

In all these cases, when modelling adsorption in layered frameworks, it is necessary to employ a 'supercell' with a higher number of layers and ideally allow relaxation of the lattice parameters, which are likely to be altered by adsorption/desorption processes.[121]

Isomerisation over more than one unit cell. Employing a supercell in modelling crystalline framework materials leads to a further opportunity for isomerism and highlights a limitation of describing frameworks' structures using their net. The well-known MOF-5 (IRMOF-1) structure, shown in Figure 2.13, provides an example of this type of isomerism: MOF-5 crystallises in the $Fm\bar{3}m$ space group and the unit cell contains two pores, a small pore and a large pore, created by the alternate orientation of the benzene dicarboxylate (BDC) linkers (inward/outward facing). Topological analysis (*e.g.* with ToposPro[68]) does not capture this detail and assigns the framework a primitive cubic, **pcu** net ($Pm\bar{3}m$ space group), consisting of one octahedral building block and three linear building blocks. Thus, as shown in Figure 2.13c, recreating this deceptively simple MOF requires firstly a 2×2×2 supercell and secondly, controlling the orientation of individual linkers in order to satisfactorily reproduce properties such as pore and window sizes

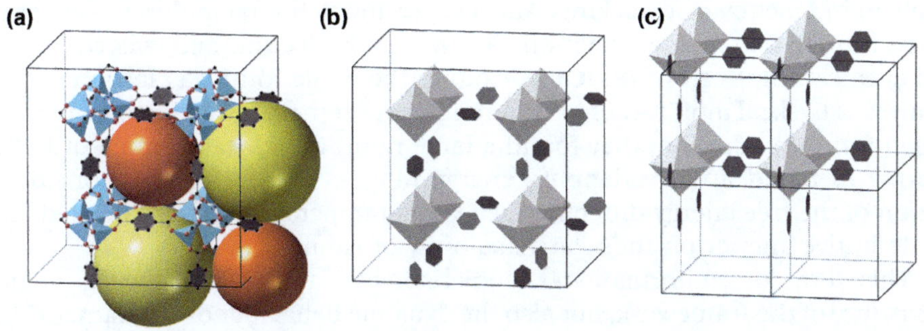

Figure 2.13 (a) The unit cell of MOF-5 shown with Zn_4O octahedra in blue and benzene dicarboxylate (BDC) linkers shown in grey. Consecutive BDC linkers in each dimension face in alternate directions creating a small pore where all linkers point into the pore, adjacent to a large pore, where all linkers point out of the pore. Yellow spheres indicate the large pore, and orange spheres indicate the small pore. (b) The unit cell of MOF-5 shown as pale grey octahedra and hexagonal linkers. (c) A 2×2×2 supercell of MOF-5 (**pcu**) as generated by AuToGraFS.[83]

and consequently dynamic behaviour of the framework. The two structures (shown in Figure 2.13b and c) represent supramolecular *conformational isomers* as they both possess the same connectivity but cannot be interconverted without breaking bonds. The effect of the linker flexibility on the conformational isomerism of several actual and hypothetical members of the IRMOF series has been extensively investigated.[122,123]

MOF-5 has the advantage that the BDC linkers are small and highly symmetric. The symmetry and rigidity of these linkers affects not only the linkers themselves, but has a broader effect on the crystal order, by dictating the orientation of the neighbouring Zn_4O building blocks: the two carboxylate units in BDC are almost always coplanar, therefore the two adjacent Zn_4O clusters are mirror images of each each other (alternately adjacent Zn_4O may be considered to be rotated 90° with respect to each other). In the case of a more flexible linker – *e.g.* the 1,4-dipropoxybenzene-2,5-dicarboxylate linker of IRMOF-4 – the two carboxylate groups are no longer necessarily coplanar and therefore, the constraint on the relative orientation of the Zn_4O building blocks is removed. As a result, in addition to disorder of the linker itself, there may be long-range disorder of the connecting building blocks.[124]

This effect is also seen in IRMOF-7, where repulsion between the carboxylic oxygen and a hydrogen atom from the naphthalene ring makes a planar structure of the linker energetically unfavourable. Disfavouring the planar conformer means that the two carboxylate groups may be either coplanar or orthogonal to each other. As a consequence, the single crystal structure of IRMOF-7 corresponds to a 1×1×1 cell with disorder in both the linkers and the Zn_4O building blocks. In large unit cells, moderate sized linkers may have the space to rotate freely; however, in small unit cells such as the unit cell of IRMOF-7 ($a = b = c = 12.914$ Å), the naphthalene linkers are able to interact by π–π stacking. Amirjalayer and Schmid employed a genetic algorithm to search conformational isomers of both cubic and twisted forms of a 2×2×2 unit cell of IRMOF-7, finding the cubic form was favoured but only a 2.04 kcal mol^{-1} energy difference between the global minimum of each form. In addition, they found a large number of local minima of both forms in broad and overlapping energy ranges, suggesting that consideration of the free energy difference at a given temperature may be needed for a definitive answer on the relative stability of isomers.

This form of conformational isomerism can affect not only the static structure of the framework, but also the dynamic behaviour of the framework. The breathing framework, DUT-8(Ni),[125] is a square layer pillar MOF that uses the same naphthalene linker as IRMOF-7 above, connected by nickel paddlewheels and each layer is separated by 1,4-diazabicyclo[2.2.2]octane (dabco) pillars. The same linker-based conformational isomerism, where the four linkers connected to each paddlewheel may be any combination of 'up' and 'down', leads to several possible conformational isomers. Petkov *et al.* considered two conformational isomers of DUT-8(Ni): firstly where all naphthalene linkers pointed the same way on each paddlewheel, and secondly where linkers alternated up/down on each paddlewheel, illustrated in Figure 2.14a (top two isomers). These isomers have very similar PXRD patterns in the open form, but strikingly different PXRD patterns in the closed form. When the framework closes, following a 'wine rack' motion, the first isomer produces an equivalent closed structure regardless of whether the framework is compressed left–right or top–bottom. However, in the second isomer, the alternating linker orientations mean that the framework has two inequivalent closed forms, where the naphthalene linkers in close contact may either point in the same direction, resulting in a high stabilisation due to dispersion, or they may point in opposing directions, in which case the stabilisation energy is low. Consequently, the closed state may or may not be a free-energy minimum, and therefore, the observation of collective flexibility in the framework depends quite sensitively on the isomer in question.

In addition to orientation, the identity of the building blocks may be changed. So-called multivariate MOFs are attractive for applications such as catalysis, where the incorporation of different functional groups can provide multiple functionalities to the pore. Multivariate MOFs are typically synthesised using a mixture of similarly-sized linkers – often the same linker with different functional groups,[127] and/or a mixture of metal salts. Given synthesis with a particular ratio of linkers, a crystalline material with a random distribution of those linkers is typically assumed to form as illustrated in Figure 2.15. However, while it is possible to find both short- and long-range order in multivariate MOFs,[128] it is not currently possible to synthesise a framework structure with a specific arrangement of functional groups.

Rather than substituting the linker or inorganic building block, changing the synthesis conditions can result in defect structures being produced. The two main types of defects not already covered are missing linkers and missing clusters (inorganic building blocks). These types of defects are most

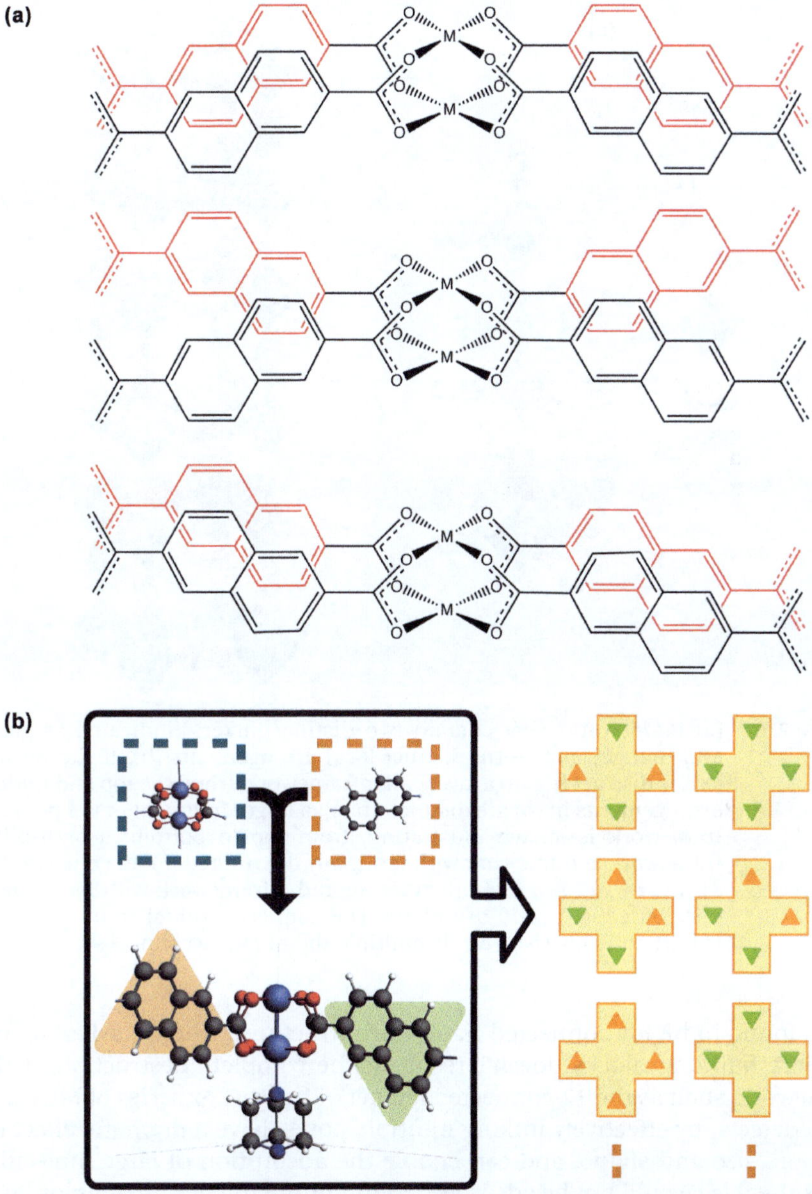

Figure 2.14 (a) Metal paddlewheel unit of DUT-8 with attached naphthalene linkers showing three different conformational isomers. (b) Schematic of how by considering different combinations of an upward pointing linker (orange triangle) and downward pointing linker (green triangle) over multiple unit cells (in each crystallographic dimension) results in multiple conformational isomers.
Adapted from ref. 126 with permission from the Royal Society of Chemistry.

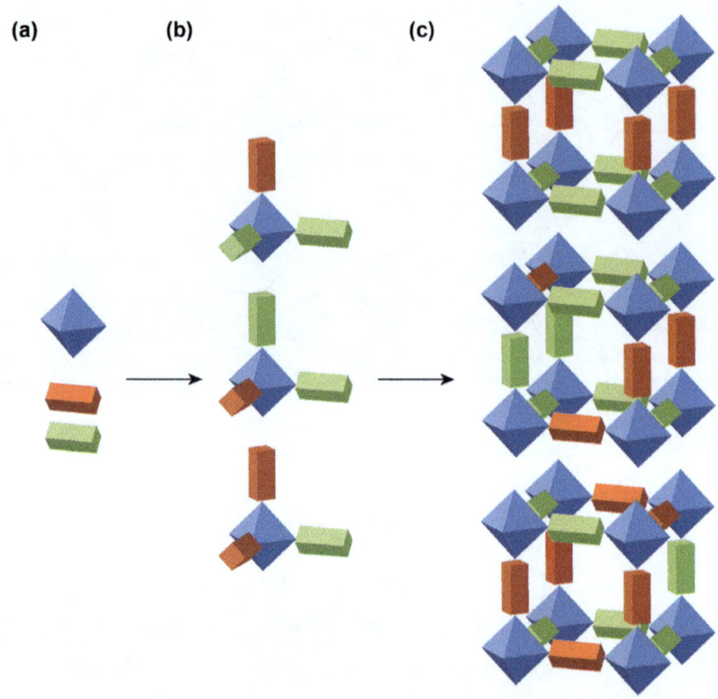

Figure 2.15 (a) Frameworks that contain two distinct linkers (indicated as green and orange), may form distinct local arrangements (b). If the vertical axis is distinct (*e.g.* in a layer-pillar framework), then the top and middle arrangements in (b) are distinct. In (c), a larger fragment of the periodic framework is shown illustrating, from top to bottom: a layer-pillar framework, a framework with randomly distributed linkers but with the same 2 : 1 ratio as the top structure and a framework with a randomly distributed 1 : 1 ratio of linkers. The connector (blue) could also contribute to isomerism–*e.g.* if multiple metal ions were present.

often found in highly connected framework structures, where the loss of one or more building blocks doesn't result in the complete destruction of the framework, such as the 12-connected **fcu** net of the UiO-6x series of MOFs.[129] Such defects, by effectively linking multiple pores, have a dramatic effect on the pore size and shape, and can enable the adsorption of large molecules that otherwise would not be adsorbed. Highlighting the need to employ large supercells in modelling defect structures, missing cluster defects have been shown to correlate in UiO-66, where regions of defects now possess the **reo** net form in amongst the regular **fcu** net of UiO (see Figure 2.16).[130,131]

2.2.3 Structure Generation of Zeolites

Zeolites are a special case of crystalline microporous framework. They are aluminosilicate frameworks composed of linked tetrahedra, connected by

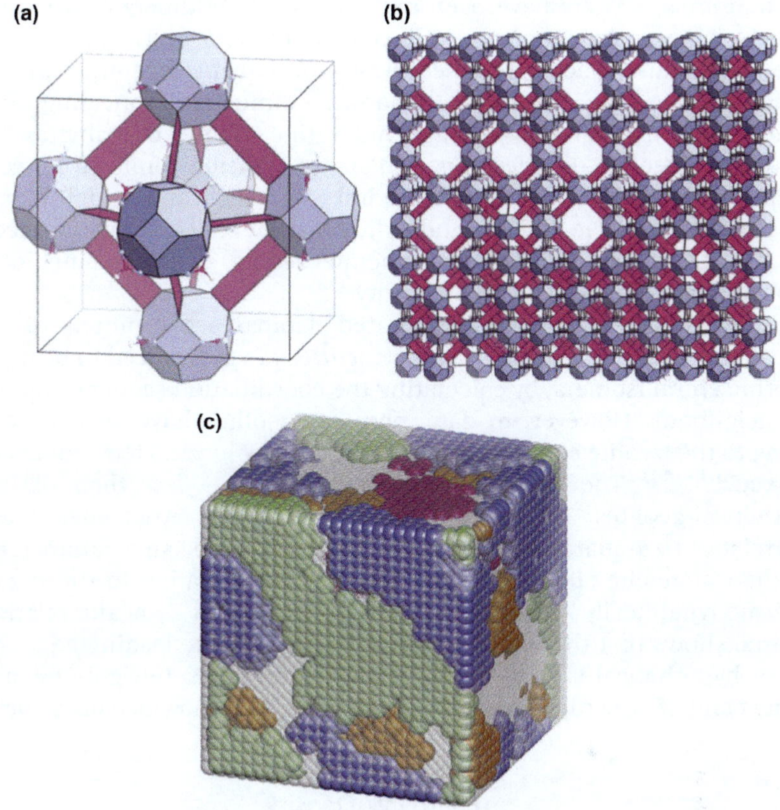

Figure 2.16 Structural description of defect nanoregions in UiO-66(Hf). (a) A polyhedral representation of a single-unit cell of the (ordered) **reo** defect structure. In this model, 25% of the cluster sites are vacant and the cluster coordination number is reduced from 12 to 8. The clusters occupy Nb sites in the defect-vacancy structure of the transition metal monoxide NbO. (b) A representative section of the supercell atomistic model. Defect-rich nanoregions are dispersed throughout a matrix of the defect-free **fcu** framework. (c) The defect concentration and domain size parameters determined experimentally can be used to generate atomistic representations of defect nanoregions in UiO-66(Hf). The representation given here is for a 6×6×6 supercell of the **fcu** unit cell (total edge length 12.4 nm), chosen for clarity to be smaller by a factor of 4^3 than the modelled configurations. There are four possible orientations of the **reo** defect structure relative to the bulk **fcu** framework, corresponding to the four cluster positions in the **fcu** unit cell. In the representation used here, individual clusters are not shown but nanodomains corresponding to each of these four orientations are represented by blue, green, orange and magenta regions. Grey-coloured regions correspond to the defect-free **fcu** framework.
Reproduced from ref. 130 with permission from Springer Nature, Copyright 2014.

oxygen atoms.[132] There have been many efforts to enumerate hypothetical zeolite structures by scanning unit cell parameters and crystallographic space groups and placing tetrahedral Si atoms using a Monte Carlo procedure, followed by molecular dynamics, simulated annealing and/or geometry optimisation to generate and refine atomic coordinates.[133–136] Zeolitic Imidazolate Frameworks (ZIFs) are typically built from corresponding zeolites, by replacing tetrahedral Si with Zn and bridging oxygen atoms with imidazole linkers and adjusting the cell parameters accordingly.[137] Other methods used to generate zeolite structures include the AASBU method[75] as well as graph theory.[138]

Together, these efforts have generated databases of millions of hypothetical zeolite structures;[48,139] these structures are guaranteed to be unique, excepting chiral isomers, by calculating the coordination sequence up to the 12th neighbour. However, to date, only 241 zeolites have been observed, leading to the 'zeolite conundrum'—namely, "Why have so few zeolites been observed?"[140,141] The subset of observed zeolites – less than 0.01% of hypothetical zeolites – lie in a well-defined region of lattice energy (calculated relative to α-quartz) *vs.* density.[142] Calculating the same parameters for hypothetical zeolite structures, and limiting consideration to those within a thermodynamically feasible window of + 30 kJ mol^{-1}, again relative to α-quartz, shows that the observed structures fall on the leading edge of the bulk of hypothetical structures, shown in Figure 2.17. Intriguingly, a low-density band of structures exists at 5 to 10 Si/Å^3. These low-density structures

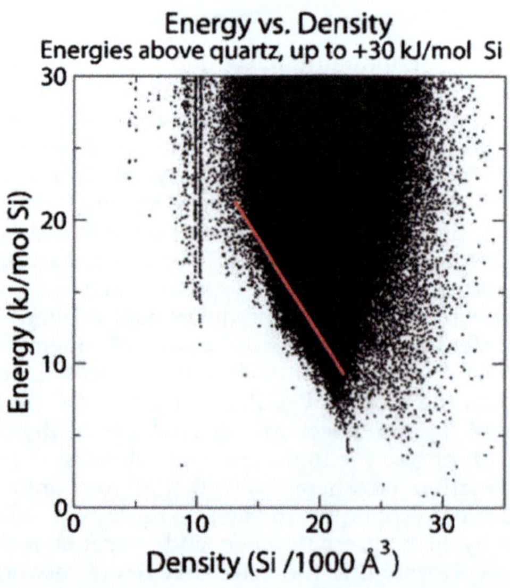

Figure 2.17 Energy *vs.* density plots of hypothetical zeolite structures. The red line indicates the least-squares fit to known zeolite structures.
Adapted from ref. 139 with permission from the Royal Society of Chemistry.

are of particular interest as they are expected to have large pores, making them ideal for adsorption and catalysis applications.

Recent efforts to resolve the zeolite conundrum have focused on high-throughput screening and characterisation[143,144] of these databases, identifying intermediate building blocks[140,145] and attempting to identify kinetic factors[146] determining the observed zeolites.

2.3 Structure Generation of Molecular Materials

Molecular cages, including Metal–Organic Polyhedra (MOPs) and Porous Organic Cages (POCs) can be computationally assembled in much the same manner as their periodic cousins. Because of the requirement to form a closed structure, the logical approach to forming porous molecular materials is a top-down approach, where building blocks are chosen for their connectivity and the geometric relationship between points of connection. Like in the nets employed when building 1-, 2- or 3-dimensional network materials, the geometric/topological design templates define the *relative* spatial positioning of molecular building blocks and their connectivity. When assembling molecular components into a closed assembly, the assembly may be a polygon, where all components are ditopic, or when one or more components are highly-coordinated (≥3), it is possible to form a polyhedral assembly. While the polygon in everyday experience is a 2D shape and a prism or a polyhedron is a 3D shape, in terms of their net/topology, they are both 0D – *i.e.* they are not extended structures. For this reason, it is perhaps clearer to speak of porous molecular materials as 0-periodic (and the crystalline networks above as 1-, 2- or 3-periodic), thus maintaining our common-sense definition of a molecular cage as a 3D object.

As previously noted in Section 2.1.3.3, the RCSR contains several 0-periodic nets, which could describe the topology of molecular cages. Where the relation to higher periodicity nets holds, the 0D nets composed of a single type of high-coordinated vertex and single type of edge (*i.e.* are vertex- and edge-transitive) would be the most likely synthetic targets. Of the 130 0-periodic nets in the RCSR, eight are composed of a single type of vertex and a single type of edge: the five regular polyhedra; tetrahedron (**tet**), cube (**cub**), octahedron (**oct**), icosahedron (**ico**) and dodecahedron (**dod**), plus two quasiregular polyhedra; cuboctahedron (**cuo**) and icosidodecahedron (**ido**). The eighth vertex- and edge-transitive net corresponds to Schoen's dodecahedron (hexagonal tiling, **hxt**) and is not typically a target for synthetic efforts. Two further polyhedra, the rhombic dodecahedron (**rdo**) and the rhombic triacontahedron (**trc**), correspond to the dual of the cuboctahedron and the icosidodecahedron, respectively; they both have two kinds of vertex, but only one kind of face. While synthetic efforts over many years have targeted these polyhedra, producing an array of truly beautiful structures, there are significant differences in forming a closed 0D cage structure compared to a structure of higher periodicity, which means that this description is limited when describing porous molecular materials. Firstly, the prohibition in a net of connecting the same pair of vertices by multiple edges excludes many possibilities for forming molecular cages. Secondly, cages may be formed in an edge-directed fashion, where building blocks serve as vertices and edges, but they may also form in a face-directed fashion, where building blocks serve as faces and vertices of the polyhedron. It is not always a clear choice to assign building blocks as vertices, edges or faces and, equally, the same geometry can often be targeted in different ways as illustrated by the triangular prism shown in Figure 2.18.

One approach for researchers to synthesise both inorganic and organic porous molecular assemblies is to analyse the angles between vertices, edges and faces in the desired polyhedron and then design building blocks to meet those geometric requirements. The angles between vertices, edges and faces for the five Platonic solids are shown in Table 2.1. While this strategy seems straightforward and has had some success in predicting and rationalising the formation of different polyhedra based on the bend angle of ditopic linkers,[147] the conformational flexibility inherent in large organic building blocks allows for significant deviation from geometrically proscribed angles. A sufficiently flexible building block may even change its shape entirely, as shown for a 6-connected dendritic building block, which could adopt conformations giving it either an octahedral or trigonal prism shape.[148] In MOPs, several small (<10°) or a a small number of large (>10°) deviations were found

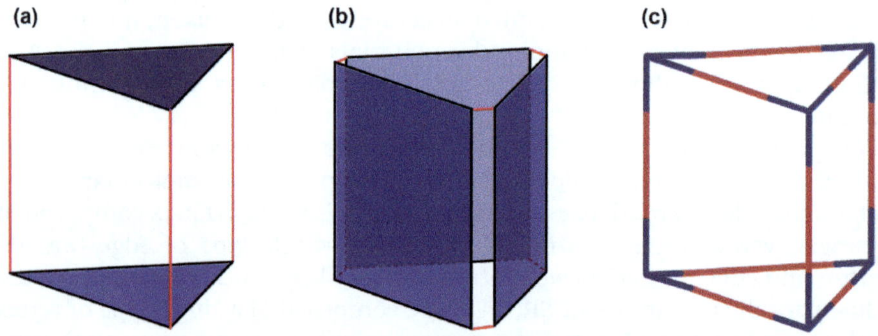

Figure 2.18 A triangular prism composed of (a) two triangular faces and three linear edges, (b) three rectangular faces linked by six linear edges and (c) six 90/90/60° vertices linked by 12 linear edges.

Table 2.1 Point group and angles between vertices (V), edges (E) and faces (F) *via* the centroid (C) for the five Platonic solids.

Polyhedron	Point group	∠VCV	∠VCE	∠VCF
Tetrahedron	*T*	109.5	54.7	70.5
Octahedron	*O*	90.0	45.0	54.7
Cube	*O*	70.5	45.0	54.7
Icosahedron	*I*	63.4	31.7	37.4
Dodecahedron	*I*	41.8	20.9	37.4

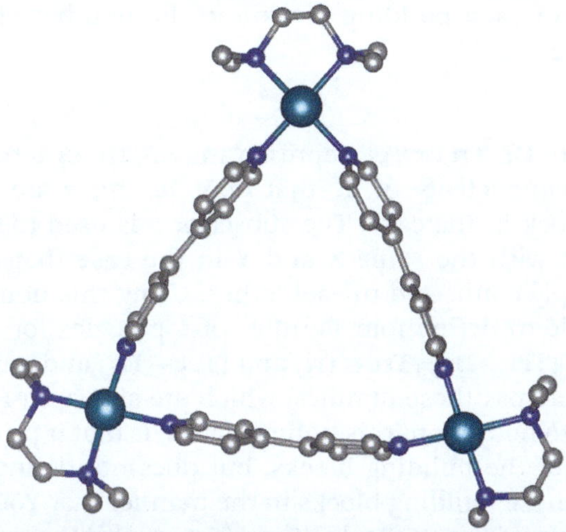

Figure 2.19 Triangular geometry of triangular assembly, CCSD refcode REZYAX,[151] composed of square planar Pd building blocks and linear bipyridine building blocks. The internal ∠N-Pd-N is 84°, significantly higher than the theoretical 60° required to form a triangle.

to be easily tolerated by deformations in other building blocks.[149] Young and Hay analysed the crystal structures of several MOPs and found that the majority did not conform to the angles expected given the target shape, such as the triangular shape of the molecule shown in Figure 2.19,[150] where the vertices are actually of square planar geometry. They also identified a common deformation motif where several building blocks underwent a concerted rotation about a symmetry axis of the molecular assembly.

Young and Hay developed a nomenclature for high-symmetry 0-periodic molecular assemblies, including those that form 2D planar polygons, prisms and those that form 3D polyhedral molecular assemblies. In the case of forming polyhedral assemblies, both edge-and face-directed assemblies are described, defined by the number of vertices (V), edges (E) and/or faces (F) that each ideal assembly possesses.[150] They noted that, in some cases, the nomenclature was not unique: for example, in defining a V_4F_4 tetrahedron, the choice of which building block is the vertex and which building block represents the face is entirely arbitrary. Similar arbitrariness in assigning building blocks results in confusion where a V_6F_8 (octahedron) could also be described as a V_8F_6 (cube); and a $V_{12}F_{20}$ (icosahedron) could also be described as a $V_{20}F_{12}$ (dodecahedron). Young and Hay were especially focused on MOPs, and so they chose to resolve the ambiguity by always defining the metal ion containing a building block as a vertex.

To describe the landscape of porous organic cages, where the choice of which building blocks represent vertices, edges and faces is again ambiguous, Jelfs and co-workers extended the nomenclature of Young and Hay for cage molecular materials formed of two components, similarly focusing on the connectivity of each building block and the number of each building block in the structure:[152]

$$\mathbf{X}_p^m\mathbf{Y}^n$$

where **X** and **Y** are **Di, Tri** or **Tet**, representing di-, tri- or tetratopic building blocks and the connectivity of **X**≥that of **Y**. *m* and *n* are the number of each building block in the cage. The subscript *p* is used to distinguish different topologies with the same **X** and **Y** in the case that two **X** building blocks are multiply connected to each other. Using this nomenclature, it is therefore possible to define four families of topologies for porous organic cages: (**Tri** + **Di**), (**Tet** + **Di**), (**Tri** + **Tri**) and (**Tet** + **Tri**) and enumerate a total of 20 topologies across these families, which are shown in Figure 2.20.

The key point to note about this nomenclature is that it primarily specifies the connectivity of the building blocks, but does not define the geometric arrangement of those building blocks in the manner that Young and Hay, or the 0-periodic nets of the RCSR do. Therefore, multiple geometric arrangements exist within the same topology. This ambiguity is illustrated in Figure 2.21, for the **Tri** + **Di** family, where $\mathbf{Tri}^4\mathbf{Di}^6$ may form a tetrahedron, an adamantoid cage or a truncated tetrahedron, depending on the geometry of the building blocks. A tritopic angle of 120° coupled with a ditopic angle of 70.5° results in an octahedron of the $\mathbf{Tri}^4\mathbf{Di}^6$ topology. Similarly, $\mathbf{Tri}^6\mathbf{Di}^{12}$ may form an octahedron or a cube and $\mathbf{Tri}^8\mathbf{Di}^{12}$ may form a cube or a cuboctahedron, as illustrated in Figure 2.22.

The ambiguity of topological *vs.* molecular shape, when considering porous molecular materials, is highlighted by one of the archetypal porous organic cages, Covalent Cage 3 (**CC3**), shown in Figure 2.23.[153] **CC3** is composed of four tritopic building blocks joined by six ditopic building blocks, *i.e.* it is a $\mathbf{Tri}^4\mathbf{Di}^6$ cage. Joining the centroids of the four tritopic building blocks reveals a tetrahedron; however, joining the centroids of the ditopic linkers reveals an octahedron, corresponding to the observed shape of the molecule. One could also describe this molecule as an octahedron formed by four alternating panels (faces) supplied by the trivalent building blocks and four 'empty' panels. The same topology and molecular shape has also been observed for a MOP composed of Mo paddlewheels joined by benzene tricarboxylate (BTC) linkers.[154]

2.3.1 Software

While several of the software packages designed and written to generate crystalline network materials can generate 0D structures using the same net definitions as supplied for 1-, 2- and 3-periodic nets, the supramolecular-toolkit, stk,[155] is designed to build structures of supramolecular assemblies, especially molecular cages, though it can also build linear polymers, macrocycles and 2D covalent organic frameworks. Unlike many of the software packages devised for periodic structures, where building blocks are typically not synthetic analogues, and connect *via* dummy atoms, which are the only

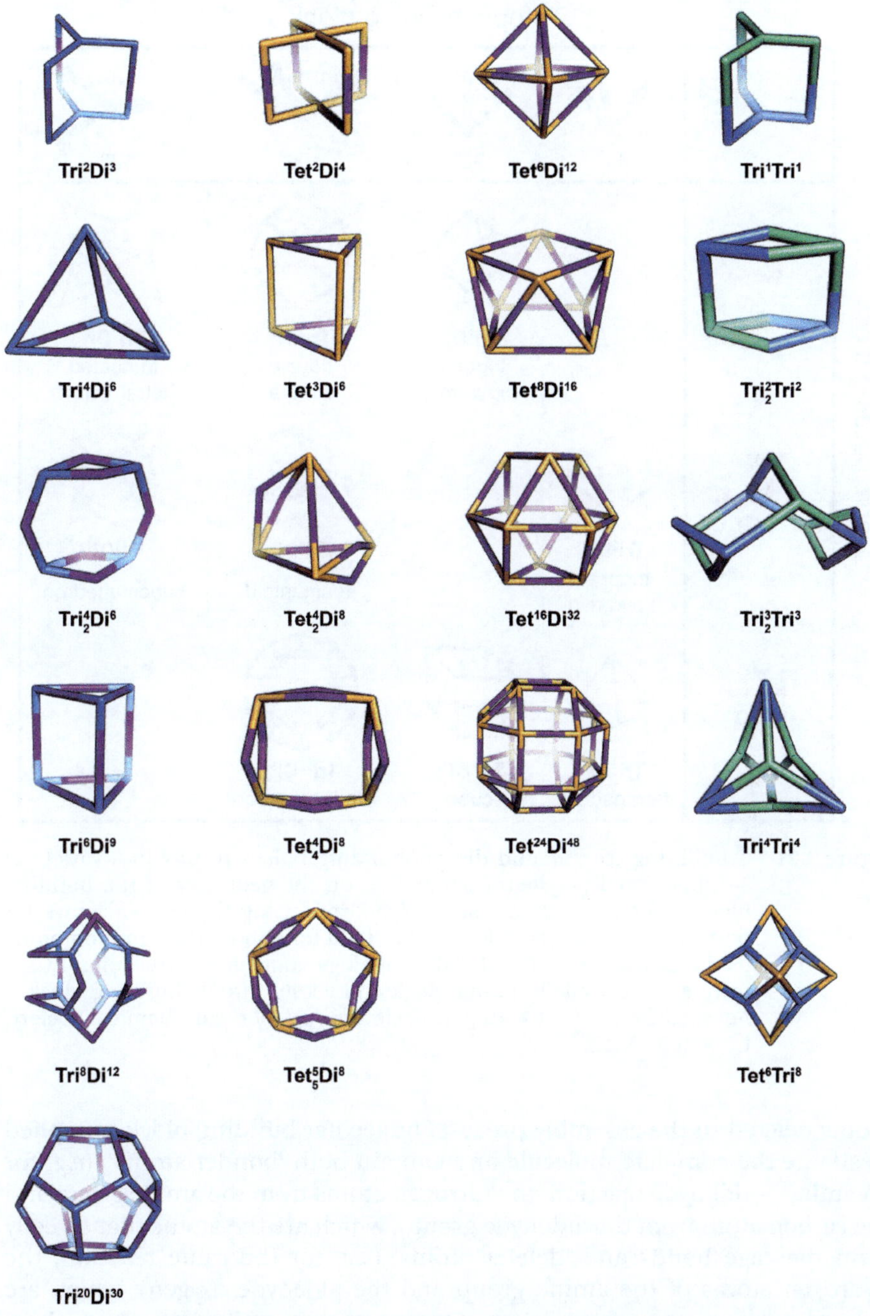

Figure 2.20 Twenty topologies for porous organic cages, ditopic building blocks are purple, tritopic building blocks are blue and cyan, and tetratopic building blocks are yellow.
Adapted from ref. 152 with permission from the Royal Society of Chemistry.

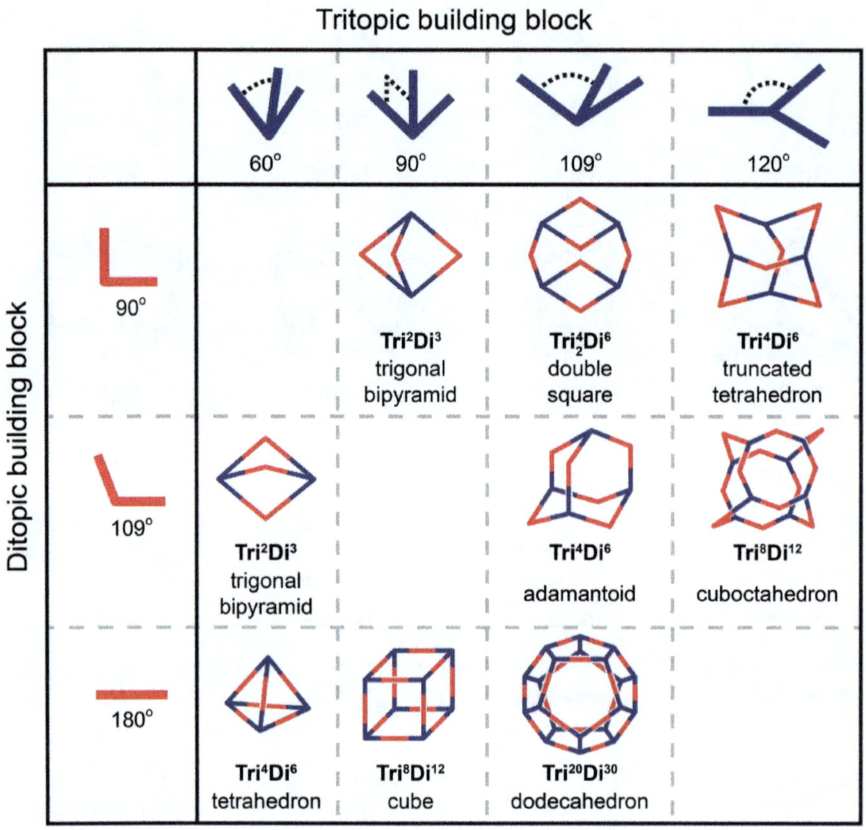

Figure 2.21 Combining tritopic and ditopic building blocks results in a variety of 3-dimensional polyhedra depending on the geometry of the building blocks. The topology of each polyhedron, using the nomenclature for porous organic cages[152] is noted in **bold** font above the common name of the polyhedron. The **Tri^{6}Di^9** topology could form a trigonal prism from a 90° tritopic building block and a 180° ditopic building block. Adapted from ref. 51 with permission from American Chemical Society, Copyright 2012.

atoms deleted in the assembly process, molecular building blocks supplied to stk are the complete molecule and contain both 'bonder atoms', (*e.g.* for an amine + aldehyde reaction, the nitrogen atoms from the amide group and the carbon atom from the aldehyde group), which are the atoms that directly form the cage bonds and 'deleter atoms' (*e.g.* for the same reaction, the hydrogen atoms of the amine group and the aldehyde oxygen), which are deleted after cage bond formation. Structures then undergo a constrained optimisation, where only the newly formed bonds are optimised, before a full geometry optimisation is carried out. Finally, the newly formed cage structure is subjected to a simulated annealing or molecular dynamics based conformer search. In the case where conformational flexibility allows the

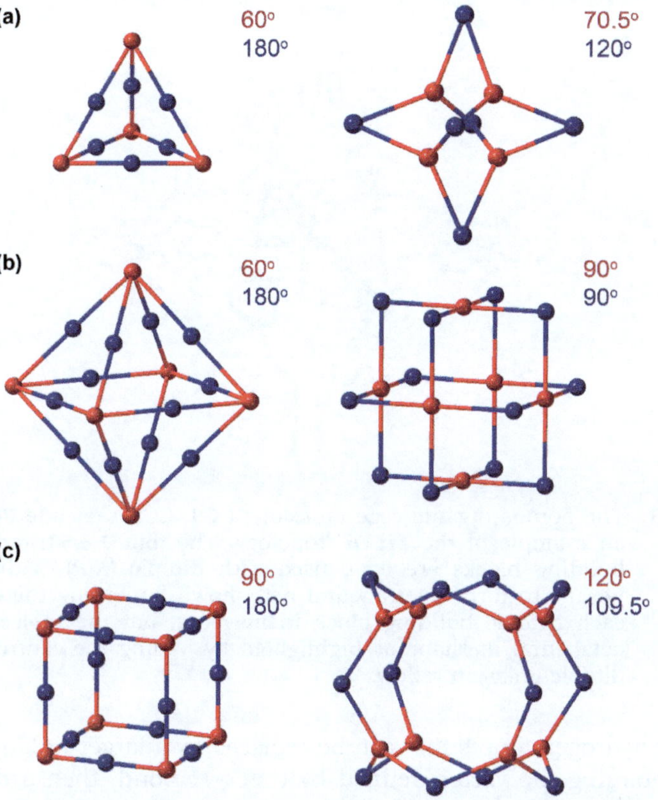

Figure 2.22 Depending on the angle between the connection points of a building block and its centroid, the same topological arrangement may form very different molecular shapes as illustrated for: (a) the $\textbf{Tri}^4\textbf{Di}^6$ topology, which may range from a tetrahedron to an octahedron, (b) the $\textbf{Tri}^6\textbf{Di}^{12}$ topology, which may form a shape between an octahedron and a cube, and (c) cage molecules possessing the $\textbf{Tri}^8\textbf{Di}^{12}$ topology that may have a molecular shape ranging from a cube to a cuboctahedron. Internal angles for each building block are shown next to each topology.

structure to collapse and become non-porous, the optimisation and conformer search may be carried out in the presence of solvent or adsorbent molecules.

HostDesigner is a software package that employs an approach more commonly used in drug discovery, called *de novo* structure-based design.[156] This approach, rather than beginning with building blocks that correspond to the molecules used in synthesis (as in stk) above, or direct analogues with bonding atoms replaced with some form of dummy or connector atom, as is commonly used in packages designed for the generation of periodic structures, begins with sub-building block size molecular fragments. Two initial fragments are chosen and placed according to the (point group) symmetry of the desired polyhedral assembly, then possible bonding vectors are identified

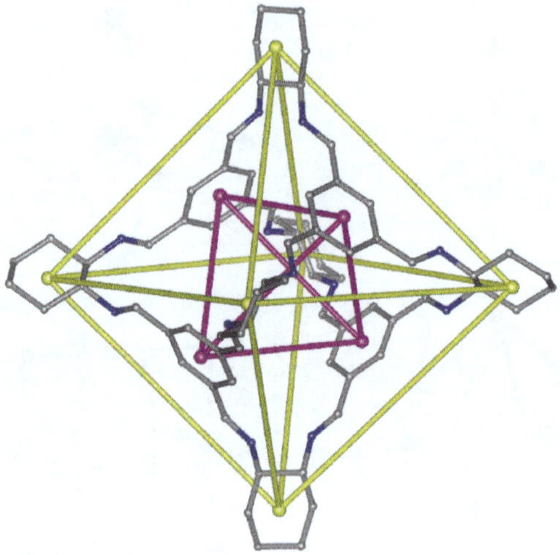

Figure 2.23 The porous organic cage molecule, **CC3**, CCSD refcode PUDXES[153] is an example of the **Tri^{4}Di^6** topology. The four 1,3,5-triformylbenzene building blocks are condensed with ditopic (*R*;*R*)-1,2-diaminocyclohexane to form a tetrahedral net, shown by joining the centroids of each tritopic building block in magenta, but the molecule itself is octahedral in shape as highlighted by joining the centroids of each ditopic linker in yellow.

by noting hydrogen atoms that can be replaced by a larger linking fragment and maintaining the vector defined by the X–H bond, then a database of linking fragments is tested to identify fragments of the appropriate shape to link two of the component fragments. The size of the polyhedral assembly and the relative orientation of the initial fragments are both scanned, generating several thousand initial geometries or *poses* (using drug design terminology) on which each linking fragment is tested. Molecular cages identified in this process are then able to be optimised and/or subjected to a conformer search.

2.3.2 Solid-state Structure

The software in Section 3.2.1 is designed to predict the structure of isolated cage structures, assuming that the thermodynamic product—the cage with the lowest free energy of formation is the product formed. This assumption is ideal for applications with cages in solution,[157] such as porous liquids[158] and there has been success in predicting bulk adsorption from single molecules.[159] Predicting the cage product of a reaction, is already a challenge – a high-throughput experimental and computational study found several instances where a **Tri^{4}Di^6** cage was the desired product, but a **Tri^{2}Di^2** product formed.[39] However, unlike porous network structures, such as MOFs and COFs, where the porosity arises at the same time as the solid-state structure *via* covalent bonding, cage molecules are realised through a two-step process, where first the porous cage is formed *via* covalent bonds, then the solid-state structure arises *via* a delicate balance of weaker interactions, especially hydrogen bonding and dispersion interactions. Structural polymorphism is common, especially in porous organic cages. As in crystal structures of non-porous molecules, small changes in the molecular structure – *i.e.* by adding functional groups – can result in completely different crystal structures, meaning that reticular design of porous cages is difficult.[160,161] Cages may form catenated assembles.[162] Even without altering the molecular structure, the synthesis and crystallisation conditions can strongly affect the structure observed. Indeed, different polymorphs of covalent cage 1 (**CC1**),[163] **CC3**[164] and **CC4**[165] have all been obtained by crystallisation in different solvent conditions. Therefore, prediction of the solid-state structures of porous cages requires crystal structure prediction (CSP).

Crystal structure prediction has an extensive history, especially in the pharmaceutical industry,[166–168] and the increasing availability of large-scale high-performance computing has seen many successes in an increasing range of structure types including flexible molecules, hydrates and salts.[169–171] A variety of CSP methods exist, employing different force fields and/or DFT combined with approaches such as simulated annealing and genetic algorithms to determine the relative energies of 10^3–10^6 possible crystal structures.

Typically, the results of a CSP search are presented as plots of the lattice energy *vs.* the density of the predicted structures and the experimentally observed polymorphs are seen on the leading edge of the plot. Adding in a third dimension, by colouring each point according to a property, such as pore-limiting diameter,[172] methane capacity or surface area, creates an energy-structure-function (ESF) map,[173] shown for the non-cage, but porous structure forming molecule, benzimidazolone, as can be seen in Figure 2.24.

For predicting the solid-state assemblies of porous cage molecules using CSP, two particular challenges exist: firstly, the large number of structures required, combined with the fact that cage molecules are typically very large (high number of atoms) compared to many 'regular' targets for CSP, means that solid-state DFT methods are too expensive for routine use and force fields are the primary computational method used, making their accuracy crucial in obtaining accurate results.[174] In particular, vibrational contributions to lattice free energies have been shown to have a significant effect on thermodynamic stability, re-ranking up to 10% of predicted polymorphs; however, the effect of thermal expansion of the free energy differences of polymorphs was found to only be $\approx$ 1–2 kJ mol^{-1}.[175,176]

A second challenge, specific to CSP of porous cages, is that guest molecules, including solvent used in synthesis, may determine the observed structure of the porous cage[164] and complicates structural prediction. Day and co-workers added an additional Monte Carlo step to their CSP procedure, where the pores of the porous organic cage, **CC1**, were stochastically filled with solvent molecules, and were therefore able to explain why the previously predicted global minimum structure had not been observed experimentally.[177]

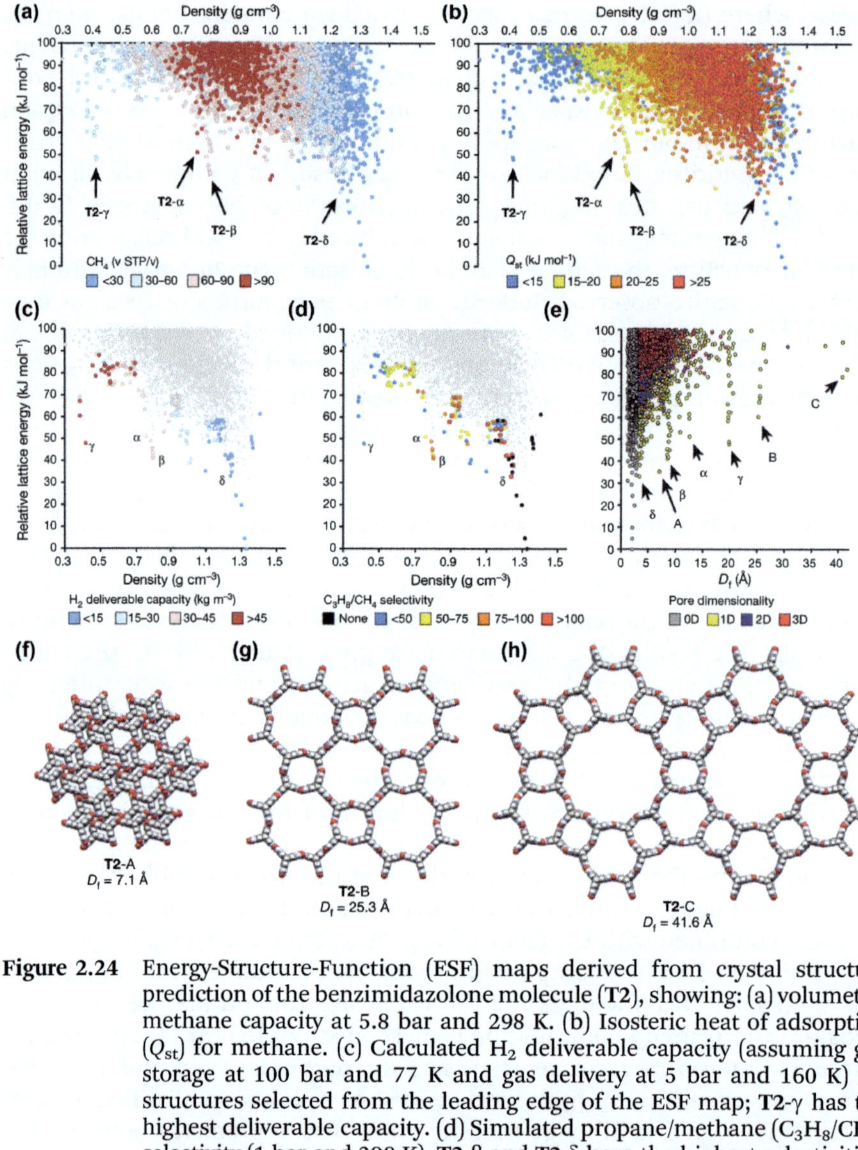

Figure 2.24 Energy-Structure-Function (ESF) maps derived from crystal structure prediction of the benzimidazolone molecule (**T2**), showing: (a) volumetric methane capacity at 5.8 bar and 298 K. (b) Isosteric heat of adsorption (Q_{st}) for methane. (c) Calculated H_2 deliverable capacity (assuming gas storage at 100 bar and 77 K and gas delivery at 5 bar and 160 K) for structures selected from the leading edge of the ESF map; **T2**-γ has the highest deliverable capacity. (d) Simulated propane/methane (C_3H_8/CH_4) selectivity (1 bar and 298 K); **T2**-β and **T2**-δ have the highest selectivities. (e) Relative lattice energy *versus* D_f, the largest free sphere, which is a measure of pore size; symbols are coloured by pore dimensionality. (f)–(h), Selected hypothetical structures from the leading edge. Reproduced from ref. 173 with permission from Springer Nature, Copyright 2017.

Moving towards the reticular synthesis of porous network structures from porous cage molecules requires cages that can act similarly to the building blocks of network materials such as MOFs and COFs. MOPs are readily employed as building blocks in MOF synthesis, creating both 3-periodic[178,179] and 2-periodic (layer) MOFs,[180] typically by substituting axial or capping ligands in the MOPs with linker building blocks. However, for porous organic cages without such (convenient) connection points, and relying on non-covalent interactions to guide self-assembly, some success has been obtained. Cooper and co-workers predicted and synthesised mixtures of four different tetrahedral covalent cages, firstly using DFT calculations on isolated cage pairs, which showed that the interface of two opposite chirality cages was nearly 20 kJ mol^{-1} more stable than the interface of homochiral cages and therefore directed window-to-window packing of the cages. This DFT prediction for heterochiral pairing, leading to a (quasi)racemic structure, was supported by crystal structure prediction, and indeed was experimentally observed.[163] Further underscoring the difficulty of structural prediction of porous organic cages, the same study observed that an eqimolar solution of the *R* enantiomers of two porous cages resulted in a quasiracemic crystal, and variable-temperature NMR measurements showed that the cages were capable of converting between *R* and *S* enantiomers in solution. The same principle of optimising the interfacial interaction in order to direct the solid-state structure was later applied to tubular (**Tet^3Di^6**) cages, where heterochiral packing of cages resulted in columns of 1D nanotubes.[160]

2.4 Structure Generation of Amorphous Materials

Porous network materials, cages[181,182] and linear polymers[183] (Figure 2.25) can also form amorphous porous phases, depending on the conditions applied, *i.e* high pressure or temperature. Amorphous porous phases can be challenging to model as they bring the possibility of porosity arising from both existing pore voids (intrinsic porosity) and from void space arising between poorly packed molecular building blocks (extrinsic porosity), as illustrated for a porous organic cage, **CC3**, in Figure 2.26. Unlike porous network materials, where single crystal or powder X-ray diffraction data is usually available, amorphous materials, by definition, do not exhibit long-range order and are therefore substantially more difficult to characterise. Characterisation of amorphous materials is often indirect, deriving properties like surface area or pore space distribution from gas adsorption isotherms. Molecular simulation of amorphous molecular materials, therefore, is invaluable to complement experimental characterisation, providing insight into molecular-level structure and the properties that arise from that structure.

The difficulty, and hence uncertainty in characterisation, means that modelling amorphous materials requires an ensemble of structures that provide a statistical description of the properties of the material. In addition, the need to model comparatively large length scales means that structural prediction of amorphous porous materials almost entirely relies on force fields, which may be either atomistic or coarse-grained. The lack of available force fields, especially for (a) inorganic systems and (b) reactive force fields, that can account for making and breaking bonds, has so far limited the systems that may be modelled.

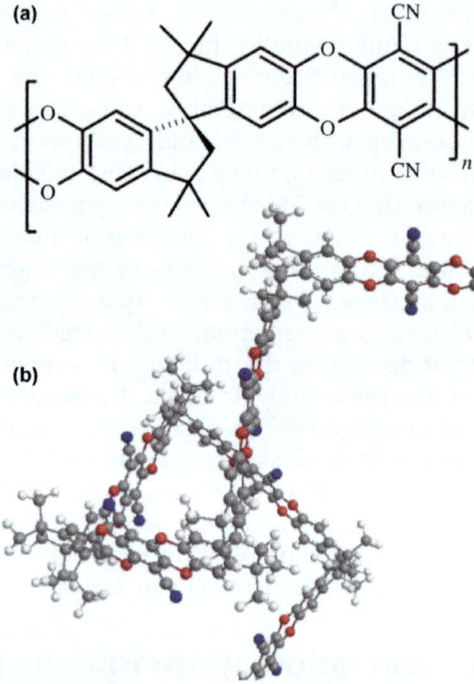

Figure 2.25 (a) The chemical structure and (b) molecular model of a random fragment of a Polymer of Intrinsic Microporosity (PIM), showing how the rigid links form intrinsic voids. Each spiro-centre twists randomly, resulting in poor packing of the PIM.
Adapted from ref. 184 with permission from the Royal Society of Chemistry.

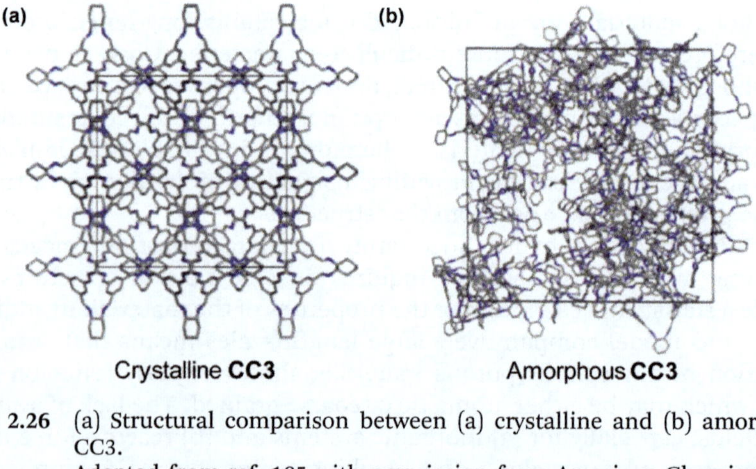

Figure 2.26 (a) Structural comparison between (a) crystalline and (b) amorphous CC3.
Adapted from ref. 185 with permission from American Chemical Society, Copyright 2013.

By definition, there is no net to guide the assembly of an amorphous structure and therefore algorithms for structural prediction of amorphous materials fall into the class of bottom-up methods:

Reverse Monte Carlo (RMC) methods[186] have been used to determine the structure of a wide variety of complex materials[187] including glasses,[188] amorphous porous carbons,[189,190] silicon[191] and to molecular adsorption in amorphous carbon.[192] A Reverse Monte Carlo method works similarly to traditional (Metropolis) Monte Carlo methods, beginning with some structure (often a crystalline one) and then generating random 'moves' in order to distort the structure. Where a Metropolis Monte Carlo method uses the energy of the newly generated configuration in order to accept or reject each randomly generated move, an RMC method uses structural information: typically either the Radial Distribution Function (RDF) or the Structure Factor (SF), to score each move. RMC simulations therefore yield structures that *on average* give the same structural information (*i.e.* the same radial distribution function) as the input/target function. However, this requirement for input data (density, RDF/SF) means that it is not possible to use RMC to predict the structures of entirely unknown phases. While RMC has been applied to ZIFs,[193] the need to account for bond breaking in moving from a crystalline phase to an amorphous one has limited its wider application to the determination of amorphous phases of porous network materials such as MOFs.

More robust approaches to generating amorphous structures of porous materials take advantage of the increasing availability of high performance computing and employ Molecular Dynamics (MD). Directly employing MD with a reactive force field allows computational modelling of melt-quenching, where the crystal structure is heated to above its melting point and then cooled. While this is an attractive analogue to the process of forming a MOF glass, it is restricted in applicability by the availability of reactive force field parameters.[194] *Simulated polymerisation* techniques gradually build structures though placing building blocks in a periodic box and 'polymerising' them over several cycles of molecular dynamics. Simulated polymerisation can be applied to 1-, 2- and 3-dimensional polymers. The essential steps in a simulated polymerisation are the following:

1. Place building blocks randomly in periodic box with a low density. A larger volume simulation cell is employed in order to have sufficient space to place building blocks without placing them unphysically close to each other.
2. Run MD with an NVT ensemble, *i.e.* keeping the number of particles (N), volume (V) and temperature (T) constant. This step allows building blocks to coalesce (polymerise).
3. From the output of the NVT MD, run a second MD with a NPT ensemble, *i.e.* keeping the number of particles (N), pressure (P) and temperature (T) constant. This step compresses the simulation cell.

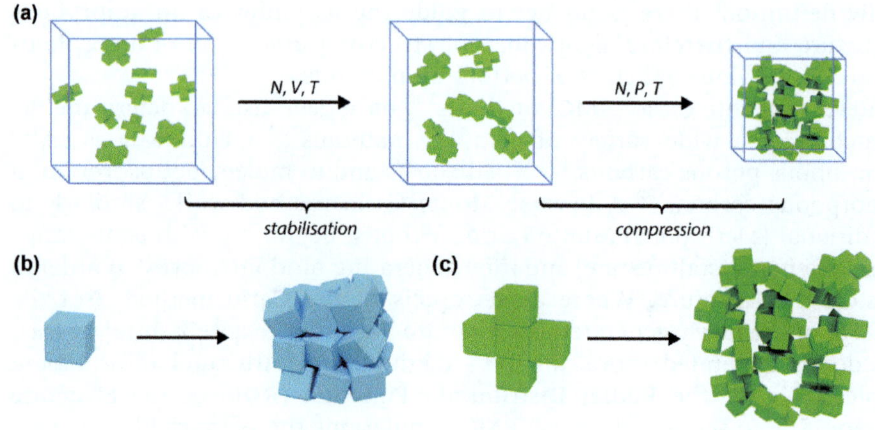

Figure 2.27 (a) The key simulation steps used for the production of amorphous porous molecular material structures. The choice of periphery functionality has been shown to significantly affect the resulting amorphous structure with (b) compact functionality leading to a dense non-porous structure and (c) bulky functionality producing a porous structure. Reproduced from ref. 195 with permission from the Royal Society of Chemistry.

These steps are also shown in Figure 2.27. A single structure prediction will employ several cycles of coalescence (stabilisation) and compression in order to reach a reasonable structure. As with the structural prediction of networked molecular frameworks, the building block libraries are generally specific to individual programs as they make different choices for the treatment of both intra-fragment bonding and especially, the treatment of inter-fragment bonding. It is important to note that simulated polymerisation methods differ from reactive force fields in that they are not directly simulating the breaking or formation of chemical bonds.

2.4.1 Software

The *Polymatic* program[196] begins with building blocks that are exactly the components of the desired polymer, with the atoms that will form a bond in the polymerisation step marked as 'reactive atoms'. Reactive atoms in building blocks are therefore typically under-coordinated and may have artificial charges added to increase the probability of reactive atoms attracting each other. Polymatic searches for pairs of bonding atoms in close proximity bonds them and optimises the newly formed bond, removing any artificial charges from the two bonding atoms and removing their bonding atom status. Several bonds may be formed in this way before beginning molecular dynamics. A total of 21 MD (NVT, NPT) steps are undertaken at different temperatures and pressures, as shown in Figure 2.28, to arrive at a final structure.

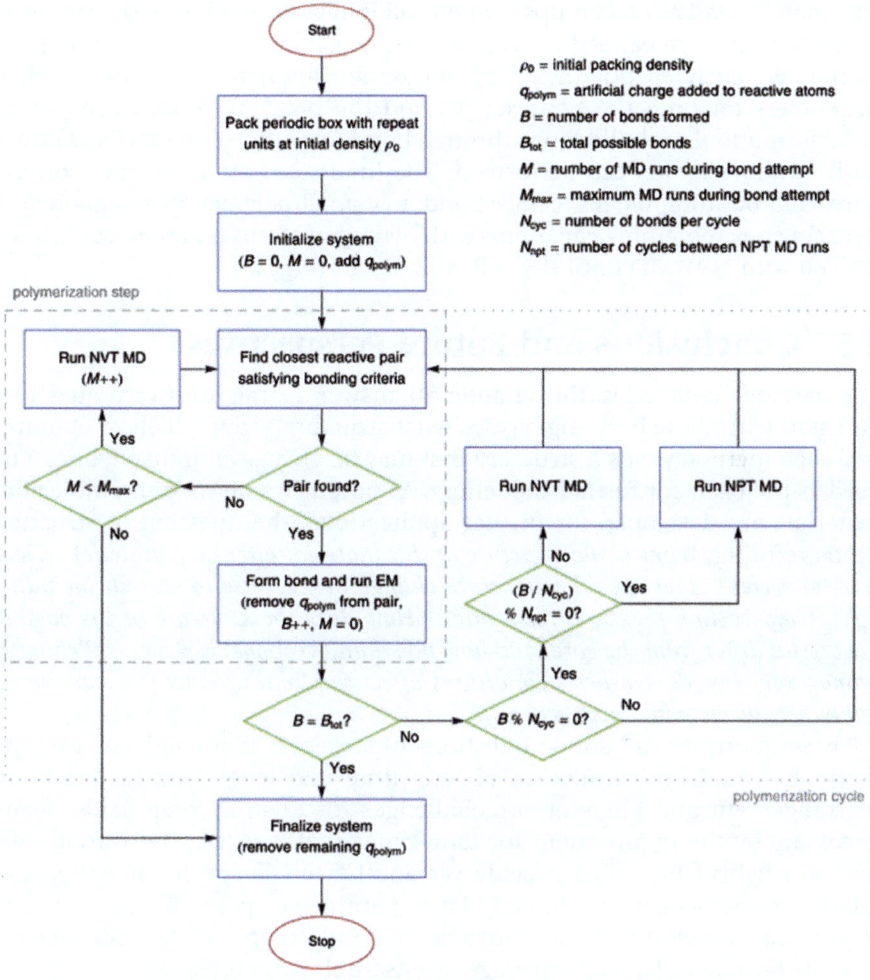

Figure 2.28 Flowchart of the polymatic simulated polymerisation algorithm. The start and stop steps are displayed in ovals, processing steps in rectangles, and conditional steps in diamonds. A polymerisation step is enclosed in a dashed box, while a polymerisation cycle is outlined in a dotted box. EM = energy minimisation, MD = molecular dynamics. Reproduced from ref. 196 with permission from Springer Nature, Copyright 2013.

Another approach for generating amorphous porous materials is to computationally mimic their synthesis. The *Ambuild* program[197] can undertake artificial synthesis of the Suzuki, Sonogashira–Hagihara, Yamamoto, Schiff Base and cyclotrimerisation reactions. In artificial synthesis, the simulation cell is seeded with not only the building blocks that comprise the final amorphous structure, but also the catalyst and solvent. Building blocks are rigid – *i.e.* their internal coordinates are held fixed throughout the simulation, and contain 'end atoms/groups', which will join to each other in the structure generation and are capped by 'cap atoms', which are deleted after bond formation. For each reaction the Ambuild program simulates, there are specified parameters for both the bond length and the bond angle that can permit bond formation. Ambuild loops through bond formation and MD (NVT) steps until no more bonds can be formed. The final steps are to remove solvent, unreacted building blocks, catalyst and any small polymer fragments before replacing any remaining cap atoms with hydrogen atoms and then equilibrate the cell with NPT MD until the cell volume converges.

2.5 Conclusions and Future Perspectives

The methods detailed in this chapter all answer a straightforward question, "If I start with these building blocks, what structure(s) am I likely to obtain?" and each method yields a structure that may be an answer in itself, or can be used as the basis for further modelling. As porous framework and molecular materials are developed for further applications, the question of structure needs refining: *What is the structure of this material after* 10 *(catalytic) cycles?* ... 100 *cycles? How does the structure change in response to stimuli including light, temperature, pressure, adsorption? How does the structure at the edge of the crystal differ from the core, and how do stimuli-responsive structural changes propagate? How do synthesis conditions affect the final product (i.e. can we reliably obtain metastable structures)?*

These questions are subtle questions of structure that challenge existing methods, requiring simulation of long time and length scales. Yet bond rearrangement and breaking are challenges for existing force fields, made necessary by the requirement for long time/length scales. Multivariate, defect and hybrid materials provide yet another challenge in that they add another dimension to an already large parameter space. These problems require the development of statistical methods that can describe the behaviour of materials, using average or ensemble structures.

High-throughput computational studies have an increasing role in materials screening, and have had several successes in identifying materials that, when synthesised, possessed record-breaking properties. Computational predictions of first the structure, then the properties of porous materials, are key in moving from materials *screening* to materials *design*. Design rules for porous materials require knowledge of *what* and *to what extent* the observable properties of materials arise from the *physical* structure (*i.e.* geometry/topology) of the material *vs.* the *chemical* structure of the material. High-throughput calculations and synthesis have both led to large databases of predicted structures. In light of needing to model structures over time and in response to various stimuli, the key challenge moving forward is firstly to define, then identify what constitutes duplicate structures.

Porous materials are a rapidly growing field with key applications in gas storage and separation, sensing and catalysis that, when realised, stand poised to change the world.[198] The methods outlined here, in conjunction with developments in data management, increased accuracy in simulations of adsorption and diffusion, and efficient, accurate and general force fields, will enable true computational design of materials, in advance of synthesis, lessening the role of serendipity in materials discovery and providing deep understanding of the microscopic origins of macroscopic material behaviour.

References

1. A. Kirchon, L. Feng, H. F. Drake, E. A. Joseph and H.-C. Zhou, *Chem. Soc. Rev.*, 2018, **47**, 8611–8638.
2. B. F. Abrahams, A. D. Dharma, P. S. Donnelly, T. A. Hudson, C. J. Kepert, R. Robson, P. D. Southon and K. F. White, *Chemistry*, 2017, **23**, 7871–7875.
3. B. Abrahams, S. Batten, H. Hamit, B. Hoskins and R. Robson, *Angew. Chem., Int. Ed.*, 1996, **35**, 1690–1692.
4. B. Abrahams, J. Coleiro, B. Hoskins and R. Robson, *Chem. Commun.*, 1996, 603–604.
5. S. L. James, *Chem. Soc. Rev.*, 2003, **32**, 276–288.
6. J. R. Long and O. M. Yaghi, *Chem. Soc. Rev.*, 2009, **38**, 1213–1214.
7. X. Feng, X. Ding and D. Jiang, *Chem. Soc. Rev.*, 2012, **41**, 6010–6022.
8. A. P. Côté, A. I. Benin, N. W. Ockwig, M. O'Keeffe, A. J. Matzger and O. M. Yaghi, *Science*, 2005, **310**, 1166.
9. J. Luo, J.-W. Wang, J.-H. Zhang, S. Lai and D.-C. Zhong, *CrystEngComm*, 2018, **20**, 5884–5898.
10. A. Karmakar, R. Illathvalappil, B. Anothumakkool, A. Sen, P. Samanta, A. V. Desai, S. Kurungot and S. K. Ghosh, *Angew. Chem., Int. Ed.*, 2016, **55**, 10667–10671.
11. B. Wang, R.-B. Lin, Z. Zhang, S. Xiang and B. Chen, *J. Am. Chem. Soc.*, 2020, **142**, 14399–14416.
12. C. J. H. Jacobsen, C. Madsen, J. Houzvicka, I. Schmidt and A. Carlsson, *J. Am. Chem. Soc.*, 2000, **122**, 7116–7117.
13. X. Huang, J. Zhang and X. Chen, *Chin. Sci. Bull.*, 2003, **48**, 1531–1534.
14. B. R. Pimentel, A. Parulkar, E.-K. Zhou, N. A. Brunelli and R. P. Lively, *ChemSusChem*, 2014, **7**, 3202–3240.
15. M. Eddaoudi, J. Kim, J. B. Wachter, H. K. Chae, M. O'Keeffe and O. M. Yaghi, *J. Am. Chem. Soc.*, 2001, **123**, 4368–4369.
16. S. Hong, M. R. Rohman, J. Jia, Y. Kim, D. Moon, Y. Kim, Y. H. Ko, E. Lee and K. Kim, *Angew. Chem., Int. Ed.*, 2015, **54**, 13241–13244.
17. M. E. Briggs and A. I. Cooper, *Chem. Mater.*, 2017, **29**, 149–157.
18. J. Tian, P. Thallapally, S. Dalgarno, P. McGrail and J. Atwood, *Angew. Chem., Int. Ed.*, 2009, **48**, 5492–5495.
19. Z. Guo, P. Sun, X. Zhang, J. Lin, T. Shi, S. Liu, A. Sun and Z. Li, *Chem. – Asian J.*, 2018, **13**, 2046–2053.
20. J. Weber, Q. Su, M. Antonietti and A. Thomas, *Macromol. Rapid Commun.*, 2007, **28**, 1871–1876.

21. X. Wang, Y. Liu, X. Ma, S. K. Das, M. Ostwal, I. Gadwal, K. Yao, X. Dong, Y. Han, I. Pinnau, K.-W. Huang and Z. Lai, *Adv. Mater.*, 2017, **29**, 1605826.
22. P. M. Budd, B. S. Ghanem, S. Makhseed, N. B. McKeown, K. J. Msayib and C. E. Tattershall, *Chem. Commun.*, 2004, 230–231.
23. J. J. Perry IV, J. A. Perman and M. J. Zaworotko, *Chem. Soc. Rev.*, 2009, **38**, 1400–1417.
24. Q. Zhu, X. Wang, R. Clowes, P. Cui, L. Chen, M. A. Little and A. I. Cooper, *J. Am. Chem. Soc.*, 2020, **142**, 16842–16848.
25. F. Wang, Z.-S. Liu, H. Yang, Y.-X. Tan and J. Zhang, *Angew. Chem., Int. Ed.*, 2011, **50**, 450–453.
26. Y. Xia, Z. Yang and Y. Zhu, *J. Mater. Chem. A*, 2013, **1**, 9365–9381.
27. N. Balahmar and R. Mokaya, *J. Mater. Chem. A*, 2019, **7**, 17466–17479.
28. O. D. Velev, T. A. Jede, R. F. Lobo and A. M. Lenhoff, *Chem. Mater.*, 1998, **10**, 3597–3602.
29. O. D. Velev, T. A. Jede, R. F. Lobo and A. M. Lenhoff, *Nature*, 1997, **389**, 447–448.
30. W. Morris, C. J. Doonan and O. M. Yaghi, *Inorg. Chem.*, 2011, **50**, 6853–6855.
31. K. K. Tanabe and S. M. Cohen, *Chem. Soc. Rev.*, 2011, **40**, 498–519.
32. K. K. Tanabe, Z. Wang and S. M. Cohen, *J. Am. Chem. Soc.*, 2008, **130**, 8508–8517.
33. W. M. Bloch, A. Burgun, C. J. Coghlan, R. Lee, M. L. Coote, C. J. Doonan and C. J. Sumby, *Nat. Chem.*, 2014, **6**, 906–912.
34. Y. Inokuma, M. Kawano and M. Fujita, *Nat. Chem.*, 2011, **3**, 349–358.
35. M. Yoshizawa, J. Klosterman and M. Fujita, *Angew. Chem., Int. Ed.*, 2009, **48**, 3418–3438.
36. M. Otte, *ACS Catal.*, 2016, **6**, 6491–6510.
37. N. Stock, *Microporous Mesoporous Mater.*, 2010, **129**, 287–295.
38. H. Reinsch, M. A. van der Veen, B. Gil, B. Marszalek, T. Verbiest, D. de Vos and N. Stock, *Chem. Mater.*, 2013, **25**, 17–26.
39. R. L. Greenaway, V. Santolini, M. J. Bennison, B. M. Alston, C. J. Pugh, M. A. Little, M. Miklitz, E. G. B. Eden-Rump, R. Clowes, A. Shakil, H. J. Cuthbertson, H. Armstrong, M. E. Briggs, K. E. Jelfs and A. I. Cooper, *Nat. Commun.*, 2018, **9**, 2849.
40. W. Lu, Z. Wei, Z.-Y. Gu, T.-F. Liu, J. Park, J. Park, J. Tian, M. Zhang, Q. Zhang, T. Gentle III, M. Bosch and H.-C. Zhou, *Chem. Soc. Rev.*, 2014, **43**, 5561–5593.
41. Y. J. Colon and R. Q. Snurr, *Chem. Soc. Rev.*, 2014, **43**, 5735–5749.
42. S. P. Collins, T. D. Daff, S. S. Piotrkowski and T. K. Woo, *Sci. Adv.*, 2016, **2**, e1600954.
43. Z. Shi, W. Yang, X. Deng, C. Cai, Y. Yan, H. Liang, Z. Liu and Z. Qiao, *Mol. Syst. Des. Eng.*, 2020, **5**, 725–742.
44. P. Z. Moghadam, A. Li, S. B. Wiggin, A. Tao, A. G. P. Maloney, P. A. Wood, S. C. Ward and D. Fairen-Jimenez, *Chem. Mater.*, 2017, **29**, 2618–2625.

45. Y. G. Chung, J. Camp, M. Haranczyk, B. J. Sikora, W. Bury, V. Krungleviciute, T. Yildirim, O. K. Farha, D. S. Sholl and R. Q. Snurr, *Chem. Mater.*, 2014, **26**, 6185–6192.
46. C. E. Wilmer, M. Leaf, C. Y. Lee, O. K. Farha, B. G. Hauser, J. T. Hupp and R. Q. Snurr, *Nat. Chem.*, 2012, **4**, 83–89.
47. R. Mercado, R.-S. Fu, A. V. Yakutovich, L. Talirz, M. Haranczyk and B. Smit, *Chem. Mater.*, 2018, **30**, 5069–5086.
48. C. Baerlocher and L. McCusker, *Database of Zeolite Structures*, http://www.iza-structure.org/databases.
49. F. Vermoortele, B. Bueken, G. Le Bars, B. Van de Voorde, M. Vandichel, K. Houthoofd, A. Vimont, M. Daturi, M. Waroquier, V. Van Speybroeck, C. Kirschhock and D. E. De Vos, *J. Am. Chem. Soc.*, 2013, **135**, 11465–11468.
50. V. V. Butova, A. P. Budnyk, A. A. Guda, K. A. Lomachenko, A. L. Bugaev, A. V. Soldatov, S. M. Chavan, S. Øien-Ødegaard, U. Olsbye, K. P. Lillerud, C. Atzori, S. Bordiga and C. Lamberti, *Cryst. Growth Des.*, 2017, **17**, 5422–5431.
51. T. R. Cook, Y.-R. Zheng and P. J. Stang, *Chem. Rev.*, 2013, **113**, 734–777.
52. E. Argueta, J. Shaji, A. Gopalan, P. Liao, R. Q. Snurr and D. A. Gómez-Gualdrón, *J. Chem. Theory Comput.*, 2018, **14**, 365–376.
53. M. O'Keeffe, M. Eddaoudi, H. Li, T. Reineke and O. Yaghi, *J. Solid State Chem.*, 2000, **152**, 3–20.
54. H. M. El-Kaderi, J. R. Hunt, J. L. Mendoza-Cortés, A. P. Côté, R. E. Taylor, M. O'Keeffe and O. M. Yaghi, *Science*, 2007, **316**, 268–272.
55. A. F. Wells, *Three-Dimensional Nets and Polyhedra*, Interscience, New York, 1977.
56. M. O'Keeffe, B. G. Hyde and J. S. Anderson, Philosophical Transactions of the Royal Society of London. *Ser. A, Math. Phys. Sci.*, 1980, 295, 553–618.
57. M. O'Keeffe and O. M. Yaghi, *Chem. Rev.*, 2012, **112**, 675–702.
58. O. Delgado-Friedrichs, M. D. Foster, M. O'Keeffe, D. M. Proserpio, M. M. Treacy and O. M. Yaghi, *J. Solid State Chem.*, 2005, **178**, 2533–2554.
59. J.-G. Eon, *Acta Crystallogr., Sect. A*, 2016, **72**, 268–293.
60. O. M. Yaghi, *ACS Cent. Sci.*, 2019, **5**, 1295–1300.
61. V. A. Blatov, *Acta Crystallogr., Sect. A*, 2007, **63**, 329–343.
62. V. A. Blatov and D. M. Proserpio, *Acta Crystallogr., Sect. A*, 2009, **65**, 202–212.
63. O. Delgado-Friedrichs and M. O'Keeffe, *J. Solid State Chem.*, 2005, **178**, 2480–2485.
64. S. Biswas, M. Grzywa, H. P. Nayek, S. Dehnen, I. Senkovska, S. Kaskel and D. Volkmer, *Dalton Trans.*, 2009, 6487–6495.
65. D. Zhao, D. Yuan, D. Sun and H.-C. Zhou, *J. Am. Chem. Soc.*, 2009, **131**, 9186–9188.
66. F. Nouar, J. F. Eubank, T. Bousquet, L. Wojtas, M. J. Zaworotko and M. Eddaoudi, *J. Am. Chem. Soc.*, 2008, **130**, 1833–1835.
67. O. Delgado-Friedrichs and M. O'Keeffe, *Acta Crystallogr., Sect. A*, 2003, **59**, 351–360.

68. V. A. Blatov, A. P. Shevchenko and D. M. Proserpio, *Cryst. Growth Des.*, 2014, **14**, 3576–3586.
69. C. Baerlocher, L. B. McCusker and D. H. Olson, *Atlas of Zeolite Framework Types*, Elsevier B.V, 6th edn, 2007.
70. M. O'Keeffe, M. A. Peskov, S. J. Ramsden and O. M. Yaghi, *Acc. Chem. Res.*, 2008, **41**, 1782–1789.
71. S. J. Ramsden, V. Robins and S. T. Hyde, *EPINET: Euclidean Patterns In Non-Euclidean Tilings*, http://epinet.anu.edu.au.
72. S. J. Ramsden, V. Robins and S. T. Hyde, *Acta Crystallogr., Sect. A*, 2009, **65**, 81–108.
73. V. A. Blatov, O. Delgado-Friedrichs, M. O'Keeffe and D. M. Proserpio, *Acta Crystallogr., Sect. A*, 2007, **63**, 418–425.
74. O. Delgado Friedrichs and D. H. Huson, *Discrete Comput. Geom.*, 1999, **21**, 299–315.
75. C. Mellot Draznieks, J. M. Newsam, A. M. Gorman, C. M. Freeman and G. Férey, *Angew. Chem., Int. Ed.*, 2000, **39**, 2270–2275.
76. C. Mellot-Draznieks, J. Dutour and G. Férey, *Angew. Chem., Int. Ed.*, 2004, **43**, 6290–6296.
77. M. Wahiduzzaman, S. Wang, B. J. Sikora, C. Serre and G. Maurin, *Chem. Commun.*, 2018, **54**, 10812–10815.
78. B. J. Sikora, R. Winnegar, D. M. Proserpio and R. Q. Snurr, *Microporous Mesoporous Mater.*, 2014, **186**, 207–213.
79. G. Ferey, C. Mellot-Draznieks, C. Serre, F. Millange, J. Dutour, S. Surblé and I. Margiolaki, *Science*, 2005, **309**, 2040–2042.
80. https://mof.tech.northwestern.edu/databases.
81. S. Bureekaew and R. Schmid, *CrystEngComm*, 2013, **15**, 1551–1562.
82. R. L. Martin and M. Haranczyk, *Cryst. Growth Des.*, 2014, **14**, 2431–2440.
83. M. A. Addicoat, D. E. Coupry and T. Heine, *J. Phys. Chem. A*, 2014, **118**, 9607–9614.
84. Y. J. Colón, D. A. Gómez-Gualdrón and R. Q. Snurr, *Cryst. Growth Des.*, 2017, **17**, 5801–5810.
85. O. Delgado-Friedrichs, M. O'Keeffe and O. M. Yaghi, *Phys. Chem. Chem. Phys.*, 2007, **9**, 1035–1043.
86. N. W. Ockwig, O. Delgado-Friedrichs, M. O'Keeffe and O. M. Yaghi, *Acc. Chem. Res.*, 2005, **38**, 176–182.
87. E. V. Alexandrov, V. A. Blatov, A. V. Kochetkov and D. M. Proserpio, *CrystEngComm*, 2011, **13**, 3947–3958.
88. M. Li, D. Li, M. O'Keeffe and O. M. Yaghi, *Chem. Rev.*, 2014, **114**, 1343–1370.
89. J. Keupp and R. Schmid, *Faraday Discuss.*, 2018, **211**, 79–101.
90. Y. Belmabkhout, R. S. Pillai, D. Alezi, O. Shekhah, P. M. Bhatt, Z. Chen, K. Adil, S. Vaesen, G. De Weireld, M. Pang, M. Suetin, A. J. Cairns, V. Solovyeva, A. Shkurenko, O. El Tall, G. Maurin and M. Eddaoudi, *J. Mater. Chem. A*, 2017, **5**, 3293–3303.
91. M. A. Addicoat, N. Vankova, I. F. Akter and T. Heine, *J. Chem. Theory Comput.*, 2014, **10**, 880–891.

92. D. E. Coupry, M. A. Addicoat and T. Heine, *J. Chem. Theory Comput.*, 2016, **12**, 5215–5225.
93. P. G. Boyd and T. K. Woo, *CrystEngComm*, 2016, **18**, 3777–3792.
94. B. Moulton and M. J. Zaworotko, *Chem. Rev.*, 2001, **101**, 1629–1658.
95. S. Barthel, E. V. Alexandrov, D. M. Proserpio and B. Smit, *Cryst. Growth Des.*, 2018, **18**, 1738–1747.
96. S. L. Gould, D. Tranchemontagne, O. M. Yaghi and M. A. Garcia-Garibay, *J. Am. Chem. Soc.*, 2008, **130**, 3246–3247.
97. J. Navarro-Sánchez, I. Mullor-Ruíz, C. Popescu, D. Santamaría-Pérez, A. Segura, D. Errandonea, J. González-Platas and C. Martí-Gastaldo, *Dalton Trans.*, 2018, **47**, 10654–10659.
98. A. P. Katsoulidis, D. Antypov, G. F. S. Whitehead, E. J. Carrington, D. J. Adams, N. G. Berry, G. R. Darling, M. S. Dyer and M. J. Rosseinsky, *Nature*, 2019, **565**, 213–217.
99. E. J. Carrington, R. Pétuya, R. K. Hylton, Y. Yan, D. Antypov, G. R. Darling, M. S. Dyer, N. G. Berry, A. P. Katsoulidis and M. J. Rosseinsky, *Cryst. Growth Des.*, 2019, **19**, 5604–5618.
100. E. S. Grape, H. Xu, O. Cheung, M. Calmels, J. Zhao, C. Dejoie, D. M. Proserpio, X. Zou and A. K. Inge, *Cryst. Growth Des.*, 2020, **20**, 320–329.
101. H. Furukawa, N. Ko, Y. B. Go, N. Aratani, S. B. Choi, E. Choi, A. Ö. Yazaydin, R. Q. Snurr, M. O'Keeffe, J. Kim and O. M. Yaghi, *Science*, 2010, **329**, 424–428.
102. F. J. Uribe-Romo, J. R. Hunt, H. Furukawa, C. Klöck, M. O'Keeffe and O. M. Yaghi, *J. Am. Chem. Soc.*, 2009, **131**, 4570–4571.
103. T. Ma, J. Li, J. Niu, L. Zhang, A. S. Etman, C. Lin, D. Shi, P. Chen, L.-H. Li, X. Du, J. Sun and W. Wang, *J. Am. Chem. Soc.*, 2018, **140**, 6763–6766.
104. P. V. Dau, M. Kim, S. J. Garibay, F. H. L. Münch, C. E. Moore and S. M. Cohen, *Inorg. Chem.*, 2012, **51**, 5671–5676.
105. L. Liu, X. Wang, Q. Zhang, Q. Li and Y. Zhao, *CrystEngComm*, 2013, **15**, 841–844.
106. R. J. Marshall, C. T. Lennon, A. Tao, H. M. Senn, C. Wilson, D. Fairen-Jimenez and R. S. Forgan, *J. Mater. Chem. A*, 2018, **6**, 1181–1187.
107. M. I. Gonzalez, E. D. Bloch, J. A. Mason, S. J. Teat and J. R. Long, *Inorg. Chem.*, 2015, **54**, 2995–3005.
108. Y. Liu, C. S. Diercks, Y. Ma, H. Lyu, C. Zhu, S. A. Alshmimri, S. Alshihri and O. M. Yaghi, *J. Am. Chem. Soc.*, 2019, **141**, 677–683.
109. Y. Liu, Y. Ma, Y. Zhao, X. Sun, F. Gándara, H. Furukawa, Z. Liu, H. Zhu, C. Zhu, K. Suenaga, P. Oleynikov, A. S. Alshammari, X. Zhang, O. Terasaki and O. M. Yaghi, *Science*, 2016, **351**, 365.
110. S. R. Batten and R. Robson, *Angew. Chem., Int. Ed.*, 1998, **37**, 1460–1494.
111. S. R. Batten, *CrystEngComm*, 2001, **3**, 67–72.
112. S. T. Hyde, B. Chen and M. O'Keeffe, *CrystEngComm*, 2016, **18**, 7607–7613.
113. E. V. Alexandrov, V. A. Blatov and D. M. Proserpio, *CrystEngComm*, 2017, **19**, 1993–2006.

114. B. Lukose, A. Kuc and T. Heine, *Chem. – Eur. J.*, 2011, **17**, 2388–2392.
115. E. L. Spitler, B. T. Koo, J. L. Novotney, J. W. Colson, F. J. Uribe-Romo, G. D. Gutierrez, P. Clancy and W. R. Dichtel, *J. Am. Chem. Soc.*, 2011, **133**, 19416–19421.
116. F. Auras, L. Ascherl, A. H. Hakimioun, J. T. Margraf, F. C. Hanusch, S. Reuter, D. Bessinger, M. Döblinger, C. Hettstedt and K. Karaghiosoff, *et al.*, *J. Am. Chem. Soc.*, 2016, **138**, 16703–16710.
117. L. Ascherl, T. Sick, J. T. Margraf, S. H. Lapidus, M. Calik, C. Hettstedt, K. Karaghiosoff, M. Döblinger, T. Clark, K. W. Chapman, F. Auras and T. Bein, *Nat. Chem.*, 2016, **8**, 310–316.
118. A. Halder, M. Ghosh, A. Khayum M, S. Bera, M. Addicoat, H. S. Sasmal, S. Karak, S. Kurungot and R. Banerjee, *J. Am. Chem. Soc.*, 2018, **140**, 10941–10945.
119. Y.-L. Li, E. V. Alexandrov, Q. Yin, L. Li, Z.-B. Fang, W. Yuan, D. M. Proserpio and T.-F. Liu, *J. Am. Chem. Soc.*, 2020, **142**, 7218–7224.
120. S. R. Batten, B. F. Hoskins, B. Moubaraki, K. S. Murray and R. Robson, *Chemical Communications*, 2000, 1095–1096.
121. C. J. Doonan, D. J. Tranchemontagne, T. G. Glover, J. R. Hunt and O. M. Yaghi, *Nat. Chem.*, 2010, **2**, 235–238.
122. S. Amirjalayer and R. Schmid, *J. Phys. Chem. C*, 2008, **112**, 14980–14987.
123. A. Kuc, A. Enyashin and G. Seifert, *J. Phys. Chem. B*, 2007, **111**, 8179–8186.
124. Y. H. Jhon, M. Cho, H. R. Jeon, I. Park, R. Chang, J. L. C. Rowsell and J. Kim, *J. Phys. Chem. C*, 2007, **111**, 16618–16625.
125. N. Klein, C. Herzog, M. Sabo, I. Senkovska, J. Getzschmann, S. Paasch, M. R. Lohe, E. Brunner and S. Kaskel, *Phys. Chem. Chem. Phys.*, 2010, **12**, 11778–11784.
126. P. S. Petkov, V. Bon, C. L. Hobday, A. B. Kuc, P. Melix, S. Kaskel, T. Düren and T. Heine, *Phys. Chem. Chem. Phys.*, 2019, **21**, 674–680.
127. L. K. Cadman, J. K. Bristow, N. E. Stubbs, D. Tiana, M. F. Mahon, A. Walsh and A. D. Burrows, *Dalton Trans.*, 2016, **45**, 4316–4326.
128. Z. Ji, T. Li and O. M. Yaghi, *Science*, 2020, **369**, 674–680.
129. J. H. Cavka, S. Jakobsen, U. Olsbye, N. Guillou, C. Lamberti, S. Bordiga and K. P. Lillerud, *J. Am. Chem. Soc.*, 2008, **130**, 13850–13851.
130. M. J. Cliffe, W. Wan, X. Zou, P. A. Chater, A. K. Kleppe, M. G. Tucker, H. Wilhelm, N. P. Funnell, F.-X. Coudert and A. L. Goodwin, *Nat. Commun.*, 2014, **5**, 4176.
131. L. Liu, Z. Chen, J. Wang, D. Zhang, Y. Zhu, S. Ling, K.-W. Huang, Y. Belmabkhout, K. Adil, Y. Zhang, B. Slater, M. Eddaoudi and Y. Han, *Nat. Chem.*, 2019, **11**, 622–628.
132. A. Sartbaeva, S. A. Wells, M. M. J. Treacy and M. F. Thorpe, *Nat. Mater.*, 2006, **5**, 962–965.
133. M. W. Deem, R. Pophale, P. A. Cheeseman and D. J. Earl, *J. Phys. Chem. C*, 2009, **113**, 21353–21360.
134. Y. Li, J. Yu and R. Xu, *J. Appl. Crystallogr.*, 2012, **45**, 855–861.
135. I. Rivin, *Nat. Mater.*, 2006, **5**, 931–932.

136. X. Liu, S. Valero, E. Argente and G. Sastre, *Faraday Discuss.*, 2018, **211**, 103–115.
137. D. W. Lewis, A. R. Ruiz-Salvador, A. Gómez, L. M. Rodriguez-Albelo, F.-X. Coudert, B. Slater, A. K. Cheetham and C. Mellot-Draznieks, *CrystEngComm*, 2009, **11**, 2272–2276.
138. M. Treacy, K. Randall, S. Rao, J. Perry and D. Chadi, *Z. Kristallogr.*, 1997, **212**, 768–791.
139. R. Pophale, P. A. Cheeseman and M. W. Deem, *Phys. Chem. Chem. Phys.*, 2011, **13**, 12407–12412.
140. V. A. Blatov, G. D. Ilyushin and D. M. Proserpio, *Chem. Mater.*, 2013, **25**, 412–424.
141. Y. G. Bushuev and G. Sastre, *J. Phys. Chem. C*, 2010, **114**, 19157–19168.
142. D. Majda, F. A. A. Paz, O. D. Friedrichs, M. D. Foster, A. Simperler, R. G. Bell and J. Klinowski, *J. Phys. Chem. C*, 2008, **112**, 1040–1047.
143. E. L. First, C. E. Gounaris, J. Wei and C. A. Floudas, *Phys. Chem. Chem. Phys.*, 2011, **13**, 17339–17358.
144. J. L. Salcedo Perez, M. Haranczyk and N. E. R. Zimmermann, *Z. Krystallog.*, 2019, **234**, 437–450.
145. Y. Li, X. Li, J. Liu, F. Duan and J. Yu, *Nat. Commun.*, 2015, **6**, 8328.
146. E. D. Kuznetsova, O. A. Blatova and V. A. Blatov, *Chem. Mater.*, 2018, **30**, 2829–2837.
147. J. Bunzen, J. Iwasa, P. Bonakdarzadeh, E. Numata, K. Rissanen, S. Sato and M. Fujita, *Angew. Chem., Int. Ed.*, 2012, **51**, 3161–3163.
148. J. Jia, F. Sun, H. Ma, L. Wang, K. Cai, Z. Bian, L. Gao and G. Zhu, *J. Mater. Chem. A*, 2013, **1**, 10112–10115.
149. D. J. Tranchemontagne, Z. Ni, M. O'Keeffe and O. M. Yaghi, *Angew. Chem., Int. Ed.*, 2008, **47**, 5136–5147.
150. N. J. Young and B. P. Hay, *Chem. Commun.*, 2013, **49**, 1354–1379.
151. K. Uehara, K. Kasai and N. Mizuno, *Inorg. Chem.*, 2007, **46**, 2563–2570.
152. V. Santolini, M. Miklitz, E. Berardo and K. E. Jelfs, *Nanoscale*, 2017, **9**, 5280–5298.
153. T. Tozawa, J. T. A. Jones, S. I. Swamy, S. Jiang, D. J. Adams, S. Shakespeare, R. Clowes, D. Bradshaw, T. Hasell, S. Y. Chong, C. Tang, S. Thompson, J. Parker, A. Trewin, J. Bacsa, A. M. Z. Slawin, A. Steiner and A. I. Cooper, *Nat. Mater.*, 2009, **8**, 973–978.
154. F. Albert Cotton, L. M. Daniels, C. Lin, C. A. Murillo and C. A. Murillo, *Chem. Commun.*, 1999, 841–842.
155. L. Turcani, E. Berardo and K. E. Jelfs, *J. Comput. Chem.*, 2018, **39**, 1931–1942.
156. B. P. Hay and T. K. Firman, *Inorg. Chem.*, 2002, **41**, 5502–5512.
157. T. Hasell and A. I. Cooper, *Nat. Rev. Mater.*, 2016, **1**, 16053.
158. N. Giri, M. G. Del Pópolo, G. Melaugh, R. L. Greenaway, K. Rätzke, T. Koschine, L. Pison, M. F. C. Gomes, A. I. Cooper and S. L. James, *Nature*, 2015, **527**, 216–220.
159. M. A. Zwijnenburg, E. Berardo, W. J. Peveler and K. E. Jelfs, *J. Phys. Chem. B*, 2016, **120**, 5063–5072.

160. A. G. Slater, M. A. Little, A. Pulido, S. Y. Chong, D. Holden, L. Chen, C. Morgan, X. Wu, G. Cheng, R. Clowes, M. E. Briggs, T. Hasell, K. E. Jelfs, G. M. Day and A. I. Cooper, *Nat. Chem.*, 2017, **9**, 17–25.
161. A. G. Slater, P. S. Reiss, A. Pulido, M. A. Little, D. L. Holden, L. Chen, S. Y. Chong, B. M. Alston, R. Clowes, M. Haranczyk, M. E. Briggs, T. Hasell, G. M. Day and A. I. Cooper, *ACS Cent. Sci.*, 2017, **3**, 734–742.
162. G. Zhang, O. Presly, F. White, I. M. Oppel and M. Mastalerz, *Angew. Chem., Int. Ed.*, 2014, **53**, 5126–5130.
163. J. T. A. Jones, T. Hasell, X. Wu, J. Bacsa, K. E. Jelfs, M. Schmidtmann, S. Y. Chong, D. J. Adams, A. Trewin, F. Schiffman, F. Cora, B. Slater, A. Steiner, G. M. Day and A. I. Cooper, *Nature*, 2011, **474**, 367–371.
164. M. A. Little, S. Y. Chong, M. Schmidtmann, T. Hasell and A. I. Cooper, *Chem. Commun.*, 2014, **50**, 9465–9468.
165. E. O. Pyzer-Knapp, H. P. G. Thompson, F. Schiffmann, K. E. Jelfs, S. Y. Chong, M. A. Little, A. I. Cooper and G. M. Day, *Chem. Sci.*, 2014, **5**, 2235–2245.
166. S. L. Price, *Chem. Soc. Rev.*, 2014, **43**, 2098–2111.
167. S. M. Woodley and R. Catlow, *Nat. Mater.*, 2008, **7**, 937–946.
168. A. R. Oganov, C. J. Pickard, Q. Zhu and R. J. Needs, *Nat. Rev. Mater.*, 2019, **4**, 331–348.
169. A. M. Reilly, R. I. Cooper, C. S. Adjiman, S. Bhattacharya, A. D. Boese, J. G. Brandenburg, P. J. Bygrave, R. Bylsma, J. E. Campbell, R. Car, D. H. Case, R. Chadha, J. C. Cole, K. Cosburn, H. M. Cuppen, F. Curtis, G. M. Day, R. A. DiStasio Jr, A. Dzyabchenko, B. P. van Eijck, D. M. Elking, J. A. van den Ende, J. C. Facelli, M. B. Ferraro, L. Fusti-Molnar, C.-A. Gatsiou, T. S. Gee, R. de Gelder, L. M. Ghiringhelli, H. Goto, S. Grimme, R. Guo, D. W. M. Hofmann, J. Hoja, R. K. Hylton, L. Iuzzolino, W. Jankiewicz, D. T. de Jong, J. Kendrick, N. J. J. de Klerk, H.-Y. Ko, L. N. Kuleshova, X. Li, S. Lohani, F. J. J. Leusen, A. M. Lund, J. Lv, Y. Ma, N. Marom, A. E. Masunov, P. McCabe, D. P. McMahon, H. Meekes, M. P. Metz, A. J. Misquitta, S. Mohamed, B. Monserrat, R. J. Needs, M. A. Neumann, J. Nyman, S. Obata, H. Oberhofer, A. R. Oganov, A. M. Orendt, G. I. Pagola, C. C. Pantelides, C. J. Pickard, R. Podeszwa, L. S. Price, S. L. Price, A. Pulido, M. G. Read, K. Reuter, E. Schneider, C. Schober, G. P. Shields, P. Singh, I. J. Sugden, K. Szalewicz, C. R. Taylor, A. Tkatchenko, M. E. Tuckerman, F. Vacarro, M. Vasileiadis, A. Vazquez-Mayagoitia, L. Vogt, Y. Wang, R. E. Watson, G. A. de Wijs, J. Yang, Q. Zhu and C. R. Groom, *Acta Crystallogr., Sect. B*, 2016, **72**, 439–459.
170. D. A. Bardwell, C. S. Adjiman, Y. A. Arnautova, E. Bartashevich, S. X. M. Boerrigter, D. E. Braun, A. J. Cruz-Cabeza, G. M. Day, R. G. Della Valle, G. R. Desiraju, B. P. van Eijck, J. C. Facelli, M. B. Ferraro, D. Grillo, M. Habgood, D. W. M. Hofmann, F. Hofmann, K. V. J. Jose, P. G. Karamertzanis, A. V. Kazantsev, J. Kendrick, L. N. Kuleshova, F. J. J. Leusen, A. V. Maleev, A. J. Misquitta,

S. Mohamed, R. J. Needs, M. A. Neumann, D. Nikylov, A. M. Orendt, R. Pal, C. C. Pantelides, C. J. Pickard, L. S. Price, S. L. Price, H. A. Scheraga, J. van de Streek, T. S. Thakur, S. Tiwari, E. Venuti and I. K. Zhitkov, *Acta Crystallogr., Sect. B*, 2011, **67**, 535–551.
171. G. M. Day, T. G. Cooper, A. J. Cruz-Cabeza, K. E. Hejczyk, H. L. Ammon, S. X. M. Boerrigter, J. S. Tan, R. G. Della Valle, E. Venuti, J. Jose, S. R. Gadre, G. R. Desiraju, T. S. Thakur, B. P. van Eijck, J. C. Facelli, V. E. Bazterra, M. B. Ferraro, D. W. M. Hofmann, M. A. Neumann, F. J. J. Leusen, J. Kendrick, S. L. Price, A. J. Misquitta, P. G. Karamertzanis, G. W. A. Welch, H. A. Scheraga, Y. A. Arnautova, M. U. Schmidt, J. van de Streek, A. K. Wolf and B. Schweizer, *Acta Crystallogr., Sect. B*, 2009, **65**, 107–125.
172. M. Bernabei, R. Pérez-Soto, I. Gómez Garca and M. Haranczyk, *Mol. Syst. Des. Eng.*, 2018, **3**, 942–950.
173. A. Pulido, L. Chen, T. Kaczorowski, D. Holden, M. A. Little, S. Y. Chong, B. J. Slater, D. P. McMahon, B. Bonillo, C. J. Stackhouse, A. Stephenson, C. M. Kane, R. Clowes, T. Hasell, A. I. Cooper and G. M. Day, *Nature*, 2017, **543**, 657–664.
174. J. Nyman, O. S. Pundyke and G. M. Day, *Phys. Chem. Chem. Phys.*, 2016, **18**, 15828–15837.
175. J. Nyman and G. M. Day, *Phys. Chem. Chem. Phys.*, 2016, **18**, 31132–31143.
176. Y. N. Heit and G. J. O. Beran, *Acta Crystallogr., Sect. B*, 2016, **72**, 514–529.
177. D. P. McMahon, A. Stephenson, S. Y. Chong, M. A. Little, J. T. A. Jones, A. I. Cooper and G. M. Day, *Faraday Discuss.*, 2018, **211**, 383–399.
178. J.-R. Li, D. J. Timmons and H.-C. Zhou, *J. Am. Chem. Soc.*, 2009, **131**, 6368–6369.
179. M. J. Prakash and M. S. Lah, *Chem. Commun.*, 2009, 3326–3341.
180. A. Mallick, B. Garai, D. D. Daz and R. Banerjee, *Angew. Chem., Int. Ed.*, 2013, **52**, 13755–13759.
181. J. D. Evans, D. M. Huang, M. R. Hill, C. J. Sumby, D. S. Sholl, A. W. Thornton and C. J. Doonan, *J. Phys. Chem. C*, 2015, **119**, 7746–7754.
182. A. Legrand, G. A. Craig, M. Bonneau, S. Minami, K. Urayama and S. Furukawa, *Chem. Sci.*, 2019, **10**, 10833–10842.
183. G. S. Larsen, P. Lin, K. E. Hart and C. M. Colina, *Macromolecules*, 2011, **44**, 6944–6951.
184. K. Althumayri, W. J. Harrison, Y. Shin, J. M. Gardiner, C. Casiraghi, P. M. Budd, P. Bernardo, G. Clarizia and J. C. Jansen, *Philos. Trans. R. Soc., A*, 2016, **374**, 20150031.
185. S. Jiang, K. E. Jelfs, D. Holden, T. Hasell, S. Y. Chong, M. Haranczyk, A. Trewin and A. I. Cooper, *J. Am. Chem. Soc.*, 2013, **135**, 17818–17830.
186. F. L. B. da Silva, W. Olivares-Rivas and P. J. Colmenares, *Mol. Simul.*, 2007, **33**, 639–647.
187. H. Y. Playford, L. R. Owen, I. Levin and M. G. Tucker, *Annu. Rev. Mater. Res.*, 2014, **44**, 429–449.
188. D. A. Keen and R. L. McGreevy, *Nature*, 1990, **344**, 423–425.

189. J. Pikunic, C. Clinard, N. Cohaut, K. E. Gubbins, J.-M. Guet, R. J. M. Pellenq, I. Rannou and J.-N. Rouzaud, *Langmuir*, 2003, **19**, 8565–8582.
190. K. T. Thomson and K. E. Gubbins, *Langmuir*, 2000, **16**, 5761–5773.
191. G. Opletal, T. C. Petersen, I. K. Snook and D. G. McCulloch, *J. Chem. Phys.*, 2007, **126**, 214705.
192. T. Iiyama, K. Hagi, T. Urushibara and S. Ozeki, *Colloids Surf., A*, 2009, **347**, 133–141.
193. T. D. Bennett, A. L. Goodwin, M. T. Dove, D. A. Keen, M. G. Tucker, E. R. Barney, A. K. Soper, E. G. Bithell, J.-C. Tan and A. K. Cheetham, *Phys. Rev. Lett.*, 2010, **104**, 115503.
194. Y. Yang, Y. K. Shin, S. Li, T. D. Bennett, A. C. T. van Duin and J. C. Mauro, *J. Phys. Chem. B*, 2018, **122**, 9616–9624.
195. J. D. Evans, K. E. Jelfs, G. M. Day and C. J. Doonan, *Chem. Soc. Rev.*, 2017, **46**, 3286–3301.
196. L. J. Abbott, K. E. Hart and C. M. Colina, *Theor. Chem. Acc.*, 2013, **132**, 1334.
197. J. M. H. Thomas, C. Mollart, L. Turner, P. Heasman, P. Fayon and A. Trewin, *J. Phys. Chem. B*, 2020, **124**, 7318–7326.
198. D. S. Sholl and R. P. Lively, *Nature*, 2016, **532**, 435.

CHAPTER 3

Atomistic Simulations of Mechanical Properties

JACK D. EVANS

Department of Inorganic Chemistry, Technische Universität Dresden, Bergstraβe 66, 01062 Dresden, Germany
Email: jack.evans@tu-dresden.de

3.1 Introduction

The fundamental strength and flexibility of materials are dictated by their mechanical properties. This generic term describes the behaviour of a material in response to combinations of stress (applied force) and strain (change in shape). In everyday life, the extremes of these properties, and their implications, are obvious, such as the flexibility exhibited by the soles of your shoes and the rigidity of a table. However, in materials science the seemingly endless combination of geometry, chemistry and electronics can provide access to mechanical properties that appear unbelievable, like a cloak that can hide the feelable-presence of an object.[1]

Mechanical properties are not simply dictated by the composition and identity of single atoms, or molecules, but instead governed by the complex interactions of large groups of atoms. This is distinctly illustrated by the extremely different properties for the allotropes of carbon (Figure 3.1).[2] Importantly, these complex interactions of large groups can span disparate scales from the vibration of atomic units to the macroscopic arrangements of crystallites.

In the understanding of mechanical properties and their implications, it is imperative to differentiate between elastic and plastic properties.

Inorganic Materials Series No. 8
Computer Simulation of Porous Materials: Current Approaches and Future Opportunities
Edited by Kim Jelfs

Published by the Royal Society of Chemistry, www.rsc.org

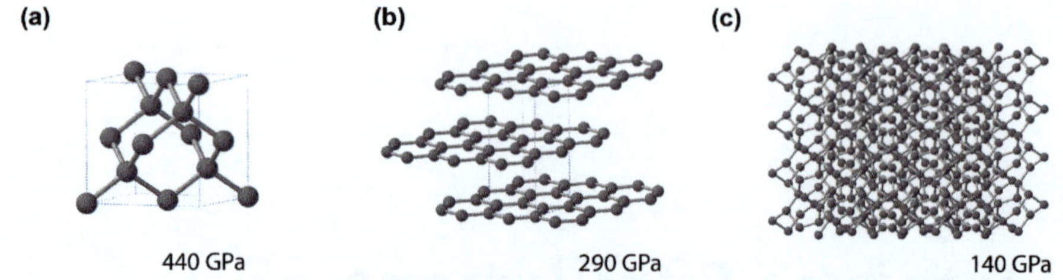

Figure 3.1 The structure of different carbon allotropes can either produce an extremely strong material or a flexible material. For example, the atomistic structure of diamond, graphite and an amorphous carbon structure (a–c, respectively) produce very different values of bulk modulus,[3–5] a simple description of flexibility.

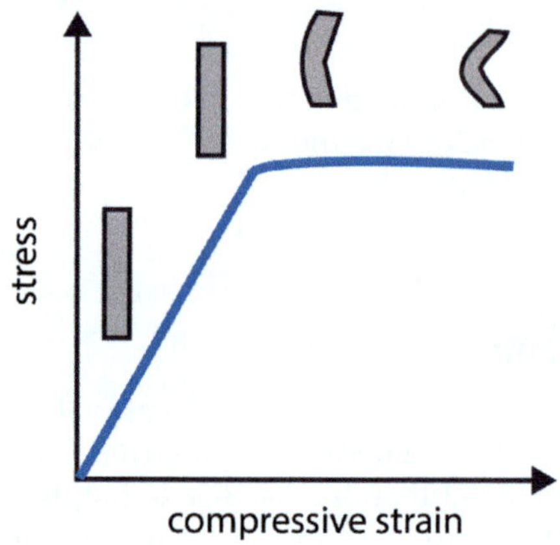

Figure 3.2 A hypothetical stress–strain curve for a steel rod under tensile strain. The axes here represent compressive strain (the length of the material) and stress (the force applied). The linear region represents the elastic region that is followed by plastic deformation, represented by bending of the rod, at large strains.

All materials, even diamond, will shrink, stretch or bend upon the application of some force. This small and reversible change in structure defines the elastic properties of a material. Beyond this elastic regime, increasing deformation produces irreversible changes in structure, such as fracture and necking. This plastic deformation is observed in the crushing of a can upon sufficient force. These regimes are illustrated in Figure 3.2, which highlights a hypothetical stress–strain curve for a steel rod under increasing compressive strain.[6] Here, we observe the elastic response and the point where plasticity occurs (yield strength).

Experimentally, the understanding of these properties is well-established, and is a fundamental requirement for building robust bridges and skyscrapers. However, simulating the mechanical behaviour of solids in response to external force and at specific temperatures requires advanced simulations. In fact, most of the practically relevant simulations for engineering and design employ finite-element approaches,[7] which generally ignore atomic or molecular effects, treating the system as a continuum model. These models can undoubtedly provide all the insight necessary for the structural design of macroscopic objects; however, complete atomistic simulation is necessary for understanding the properties of emerging materials, where their essential mechanical behaviour is unknown. This is especially important for materials that may behave in unexpected ways, as the standard experimental protocols for mechanical testing are often not amenable to micron-size crystalline materials, thus simulations are unique in providing insight into the mechanical behaviour of these structures.

In the search for new materials that display extraordinary mechanical characteristics for use in emerging applications, researchers are increasingly considering diverse families of different materials.[8] Specifically, porous materials display many interesting properties and these will be the focus of this chapter. This includes conventional dense, inorganic nanoporous materials such as metal oxides and zeolites and also crystalline porous materials, which are constructed from relatively weaker bonds (metal–ligand bonds, π–π interactions and hydrogen bonding). These softer, although crystalline, materials commonly exhibit anomalous behaviour due to their flexibility (Figure 3.3). Metal–organic frameworks (MOFs) are one class of porous materials that have seen significant attention since Yaghi and co-workers first demonstrated the immense microporosity of MOF-5, a three-dimensional framework constructed from zinc and 1,4-benzenedicarboxylate.[9] MOF materials are of particular interest as they are constructed using a modular chemical approach that uses different metal nodes and ligands to finely tune the structure and properties of the resulting framework.[10] Interestingly, the mechanical properties of MOFs, such as rigidity, are generally in between that observed for organic polymers and inorganic ceramics, such as nanoporous aluminosilicate zeolites.[11] Moreover, the properties of these conventional porous solids, such as porous silicas produced from traditional hydrothermal synthetic methods, are often difficult to control. The reactants do not usually retain their structure during the reaction, leading to a poor correlation between reactants and products. In contrast,

Figure 3.3 Atomistic structures of a conventional dense solid CsCl,[12] a porous siliceous zeolite with RHO framework type[13] and the open framework structure of bio-MOF1.[14] Atom colours are represented by: Cs, pink; Cl, green; Si, yellow; O, red; C, grey; H, white; N, blue; Zn, dark red.

MOF materials can be designed beginning with well-defined molecular building blocks that maintain their structural integrity throughout synthesis. The powerful design principles demonstrated in MOFs allow significant control over the framework structure and pore chemistry. In addition to MOFs, other crystalline frameworks such as covalent organic frameworks (COFs), porous molecular organic solids and other molecular frameworks, which have been shown to have similar properties, have also attracted attention from the research community.

Given these unique and exciting arrangements of structure and chemistry, a thorough understanding of nanoscale mechanics and the ability to quantify the mechanics of porous materials and other interesting materials using computational approaches is crucial. In this chapter, I will outline fundamental mechanical properties, the different computational approaches employed to simulate these properties and, finally, interesting consequences of these properties for different structural arrangements as observed in different families of materials.

3.2 Fundamental Mechanical Properties

The fundamental elastic properties of a material will generally determine how much it will compress or expand for a specific amount of external pressure. It is usually defined by the ratio of the change in deformation or strain for the given applied force. This is nicely illustrated in the definition of the bulk modulus. The bulk modulus is a measure of how resistant to compression a material is and is defined as the ratio of an infinitesimal pressure increase to the relative decrease in volume. Bulk modulus is often designated by K or B. This is formally defined in the equation below, where P is pressure, V is volume:

$$K = -V\frac{\mathrm{d}P}{\mathrm{d}V}$$

The bulk moduli of solids is often quoted in GPa, for example, the bulk modulus of diamond is approximately 400 GPa and the bulk modulus of steel is approximately 100 GPa.[6] This value straightforwardly allows one to understand the amount of deformation expected for a given pressure (or isotropic hydrostatic force). For example, a material with a bulk modulus of 400 GPa requires an external pressure of 4 GPa (almost 40 000 times atmospheric pressure) to produce a 1% change in volume. Bulk moduli for a selection of materials are displayed in Table 3.1.

The bulk modulus is an important property that can be straightforwardly measured experimentally. This is often achieved, for example, using *in situ* powder X-ray diffraction, where pressure is applied using a diamond-anvil cell and a pressure transducing medium to produce hydrostatic pressure.[18] However, it is a bulk property and can overlook important atomistic detail, treating the solid as a homogeneous continuous medium defined by only volume rather than as a periodic array of atoms.

Table 3.1 Reported bulk moduli for a variety of solids.

Material	Bulk modulus/GPa	Reference
Diamond	440	4
Butadiene rubber	2.4	15
Stainless steel	160	16
Ice (Ih phase)	10	17

The study of mechanics, with atomistic detail, ultimately, is rooted in established and fundamental physical concepts, Hooke's law and Newton's second law. Hooke's law states that in an elastic solid the strain is directly proportional to the stress and easily defined for the compression (or extension) of a spring. This is defined in the following equation, where the characteristic elasticity of a spring is defined by the force F applied to produce a displacement x:

$$k = \frac{F}{x}$$

Notably, and as discussed in the introduction, this treatment applies to small strains only and is not appropriate for the plastic regime. In fact, plastic deformation, which is displacement relating to permanent deformation or non-reversible shape change, is defined when Hooke's law is no longer satisfied.

3.2.1 Complete Elastic Properties

In order to completely calculate the elastic response for a three-dimensional solid, the strain (or displacement) must be defined. If we consider three orthogonal vectors $\hat{\mathbf{x}}, \hat{\mathbf{y}}, \hat{\mathbf{z}}$ of unit length embedded in an unstrained solid, for example, these could represent the periodicity in a crystal (the unit cell). After a small uniform deformation of the solid has taken place, these axes can be distorted in both orientation and in length. For a uniform deformation, each primitive cell of the crystal is deformed in the same way. The distorted axes x', y', z' can be written in terms of the old axes, as below:

$$x' = (1 + \epsilon_{xx})\hat{\mathbf{x}} + \epsilon_{xy}\hat{\mathbf{y}} + \epsilon_{xz}\hat{\mathbf{z}}$$
$$y' = \epsilon_{yx}\hat{\mathbf{x}} + (1 + \epsilon_{yy})\hat{\mathbf{y}} + \epsilon_{yz}\hat{\mathbf{z}}$$
$$z' = \epsilon_{zx}\hat{\mathbf{x}} + \epsilon_{zy}\hat{\mathbf{y}} + (1 + \epsilon_{zz})\hat{\mathbf{z}}$$

These coefficients define the strain in each direction and are dimensionless values. As demonstrated by Kittel,[19] if we consider the effect of this strain on an atom, initially at $\vec{r} = x\hat{\mathbf{x}} + y\hat{\mathbf{y}} + z\hat{\mathbf{z}}$, the new position will be $\overrightarrow{r'} = x\mathbf{x}' + y\mathbf{y}' + z\mathbf{z}'$. As a result, the displacement can be defined in a general form by the equation for $R(\vec{r})$:

$$R(\vec{r}) = \vec{r} - \overrightarrow{r'} = u(\vec{r})\hat{\mathbf{x}} + v(\vec{r})\hat{\mathbf{y}} + w(\vec{r})\hat{\mathbf{z}}$$

It is not common to use the coefficients ϵ but instead strain components e, which completely define the strain:

$$e_{xx}=\epsilon_{xx}=\frac{\partial u}{\partial x} \qquad e_{xy}=\epsilon_{yx}+\epsilon_{xy}=\frac{\partial u}{\partial y}+\frac{\partial v}{\partial x}$$

$$e_{yy}=\epsilon_{yy}=\frac{\partial v}{\partial y} \qquad e_{yz}=\epsilon_{zy}+\epsilon_{yz}=\frac{\partial v}{\partial z}+\frac{\partial w}{\partial y}$$

$$e_{zz}=\epsilon_{zz}=\frac{\partial w}{\partial z} \qquad e_{zx}=\epsilon_{zx}+\epsilon_{xz}=\frac{\partial u}{\partial z}+\frac{\partial w}{\partial x}$$

The strain components in the orthogonal directions are straightforwardly defined, while the components e_{xy}, e_{yz} and e_{zx} describe the change in angle of the axes.

We can define stress or the force acting on a unit area, in a Cartesian coordinate system, using nine distinct stress components: σ_{xx}, σ_{xy}, σ_{xz}, σ_{yx}, σ_{yy}, σ_{yz}, σ_{zx}, σ_{zy}, σ_{zz}, where the subscript indicates the direction of the force and the normal to the plane of which the force is applied as visualised in Figure 3.4.

When this definition of stress is applied, we can remove three components as the conditions $\sigma_{yz}=\sigma_{zy}$, $\sigma_{zx}=\sigma_{xz}$, $\sigma_{xy}=\sigma_{yz}$ must be true if the object is stationary, the converse produces a net torque and thus angular acceleration. These components of stress can be defined by force per unit area or alternatively energy per unit volume. In the formulation presented, and elsewhere in literature, the standard Voigt notation can be used for the indices of strains, stresses and elastic tensors. This is achieved by contracting a

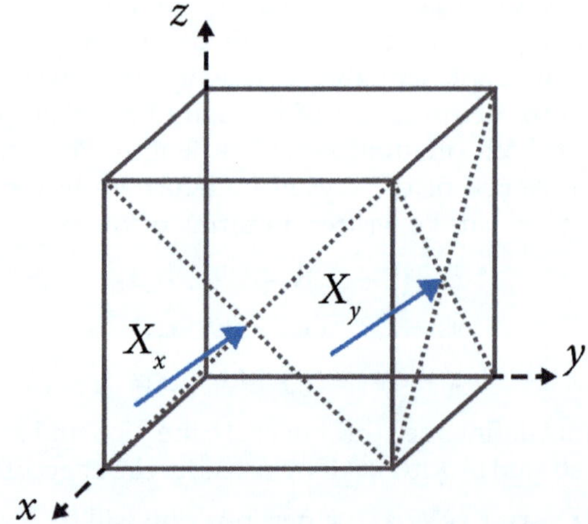

Figure 3.4 An illustration of the directional stress components on a cube in a Cartesian coordinate system. σ_{X_x} refers to a force applied in the x-direction to the area of a plane with the normal also lying in the x-direction (X_x). In contrast, σ_{X_y} refers to a force also applied in the x-direction, but to the area of a plane with normal lying in the y-direction (X_y).

pair of Cartesian indices into a single subscript integer: $xx \rightarrow 1$, $yy \rightarrow 2$, $zz \rightarrow 3$, $yz \rightarrow 4$, $xz \rightarrow 5$, $xy \rightarrow 6$.

Small strains (or deformations), as per Hooke's law, are directly proportional to the applied stress and from the presented definitions of strain and stress we can define the elastic stiffness constants, C_{ij}.

$$\sigma_1 = C_{11}e_1 + C_{12}e_2 + C_{13}e_3 + C_{14}e_4 + C_{15}e_5 + C_{16}e_6$$
$$\sigma_2 = C_{21}e_1 + C_{22}e_2 + C_{23}e_3 + C_{24}e_4 + C_{25}e_5 + C_{26}e_6$$
$$\sigma_3 = C_{31}e_1 + C_{32}e_2 + C_{33}e_3 + C_{34}e_4 + C_{35}e_5 + C_{36}e_6$$
$$\sigma_4 = C_{41}e_1 + C_{42}e_2 + C_{43}e_3 + C_{44}e_4 + C_{45}e_5 + C_{46}e_6$$
$$\sigma_5 = C_{51}e_1 + C_{52}e_2 + C_{53}e_3 + C_{54}e_4 + C_{55}e_5 + C_{56}e_6$$
$$\sigma_6 = C_{61}e_1 + C_{62}e_2 + C_{63}e_3 + C_{64}e_4 + C_{65}e_5 + C_{66}e_6$$

These constants or moduli of elasticity, as deformation is unitless, also have the dimensions of force per unit area or energy per unit volume. These constants are often presented as a matrix C_{ij}, and as a result Hooke's law can be presented as $\sigma_i = C_{ij}e_j$. Moreover, instead of describing this proportionality with respect to stress, this can also be presented with respect to the deformation, to produce $e_i = S_{ij}\sigma_j$, where the matrix S_{ij} are the elastic compliance constants.

The above 36 elements of the C_{ij} matrix completely define the elasticity of a crystalline solid; however, in practice the number of unique constants can be reduced by several considerations. For example, the C_{ij} matrix is symmetric and is thus composed of 21 independent components and by applying our understanding of necessary stability conditions this can be reduced further.[20] The elastic energy for a crystal can be approximated using the energy expression as dictated by Hooke's law:

$$E = E_0 + \frac{V_0}{2}\sum_{i,j=1}^{6} C_{ij}e_ie_j$$

where E_0 and V_0 are the equilibrium energy and volume of the crystal. Based on this harmonic energy approximation, for a stable system in the absence of external load, the elastic energy, E, must increase in response to a deformation ($E > 0, \forall\, e \neq 0$). This condition is referred to as the elastic stability criterion and was mathematically related by Born to indices of the C_{ij} by four conditions:[21]

1. The matrix C_{ij} is definite positive.
2. All eigenvalues of C_{ij} are positive.
3. All the leading principal minors of C_{ij} are positive, a property which is also known as Sylvester's criterion.
4. An arbitrary set of minors of C_{ij} are all positive.

These above formulations are the generic Born elastic stability conditions for an unstressed crystal. They are valid regardless of the symmetry of the crystal.

Table 3.2 Laue-classes and number of independent second-order elastic constants C_{ij} as reported by Mouhat and Coudert.[20] Please note the naming convention of Wallace is used to distinguish the different Laue-classes of the same crystal system.[22]

Crystal system	Laue-class	Number of independent C_{ij}
Triclinic	$\bar{1}$	21
Monoclinic	$2/m$	13
Orthorhombic	mmm	9
Tetragonal (II)	$4/m$	7
Tetragonal (I)	$4/mmm$	6
Rhombohedral (II)	$\bar{3}$	7
Rhombohedral (I)	$\bar{3}m$	6
Hexagonal (II)	$6/m$	5
Hexagonal (I)	$6/mmm$	5
Cubic (II)	$m\bar{3}$	3
Cubic (I)	$m\bar{3}m$	3

(a)

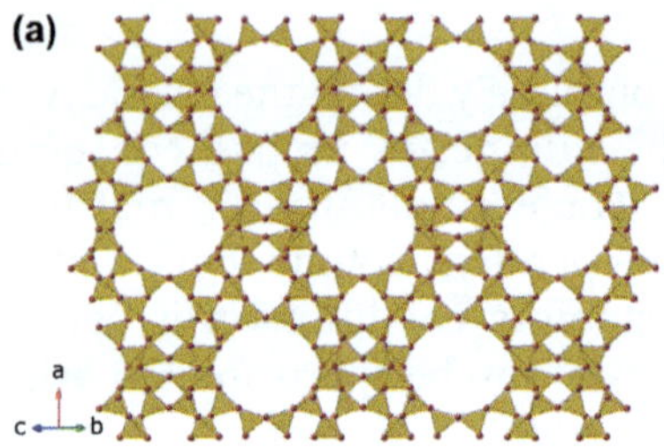

(b)

$$C_{ij} = \begin{pmatrix} 83.504 & 50.299 & 50.299 & 0 & 0 & 0 \\ 50.299 & 83.504 & 50.299 & 0 & 0 & 0 \\ 50.299 & 50.299 & 83.504 & 0 & 0 & 0 \\ 0 & 0 & 0 & 19.636 & 0 & 0 \\ 0 & 0 & 0 & 0 & 19.636 & 0 \\ 0 & 0 & 0 & 0 & 0 & 19.636 \end{pmatrix}$$

Figure 3.5 An atomistic depiction of the FAU zeolite (faujasite-type structure), which exhibits cubic high symmetry or more specifically the Laue-class $m\bar{3}m$ (a). The resulting C_{ij} matrix in units GPa computed by Coudert[24] (b). The tetrahedral units of SiO_4 are represented by yellow polyhedra and oxygen atoms by red spheres.

Mouhat and Coudert extended these necessary stability conditions for application to the 11 Laue-classes, which describe the symmetry of a crystal, as shown in Table 3.2.[20]

As a result, the number of unique C_{ij} can vary between 3 and 21, depending on the symmetry of the crystal system studied. For example, a material like the FAU zeolite[23] (Figure 3.5) with a cubic high symmetry system ($m\bar{3}m$) has only three unique elements of C_{ij} that define the harmonic elastic response of this material.

The above formulation of the elastic matrix (C_{ij}) completely defines the elastic behaviour of a material for forces applied in any number of directions. Thus, all the measured compressive, tensile and shear behaviour can be derived from the elastic matrix.

3.2.2 Young's Modulus

The most common tensile moduli discussed in materials research is Young's modulus. This simply describes a relationship between force and resulting deformation for a uniaxial deformation, either compressive or tensile (compression or extension):

$$E = \frac{\sigma(\varepsilon)}{\varepsilon}$$

where σ and ε are the stress and strain. For example, the Young's modulus can be readily computed for the deformation of a rod under a given compressive (or tensile) load and is equivalent to the initial slope of the stress–strain curve (Figure 3.2).

On the microscale, the Young's modulus can be experimentally measured by atomic force microscopy (AFM) allowing for the determination of this fundamental quantity for all different types of materials.[25] This is achieved by careful interpretation of AFM force curves, which characterise tip–sample interactions,[26] to provide mechanical and adhesive properties of the investigated surface.

The Young's modulus can also be obtained by using the elastic compliance matrix (S_{ij}), where the directions of the force are characterised by a unit vector **u**:

$$\mathbf{u} = \begin{pmatrix} \sin(\theta)\cos(\varphi) \\ \sin(\theta)\sin(\varphi) \\ \cos(\theta) \end{pmatrix}$$

As a result, the elastic compliance matrix is transformed to a new basis set represented by two angles (θ, φ) that dictate the direction of the normal stress, using points on the unit sphere:

$$E(\theta, \varphi) = \frac{1}{\sum_{ijkl} \mathbf{u}_i \mathbf{u}_j \mathbf{u}_k \mathbf{u}_l S_{ijkl}}$$

A detailed description of this is reported by Marmier *et al.*[27] and here we use notation introduced in their report.

3.2.3 Linear Compressibility

The compressibility of a material describes the relative deformation of its linear dimensions with respect to isotropic stress, for example, external pressure. The resulting deformation is straightforwardly calculated in tensorial form using the compliance matrix:

$$\beta(\theta, \varphi) = \sum_{ijk} \mathbf{u}_i \mathbf{u}_j S_{ijkk}$$

which gives the linear compressibility for any given direction, $\beta(\theta, \varphi)$. The linear compressibility averaged over all space is the compressibility, which is the inverse of the bulk modulus. Negative linear compressibility is considered an unusual mechanical property as it is rarely observed; however, several examples have been recently reported.[28]

3.2.4 Poisson's Ratio

A positive strain (tensile) in one direction can also contribute a negative strain (compressive) in the other direction. This is simply observed in the stretching of a rubber band, which when stretched in one direction becomes thinner in the other directions. This lateral deformation that accompanies a longitudinal deformation is called the Poisson effect.

The Poisson's ratio (ν) is a material property defined simply as the ratio between the applied strain to the observed strain, in the orthogonal direction. For example:

$$\nu = -\frac{\varepsilon_{\text{lateral}}}{\varepsilon_{\text{longitudinal}}}$$

The negative sign in the definition accounts for the expected behaviour of expansion in the longitudinal direction, in response to a contraction in the lateral direction. The average Poisson's ratio is a dimensionless parameter that can provide insight into the mechanical nature of the material. Distinct classes of engineered structures can be easily separated by comparing the Poisson's ratio (see Table 3.3).

For periodic atomistic structures, the Poisson's ratio can be defined for all directions, not just longitudinal and lateral, and thus depends on three angles and can be defined using a second, perpendicular unit direction **v**:

$$\mathbf{v} = \begin{pmatrix} \cos(\theta)\cos(\varphi)\cos(\chi) - \sin(\varphi)\sin(\chi) \\ \cos(\theta)\sin(\varphi)\cos(\chi) - \cos(\varphi)\sin(\chi) \\ -\sin(\theta)\cos(\chi) \end{pmatrix}$$

to give the Poisson's ratio in all directions:

$$\nu(\theta, \varphi, \chi) = -\frac{\sum_{ijkl} \mathbf{u}_i \mathbf{u}_j \mathbf{v}_k \mathbf{v}_l S_{ijkl}}{\sum_{ijkl} \mathbf{u}_i \mathbf{u}_j \mathbf{u}_k \mathbf{u}_l S_{ijkl}}$$

Table 3.3 Reported Poisson's ratio for a variety of materials.

Material	Poisson's ratio	Reference
Natural rubber	0.4999	29
Cork	0.097	30
Polystyrene	0.34	31
Silicate glasses	0.20–0.37	32

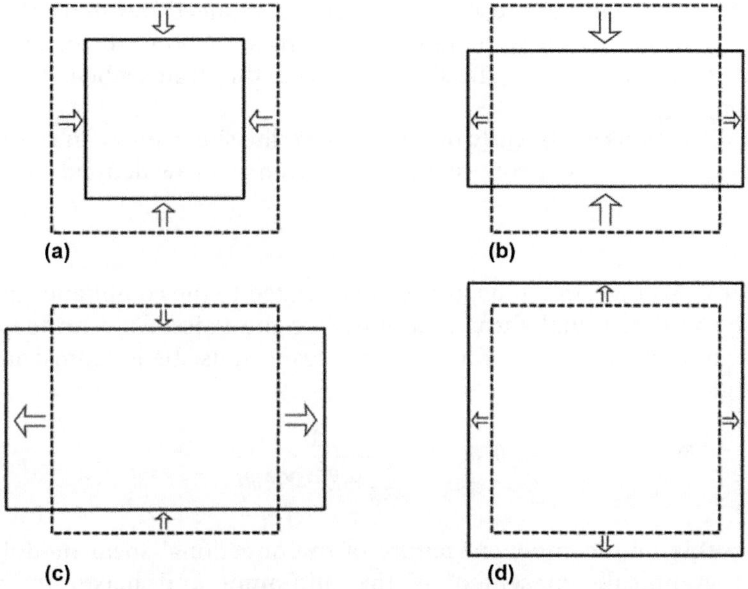

Figure 3.6 Demonstration of various auxetic types for the cross-section of a material subjected to an applied tensile load in the direction normal to the page. The original cross-section is represented by a dotted line and the final cross-section a full line. Conventional behaviour with a positive Poisson's ratio in all directions (a). A material with this axis has at least some negative Poisson's ratio (b). A material where the mean of the Poisson's ratios around this axis is negative (c). A material that shows a negative Poisson's ratio for all transverse directions (d).
Reproduced ref. 33 with permission from the Royal Society of Chemistry.

The Poisson's ratio can be graphically presented as the minimum and maximum value for a prescribed unit direction (θ, φ).

On average, most materials demonstrate a positive Poisson's ratio and materials that have approximately zero Poisson's ratio are named anepirretic materials. Surprisingly, some materials even exhibit a negative Poisson's ratio, labelled as auxetic materials. The directionality dependence of Poisson's ratio has necessitated further classification.[33] Materials can be classified in three groups and with three subgroups to describe the dependence of auxeticity on the direction of stress considered (Figure 3.6). A material can belong to several classes of auxeticity and, very rarely, materials can be classified as completely auxetic, where they will show a negative Poisson's ratio in all directions regardless of the direction of force.

3.2.5 Shear Modulus

The previously considered elastic moduli, for example, Young's moduli and bulk modulus, result from describing the structural behaviour for either compressive or tensile stress. However, not all forces that a material can encounter are a normal stress and as a result we need to extend the fundamental elastic moduli to include shear forces, or other distortional deformations. In Section 3.2.1, the shear forces are the strains labelled: σ_{xy}, σ_{xz}, σ_{yx}, σ_{yz}, σ_{zx}, σ_{zy}.

Shear distortions are linearly proportional to the shear stress, in the elastic regime, to give the shear counterpart (G) of Hooke's Law defined as:

$$\sigma_{xy} = Ge_{xy}$$

Similar to Young's modulus, this can be related to the compliance matrix; however, the directional shear modulus, like the Poisson's ratio, depends on two perpendicular directions and as a result must be is represented by three angles:

$$G(\theta, \varphi, \chi) = \frac{1}{\sum_{ijkl} \mathbf{u}_i \mathbf{v}_j \mathbf{u}_k \mathbf{v}_l S_{ijkl}}$$

Due to this multicomponent nature of the directional shear modulus, it is often graphically presented as the minimum and maximum shear modulus in a given unit direction (θ, φ). To achieve this for each θ, φ, the angle of the shear, χ, is scanned to find the minimum and maximum values in this direction.

3.2.6 Averaging Schemes for Elastic Moduli

The elastic moduli discussed above will be the same in all directions for isotropic materials; however, many materials are anisotropic and will show orientation dependent moduli. For example, Figure 3.7 demonstrates the

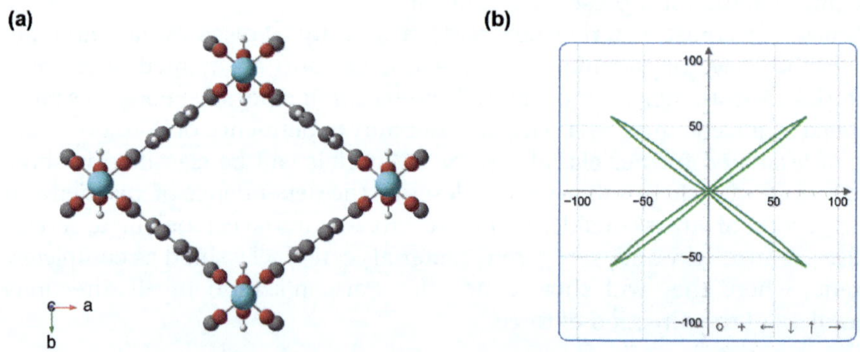

Figure 3.7 An atomistic depiction of the metal–organic framework MIL-53(Al), which exhibits orthorhombic symmetry (a).[34] The resulting spatial dependence of Young's modulus as computed by Ortiz *et al.*[35] displayed in units GPa and for the *a*, *b*-plane (b). Atom colours are represented by: Al, teal; O, red; C, grey; H, white. The mechanical properties are plotted using the ELATE software.[36]

spatial dependence of Young's modulus for the metal–organic framework MIL-53(Al).

For understanding and comparison, the elastic properties of anisotropic materials can be simplified by considering an equivalent isotropic material. There are three main schemes used to average elastic moduli: Voigt,[37] Reuss[38] and Hill.[39]

The Voigt averaging scheme uses specific elements of the elastic matrix (C_{ij}):

$$a_V = \frac{C_{11} + C_{22} + C_{33}}{3}, \quad b_V = \frac{C_{23} + C_{13} + C_{12}}{3}, \quad c_V = \frac{C_{44} + C_{55} + C_{66}}{3}$$

to give the Voigt averaged bulk modulus and shear modulus:

$$K_V = \frac{a_V - 2b_V}{3}, \quad G_V = \frac{a_V - b_V + 3c_V}{5}$$

Contrastingly, the Reuss averaging scheme uses specific elements of the compliance matrix (S_{ij}):

$$a_R = \frac{S_{11} + S_{22} + S_{33}}{3}, \quad b_R = \frac{S_{23} + S_{13} + S_{12}}{3}, \quad c_R = \frac{S_{44} + S_{55} + S_{66}}{3}$$

to give the Reuss averaged bulk modulus and shear modulus:

$$K_V = \frac{1}{3a_R + 6b_R}, \quad G_V = \frac{5}{4a_R - 4b_R + 3c_R}$$

The Hill average is subsequently defined as the arithmetic average of the Voigt and Reuss moduli.

The averaged bulk and shear moduli values can then be used to give an average Young's modulus and Poisson's ratio given the relations; these are identical for both the Voigt and Reuss schemes:

$$E = \left(\frac{1}{3G} + \frac{1}{9K}\right)^{-1}, \quad \nu = \frac{1}{2}\left(1 - \frac{3G}{3K - G}\right)$$

Alternatively, a non-analytical direct averaging scheme can be employed, based on a numerical average of the calculated moduli. This scheme must be converged with a given mesh of different directions and thus cannot be straightforwardly computed and is rarely used.

3.2.7 Anisotropy of Mechanical Properties

An important indicator of the average mechanical properties of a material is the anisotropy magnitude. For example, Oritz *et al.* suggested the importance of the anisotropy of mechanical properties for understanding the flexibility of porous crystals.[35] The anisotropy of the Young's modulus, shear modulus and Poisson's ratio can be straightforwardly determined from our definitions of these directional moduli. The value of anisotropy of each of these properties is calculated as the ratio of its maximum value to its minimum value, which varies from 1 to infinity, computed by scanning the angles of the unit direction. A value of one for this property would describe an isotropic response, whereas very directional, anisotropic materials, would have moduli with an anisotropic value much greater than one, possibly infinity.

As described above, it is straightforward to consider the anisotropy for a specific elastic modulus; however, there have been efforts to quantify elastic anisotropy using a single parameter. This was first introduced for cubic crystal systems, as detailed by Chung and Buessem as the parameter A and A^*, which use the unique elements of a complete elasticity matrix C_{ij} with cubic symmetry:[40]

$$A = \frac{2C_{44}}{C_{11} - C_{12}}$$

$$A^* = \frac{3(A-1)^2}{[3(A-1)^2 + 25A]}$$

The empirical corrections used to define A^* produces a useful metric for the relative magnitude of elastic anisotropy. For a perfectly isotropic system $A^* = 0$ and for increasing anisotropy $A^* > 0$. These empirical improvements ensure this metric is single-valued and a practical measure of anisotropy. However, A and A^* are only applicable to cubic systems and are meaningless for other crystal classes. Recently, in fact, until the work of Ranganathan and Ostoja-Starzewski[41] there was no available measure for anisotropy featuring universal applicability to different crystal symmetries.

The universal anisotropy index (A^U), introduced by Ranganathan and Ostoja-Starzewski,[41] quantifies single crystal anisotropy and, importantly, is unique in its definition and can also account for bulk contributions. This can be defined with respect to the Voigt and Reuss averaging schemes for Young's and shear modulus:

$$A^U = 5\frac{G_V}{G_R} + \frac{K_V}{K_R} - 6$$

This can be interpreted as a generalisation of the anisotropy index (A), using all the stiffness coefficients to account for the tensorial nature of elastic stiffness. A^U is zero for locally isotropic single crystals and the value defines the extent of anisotropy ($A^U > 0$) considering both shear and the bulk contributions.

3.2.8 Beyond the Elastic Regime

To capture the mechanical behaviour of materials beyond the elastic regime, which is generally represented by deformations of less than 1%, it is essential to construct pressure–volume equations of state. Equations of state describe a material's property, volume, with reference to an imposed state, in this case external force or pressure. As discussed later, in Section 3.3.1, equations of state can be fit by empirical relations to give properties such as bulk modulus.

To produce an equation of state for an atomistic system, molecular dynamics simulations are performed at a defined temperature for a series of volume, within a given range of interest. The volume of the system is kept constant, however, the cell shape should be allowed to fluctuate.[42] This should be achieved so that the ensemble-averaged stress $\langle \sigma_a \rangle$ vanishes. This is especially achieved by using the (N, V, $\sigma_a = 0$, T) ensemble, which was specifically designed because large fluctuations in the internal pressure, present in the (N, P, $\sigma_a = 0$, T) ensemble, generally lead to poor convergence. The resulting ensemble-averaged internal pressure $\langle P \rangle$, assuming mechanical equilibrium, exerted by the material at a given volume is the exerted pressure from the environment on the system at this volume. Hence, the obtained relationship corresponds to the macroscopic pressure–volume equation of state.

The pressure–volume equation of state provides the structure of the atomistic system at any given pressure. This can be graphically demonstrated by computing the intersection between the equation of state and a horizontal line at the required pressure. Importantly, all intersections with a negative slope represent thermodynamically stable structures $\left(\frac{\partial P}{\partial V} < 0\right)$ and thus can be produced experimentally. As an example, this is demonstrated for the metal–organic framework material DMOF-1(Zn). Figure 3.8 illustrates that we find that the transition pressure for instability of this material changes with temperature and a possible metastable dense phase also disappears with increasing temperature.[43]

As the equation of state provides information outside of the elastic regime this also provides information of the different phases of a system. Crucially, the transition pressure, which represents the pressures above or below a phase that is no longer stable, can directly be identified as the maxima and minima of the equation of state. As a result, this analysis can provide unsurpassed information on the stability and switchability of materials.

3.3 Simulation Approaches

The simulation of these properties can be achieved using highly accurate density functional theory or more approximate classical potentials that are used to describe the interactions between the atoms within the structure. Regardless of the method used to obtain an energy for a given deformation, there are a range of methods that can be applied.

3.3.1 Static Methods

Elastic moduli, such as bulk modulus and Young's modulus, can be calculated from their specified definitions. For example, the bulk modulus is the

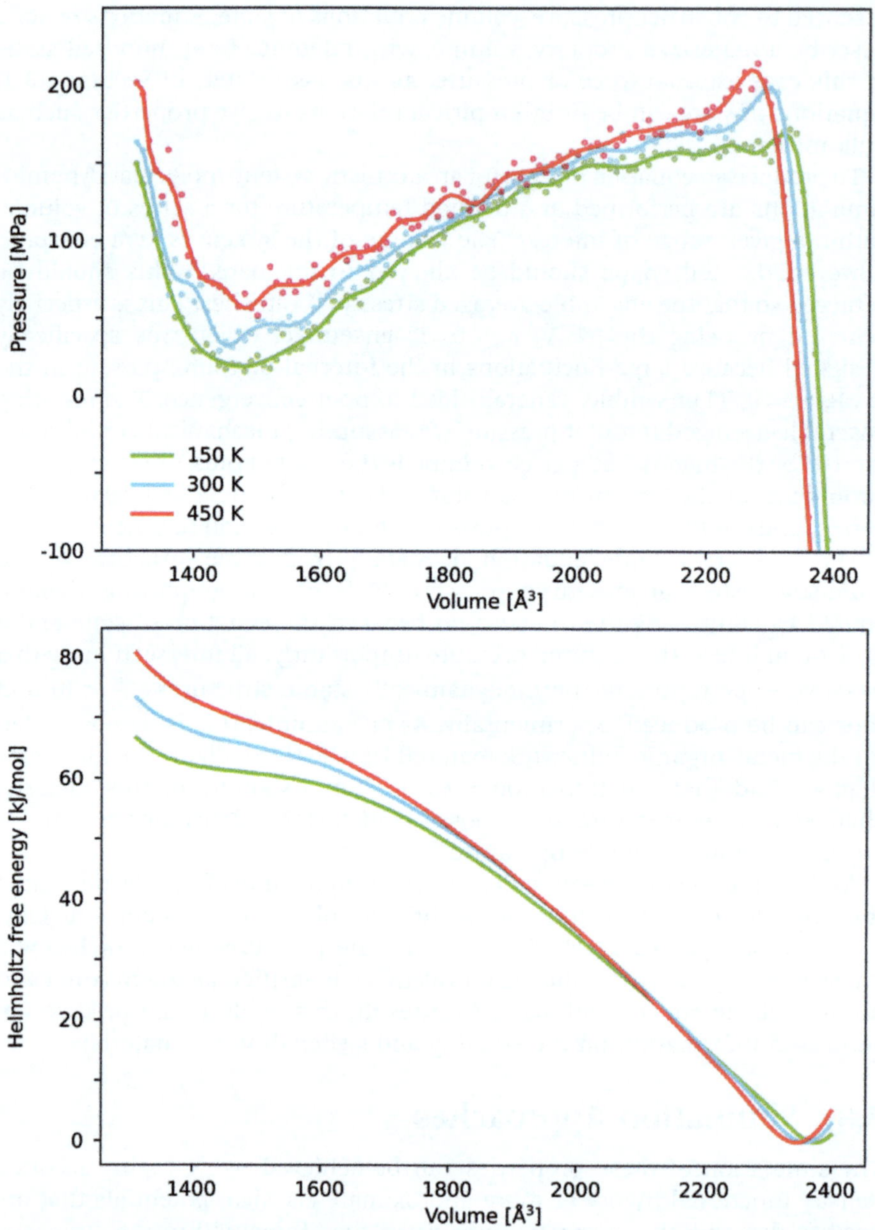

Figure 3.8 The pressure–volume equation of state and Helmholtz free energy with respect to volume obtained through force field molecular dynamics simulations of the metal–organic framework material DMOF-1(Zn) at 150 K, 300 K and 450 K.
Reproduced from ref. 43 with permission from the Royal Society of Chemistry.

relative change in volume for an applied pressure and thus can be calculated from constrained cell optimisations. Here, the residual stress for a given change in volume is computed from the equilibrium volume. Similar cell deformations and specific elements of the residual stress tensor can be simulated to compute the Young's modulus or shear modulus.

Alternatively, models can be applied to a series of simulations to find these elastic moduli in an equivalent methodology to experimental studies. This is particularly applied for calculating the bulk modulus, where the cell volume is optimised for increasing amounts of applied pressure (or alternatively the residual stress is computed for a series of constrained cell optimisations) and can be fitted to analytical expressions. One such analytical expression is the Birch–Murnaghan isothermal equation of state.[44] These analytical expressions relate the volume of a body to the applied pressure. The third-order Birch–Murnaghan isothermal equation of state is given by the expression:

$$P(V) = \frac{3B_0}{2}\left[\left(\frac{V_0}{V}\right)^{\frac{7}{3}} - \left(\frac{V_0}{V}\right)^{\frac{5}{3}}\right]\left\{1 + \frac{3}{4}(B_0' - 4)\left[\left(\frac{V_0}{V}\right)^{\frac{2}{3}} - 1\right]\right\}$$

where P is the pressure, V_0 is the equilibrium volume, B_0 is the bulk modulus, and B_0' is the first derivative of the bulk modulus, with respect to pressure. This third-order Birch–Murnaghan has the advantage over first-order equations of state, which do not include a first derivative of the bulk modulus term, as it is able to capture the change of bulk modulus with changes in pressure. Materials can show softening or stiffening following the application of pressure.[45]

The complete elastic constants can be calculated by combining a series of deformations and geometry optimisations. Specifically, starting from a relaxed structure, a set of distorted structures can be generated. As described in Section 3.2.1, the strain tensor has six independent components. Each of these strains can be applied independently with several magnitudes to produce a set of deformed structures. For example, if four distinct strain magnitudes (−0.010, −0.005, +0.005, +0.010) were considered, this would result in 24 deformed structures. For each deformed structure, constrained cell optimisation is performed to produce an optimised structure and the resulting stress tensor can be calculated. The stress tensor is the resulting force which then, under the conditions of linear elasticity, can be directly related to the strains *via* the elastic matrix ($\sigma_i = C_{ij}\, e_j$). Practically, one row (or equivalently, column) of the elastic matrix is obtained from a linear fit of the calculated stresses over the range of imposed strains. This is then repeated for each of the six independent strain components to result in all elements of the elastic modulus tensor. This complete elastic matrix can also be computed from total energy as a function of the strain. There are a number of implementations of this method, for example the methodology, used by de Jong *et al.*[46] for a high-throughput analysis of the elastic properties of inorganic compounds (Figure 3.9), is well documented in the python library pymatgen.[47] To improve

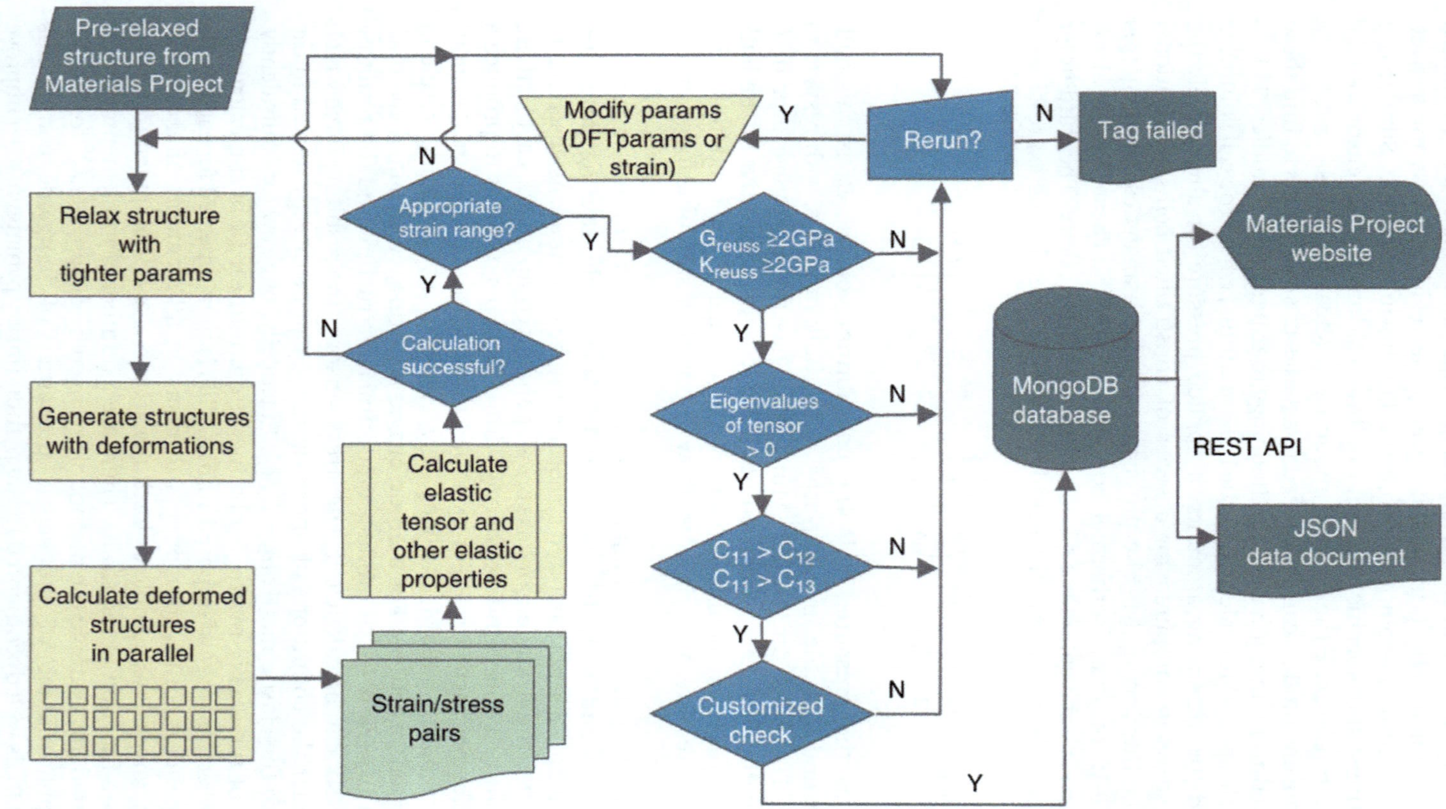

Figure 3.9 Workflow for calculating and filtering the elastic constants using the Materials Project and a high-throughput simulation methodology.
Reproduced from ref. 46, https://doi.org/10.1038/sdata.2015.9, under the terms of the CC BY 4.0 license, http://creativecommons.org/licenses/by/4.0.

computational efficiency, several deformed structures can be considered equivalent, owing to the symmetry present in the periodic structure. This methodology was reported by Perger *et al.*[48] and is used in the CRYSTAL software.[49] Symmetry can be used to limit the number of unique elements of the complete elastic matrix, and the number of deformations, to greatly improve the computational efficiency. This approach when applied to highly symmetric cubic structures, such as LiH, will result in only three deformed structures that are required to be optimised.[48]

This static approach is the established method for calculating the elastic properties for a given material at the zero-kelvin limit. It has been used successfully to chart the elastic properties for a large database of inorganic crystal structures.

3.3.2 Dynamic Methods

Elastic properties can also be calculated by using the natural deformations that occur at finite temperature. This provides a method for investigating the mechanical properties for realistic conditions of pressure and temperature. Parrinello and Rahman demonstrated that fluctuations of strain in a (N, σ, H) ensemble, using molecular dynamics, are related to adiabatic (constant entropy) compliance and thus to the elastic moduli of the system.[50] This is also applicable to (N, σ, T) or (N, P, T) ensembles.[51] The time series of cell matrices and cell volumes can be extracted from a molecular dynamics simulation at a constant pressure P and temperature T. The strain at every timestep can then be computed as:

$$\varepsilon = \frac{1}{2}[\langle \mathbf{h} \rangle^{-1} \langle \mathbf{h}\mathbf{h}^T \rangle (\langle \mathbf{h} \rangle^T)^{-1} - \mathbf{1}]$$

where $\mathbf{h}$ is the cell matrix and $\mathbf{1}$ the 3×3 identity matrix. Note, it is common convention to use the unit cell vectors a, b and c as the rows in the cell matrix $\mathbf{h}$.

The strain at each timestep is then used to compute the compliance matrix for constant pressure and temperature, as demonstrated by Oritz *et al.*:[45]

$$S_{ij} = \frac{V_0}{k_B T}\langle(\varepsilon_i - \langle \varepsilon_i \rangle)(\varepsilon_j - \langle \varepsilon_j \rangle)\rangle$$

where the reference volume, V_0, is usually computed as the average volume.

Analysis of volume fluctuations from molecular dynamics simulations can be used to accurately determine the complete elastic properties of a material (Figure 3.10). However, it is imperative that physically correct and adequate sampling is performed. Molecular dynamics trajectories employed to calculate the elastic constants often sample every timestep for a long trajectory, multiple nanoseconds. The reason for such sampling is to ensure convergence of fluctuation.[52] This is especially important for soft materials, with

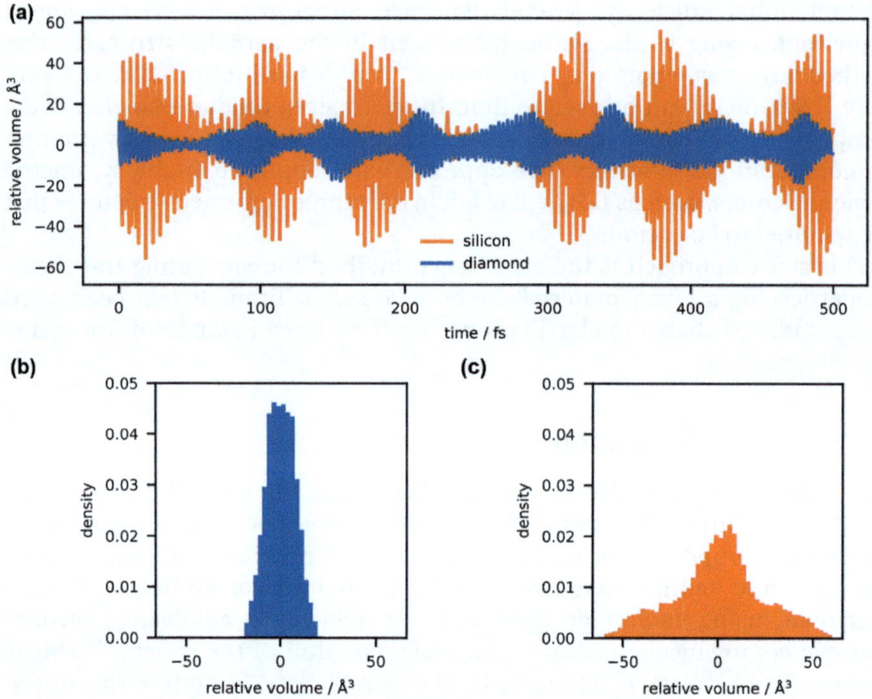

Figure 3.10 Volume fluctuations during (N, P, $\sigma_a = 0$, T) simulations for diamond and silicon (a), which have a bulk modulus of 440 GPa[4] and 98 GPa,[56] respectively, and their resulting probability distributions (b and c). Simulations were conducted using LAMMPS[57] employing a Nosé–Hoover-style thermostat and barostat set to 300 K and 0 Pa with time constants of 1 fs and 10 fs, respectively.[58] A timestep of 0.001 fs was used and Tersoff potentials[59] were used to describe the potential energy.

elastic constant which are small in magnitude that have large volume fluctuations. The importance of barostat choice has been discussed by Rogge *et al.*[53] Static properties, such as average volume, are correctly described with most barostats. However, this is not the case for properties that depend nonlinearly on pressure. The Berendsen barostat[54] is shown to produce less variance in volume as it suppresses pressure fluctuations more drastically than other barostats, such as the Martyna – Tuckerman – Tobias – Klein barostat,[55] and thus is inappropriate for computing the elastic constants.

The advantage of the fluctuation methodology, described above, is that it is able to capture changes in elastic properties resulting from changes in temperature or pressure. In particular, it has highlighted pressure and temperature induced softening processes in framework materials.[60]

Dynamic methods are also used to reliably predict and understand the behaviour of materials beyond the elastic regime. This is particularly useful for examining the phase changes exhibited by highly flexible MOFs.

Accurate simulations can include important thermodynamic considerations to provide an estimation of the free energy difference between different metastable states. A variety of free energy estimation methods can be used to construct free energy profiles as a function of volume, which exactly describe the propensity for material to demonstrate flexibility or phase changes.[61] Examples of these methods include free energy perturbation, thermodynamic integration, umbrella sampling, metadynamics and variationally enhanced sampling. Primarily, free energy calculation consists of three important criteria: an appropriate description of the potential energy of the system, a sampling methodology to generate a representative ensemble of configurations, and methods to estimate the relevant free energy differences.

In this chapter we have generally omitted the importance of an accurate description of the potential energy. However, in free energy computations it is crucial that this description is correct, as simulation requires sampling outside of equilibrium. A potential energy surface can be constructed with first-principles methods, such as DFT, to ensure energies and forces are accurate, but it is not straightforward to use DFT for free energy simulations. These simulations require large portions of the configuration space to be sampled and as a result there are many energy evaluations of the system that lead to huge computational cost. It is therefore necessary to use analytical expressions or force fields of the potential energy to allow the sampling to examine longer time and length scales. This approximation will neglect the detailed electronic structure of a system and simply provide an analytical expression for the potential energy with respect to atomic positions. Many of these force field approaches are able to accurately capture the flexibility of complex materials,[62] but it is important to test the suitability of this approximation.

Approaches that sample the phase space, with respect to volume, can use molecular dynamics simulations, Monte Carlo methods, or a combination of both. Generally, molecular dynamics approaches are used. However, sufficient sampling of configurational space can become problematic for systems that exhibit metastable phases separated by high free energy barriers. This is because the probability to overcome these barriers during a molecular dynamics trajectory can be very low. As a result, several sampling methods have been developed to enhance the sampling of usually rare regions of phase space.[63] Generally, techniques either enhance the sampling of all degrees of freedom or enhance sampling for certain degrees of freedom, also named collective variables. These variables describe a usually high-dimensional function of bond lengths, angles, dihedrals and must be chosen carefully to effectively sample interesting phases, to especially include phases separated by large free energy barriers. Subsequent sampling of collective variables can be enhanced by either directly constraining the collective variable,[64] introducing a bias potential[65] or increasing the temperature.[66] In studying flexibility, phase transitions are often characterised by volume changes and as a result volume can also be used as a collective variable.

A summary of general methods used in free energy sampling as discussed by Demuynck *et al.*[61] are depicted in Figure 3.11. Many of these methods rely on equilibrium simulations where sufficiently long simulations are performed in specific regions of the phase space. Methods that use equilibrium simulations include thermodynamic integration and umbrella sampling. Thermodynamic integration uses specified sampling regions of the phase space that correspond to different volume states. These different states are simulated with a fixed volume simulation, for example, by using the (N, V, $\sigma_a = 0$, T) ensemble, to produce the pressure–volume equation of state, as described in Section 3.2.8. Free energy differences can then be obtained by integrating over the negative of average pressure, which is the averaged

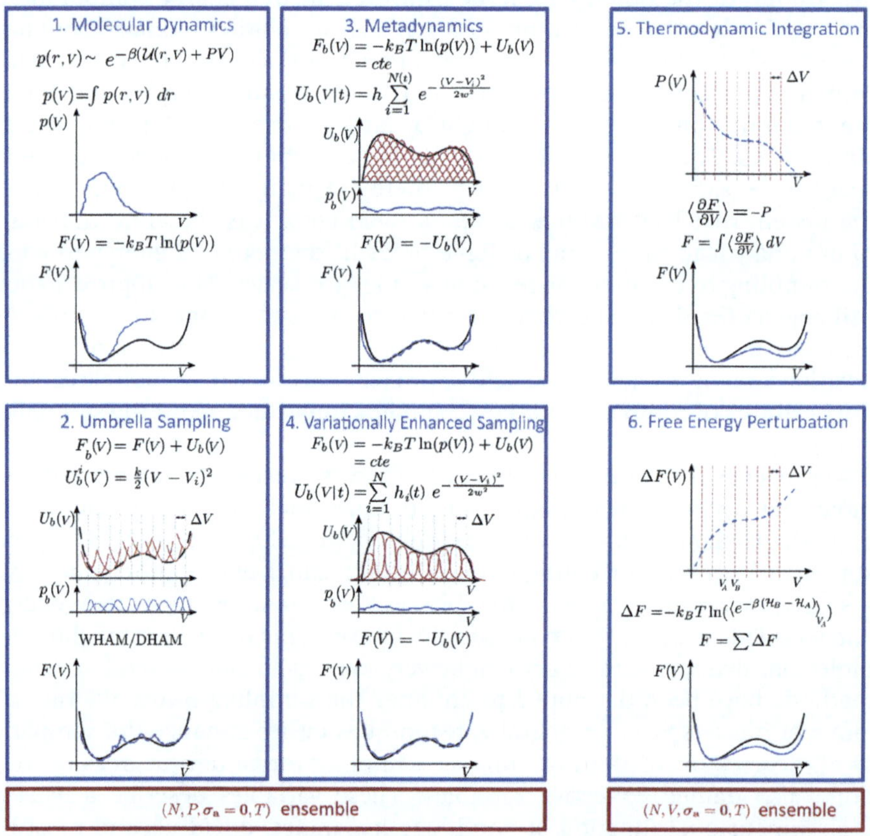

Figure 3.11 Representations of several free energy methods stylistically illustrated. A different free energy method is shown in each pane, where the top plot shows the simulation result and the bottom the resulting free energy profile.

free energy derivative, with respect to volume.[67] The umbrella sampling method, alternatively, introduces an external potential to enhance sampling in rarely visited regions.[68] This is achieved by using various simulations of the (N, P, $\sigma_a = 0$, T) ensemble where an external harmonic bias potential is included to force sampling in the phase space around a specific reference volume. The free energy differences can then be obtained with advanced analysis methods, such as the weighted histogram analysis method.[69] These methods attempt to connect this local information from biased molecular dynamics trajectories to an unbiased system.

In contrast to the above equilibrium methods, it is also possible to use a nonequilibrium approach, where equilibrium free energy can be estimated from a nonequilibrium simulation. For example, the free energy difference between two distinct configurations can be computed from the average work done to the system by forcing the transition to occur within a measured finite time period.[70] One nonequilibrium method that is employed in this respect is metadynamics.[71] Metadynamics is simply described by simulations that increasingly fill the free energy landscape to encourage trajectories to sample different phases. For example, an algorithm may use a few collective variables to describe a system of interest. During a simulation, the configuration of the system is determined using collective variables and a positive Gaussian potential is added to this energy landscape to discourage the system to return a single equilibrium configuration. For the application of examining the flexibility of materials, metadynamics can be applied by adding a history-dependent potential to the potential energy surface. For example, a number of Gaussian hills are added to a bias potential, where their locations are related to the volume states visited during a molecular dynamics simulation.[61] Subsequently, the metadynamics simulation would be considered converged when all volume states are uniformly distributed, as a result of the combination of the system potential energy and the applied bias potential. Using this negative bias potential, the free energy profile can be estimated.

Variational enhancing sampling is a relatively recently introduced sampling concept, similar to metadynamics, which is useful for sampling high-dimensional free energy landscapes and large systems that are not straightforwardly described by a few collective variables.[72] Here, the variational principle allows for the minimisation of the functional describing the bias potential in a simple way relative to the free energy surface, representing a practical, flexible and efficient sampling method. Specifically, the bias potential is constructed variationally by minimising a functional of the bias potential and this is achieved using a stochastic optimisation routine. There is a flexibility in the choice of target distribution, representing the region of sampling, and this approach has been applied to describe the free energy landscape with respect to volume for the material MIL-53(Al).[61]

The dynamic methods described above can be used to define the simple and complex features of flexibility in an atomistic system. This can be as simple as the elastic features of a system to the complete free energy landscape.

3.3.3 Abstract Methods

The computational determination of mechanical properties of atomistic materials usually involves the optimisation and energy determination of many deformation modes (static methods) or many time steps (dynamic methods). Thus, mechanical property calculations require greater computational cost than routine optimisations. These calculations are usually performed using high accuracy first-principles methods and efficient computing technologies have permitted calculations of large databases of materials. For example, de Jong *et al.* reported the elastic properties for more than 1000 inorganic solids, and this database is being continually updated.[46] However, even with current advanced computing resources, high-throughput methods can be applied only to a small fraction of realised materials. Computationally cheap classical potentials may also be employed to improve the efficiency of these simulations; however, this often results in lower and sometimes unacceptable accuracy.

Machine learning represents a potential approach to accelerate materials discovery, bypassing the need for conventional energy-based calculations entirely. This can efficiently permit the large-scale high-throughput screening of hypothetical materials. It is especially powerful in predicting properties for which molecular simulations are complex. Generally, machine learning algorithms seek to build a predictive model for a given property based on a set of features that describe the systems of interest. There are many different machine learning algorithms that can provide regression or classification for mechanical properties and importantly there are many ways to represent the features of a material.[73]

One of the first applications of machine learning for this mechanical properties was reported by de Jong *et al.*[74] In this study, Hölder means of chemical and structural descriptions were used to construct a feature set to describe a database of inorganic crystals. Multivariate local regression within a gradient boosting framework was then used to produce a predictive model for average bulk and shear moduli (Figure 3.12). The model produced a root-mean-squared-error of 0.0750 log (GPa) and enabled the accurate screening of 30 000 structures for the special property of ultra-hardness.

We also tested a machine learning approach that relied solely on geometric features, related to local geometry, structure and porosity, to predict the mechanical properties of siliceous zeolites.[75] This approach was able to predict the average bulk and shear moduli of zeolites with an accuracy exceeding that of force field approaches. Analysis of the model illustrated clear correlations between characteristic features of a zeolite and elastic moduli. Notably, this approach has been extended to search for auxeticity.[76]

Recently, this machine learning approach to predict the mechanical properties of materials was extended to metal–organic framework systems by Moghadam *et al.*[77] In their application, an artificial neural network was trained to predict the bulk modulus of the training data, where the free parameters of the network were optimised using a *tanh* function. This predicted model was then validated by cross validation. Interestingly, the best

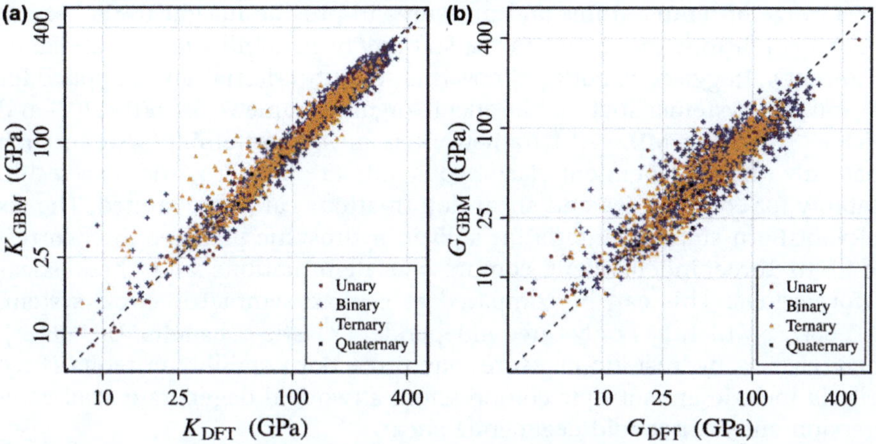

Figure 3.12 Comparison of DFT training data with machine learning predictions for bulk modulus (K) (a) and shear modulus (G) (b), for a database of k-nary inorganic crystals.
Reproduced from ref. 74, https://doi.org/10.1038/srep34256, under the terms of the CC BY 4.0 license, http://creativecommons.org/licenses/by/4.0/.

performing model, using topology as a descriptor, dramatically improved the accuracy of their model, suggesting that topological features are crucial for accurately predicting bulk modulus.

3.4 Applications of Mechanical Properties

3.4.1 Understanding Mechanical Stability or Instability

As demonstrated by Born, we can use this understanding of the mechanical properties of an atomistic system to produce a set of criteria that determine whether a system is mechanically stable. These stability criteria are obtained by requiring that the energy of the lattice increases for any infinitesimal strain imposed on the unit cell, at an equilibrium volume. We have briefly discussed in Section 3.2.1 how this stability criteria relate to the complete elastic stiffness matrix (C_{ij}) for a given system.

This Born stability criteria can be generalised to also account for systems subject to an arbitrary external stress, such as isotropic pressure or any given anisotropic stress. For a given external stress (σ), the elastic stiffness matrix (C_{ijkl}) is replaced by its stressed analogue given by:

$$B_{ijkl} = C_{ijkl} - 1/2\ (\delta_{ik}\sigma_{jl} + \delta_{jk}\sigma_{il} + \delta_{il}\sigma_{jk} + \delta_{jl}\sigma_{ik} - 2\delta_{kl}\sigma_{ij})$$

To confirm the mechanical stability of material at a given temperature and subject to a given pressure, the outlined sufficient and necessary stability conditions (Section 3.2.1) must also hold true for this stressed analogue of the stiffness matrix.[20] Notably, the elastic stiffness matrix can be computed for a range of temperatures and pressures to provide insight to the experimental conditions that result in the stability or instability for a material.

Recently, Rogge *et al.* outlined how this stability criteria can be applied for several test systems, such as the metal–organic frameworks MIL-47(V) and UiO-66(Zr).[78] The MIL-47(V) framework is an orthorhombic system, such that only nine independent elastic constants are nonzero, and as a result, stability for compressive and shear deformations can be separated. The six relevant Born stability criteria for a given hydrostatic pressure then correspond to three independent compressive deformations and three shear deformations. This can be compared to a more symmetric cubic system, UiO-66(Zr), which has only three independent elastic constants. UiO-66(Zr), for a given a hydrostatic pressure, has three Born stability criteria. These criteria include an isotropic compression, a two-fold degenerate axial compression and a three-fold degenerate shear.

These case studies exemplify that the anisotropy of the Young's and shear moduli (Figure 3.13) provides a reliable prediction of flexible (MIL-47(V)) or rigid behaviour (UiO-66(Zn)). This has also been demonstrated in other studies.[35] Importantly, however, this has proved inadequate to reveal quantitative information, with respect to the critical pressure of mechanical instability, or significant insight into the process of the instability.

To extract quantitative information of mechanical instability, the Born stability criteria with respect to changes in pressure or the complete pressure–volume equation of state must be computed. For example, to investigate the weakest mode of deformation, the Born stability criteria can be assessed for increasing values of pressure. For highly symmetric materials these deformation modes are relatively easy to interpret (MIL-47(V) and UiO-66(Zr), for example). However, for low symmetries, and systems with no apparent symmetry, this type of analysis is not tractable. Furthermore, molecular dynamics simulations in the (N, V, T) or (N, P, T) ensembles can result in large pressure fluctuations, despite using state-of-the-art barostats. These large pressure fluctuations may induce premature collapse that disrupts an accurate computation of the complete elastic stiffness matrix (C_{ij}), especially near areas of mechanical instability. This is particularly problematic for flexible 'soft' materials where there are already large fluctuations present. Simply, these problems can be resolved by computing pressure–volume equations of state. Simulations are employed to provide a converged value of pressure for a given volume to result in the pressure–volume equation of state. These profiles can provide microscopic insight into inelastic deformations and thermodynamic integration can be applied to the resulting volume equation of state to compute the relative stability of phases present. For example, a single equation of state can identify the multiple metastable phases present in MIL-47(V).[67]

Combining all the information above provides unprecedented information for understanding the mechanical stability of a material. For example, Rogge *et al.* demonstrated this in a series of metal–organic framework structures to identify the critical pressure to produce an unwanted phase transition or amorphisation or alternatively flexibility (Figure 3.14).

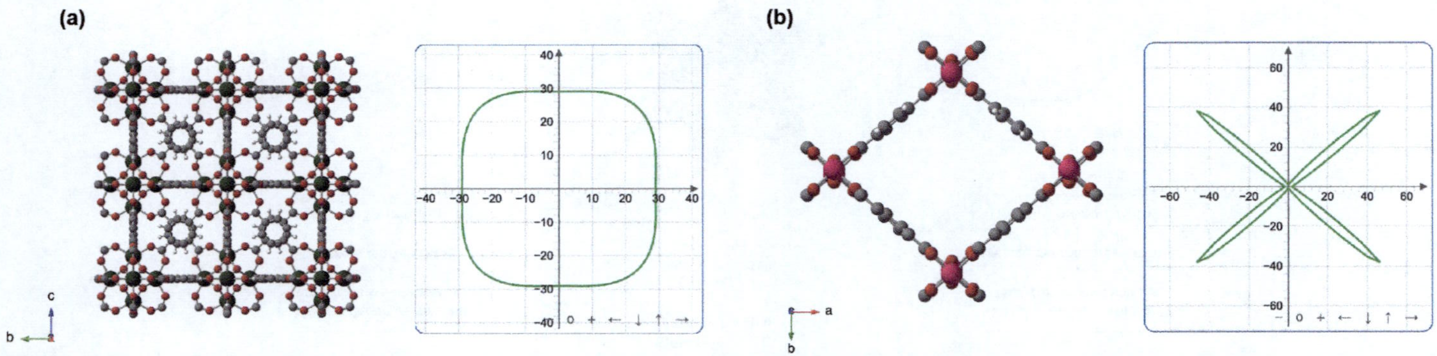

Figure 3.13 Atomistic depictions of the metal–organic frameworks UIO-66(Zr) and MIL-47(V) (a and b, respectively). The resulting spatial dependence of Young's modulus as computed by Rogge *et al.*[78] displayed in units GPa and for the depicted projection of the structures. Atom colours are represented by: Zr, green; V, magenta; O, red; C, grey; H, white. The mechanical properties are plotted using the ELATE software.[36]

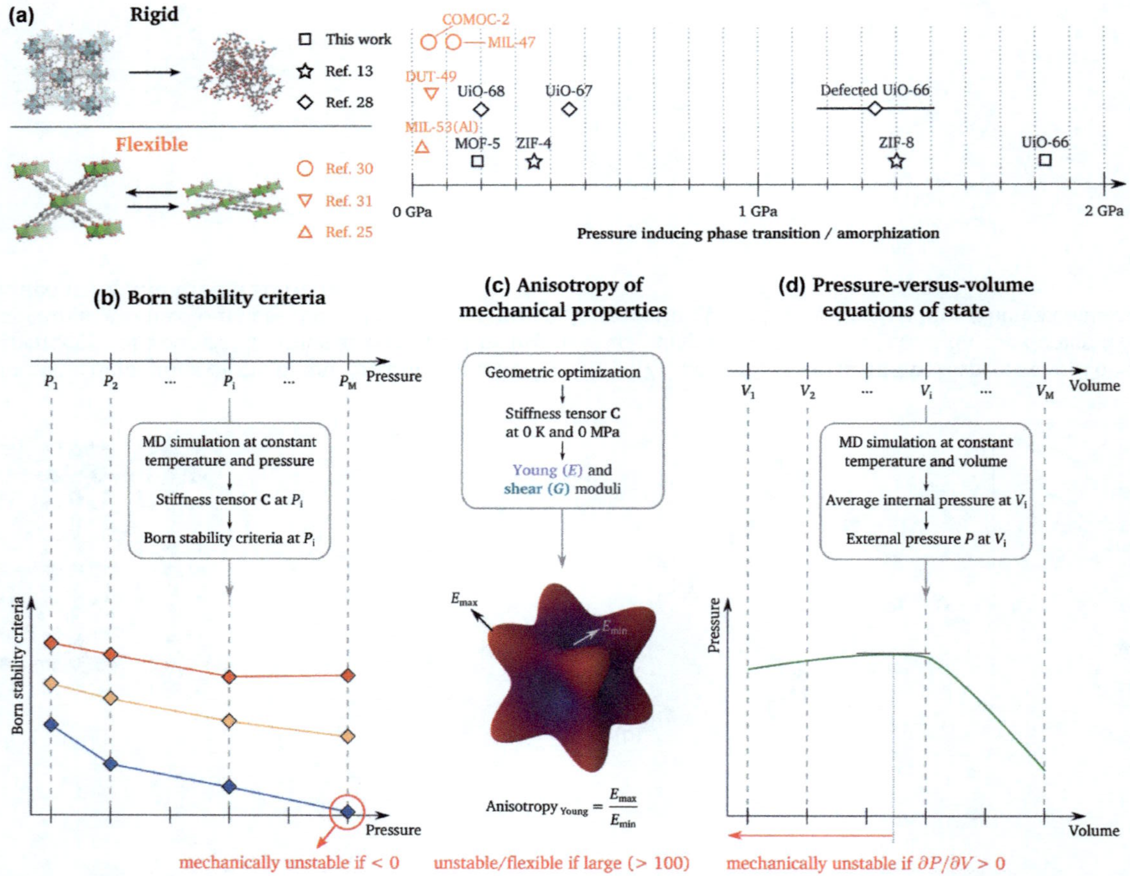

Figure 3.14 Critical pressures that can induce phase transition or amorphisation for a series of flexible and rigid metal–organic framework materials (a). Schematic illustrations of the Born stability criteria (b), anisotropy (c) and pressure–volume equations of state (d), which can be used to determine mechanical stability.
Reproduced from ref. 78, https://pubs.acs.org/doi/10.1021/acs.accounts.7b00404, with permission from American Chemical Society, Copyright 2017. Further permissions requests related to the materials excerpted should be directed to the ACS.

The above has described the ability for pressure–volume equations of state to predict the mechanical stability of a given material. However, the accuracy of this prediction is inherently determined by the level of theory used to compute the equation of state. The level of theory should be chosen to reliably reproduce the potential energy surface. This includes the correct description of all metastable phases and their relative energetics. Although *ab initio* electronic structure methods can yield an accurate description of the potential energy surface, they are computationally demanding. Moreover, given that simulations need to be carried out at finite temperatures, stability can be affected by a subtle combination of entropic and dispersion interactions,[79] this can be prohibitively expensive. Thus, there is a need for the development of accurate force fields, and researchers should consider carefully the accuracy achieved by classical methods.

3.4.2 Mechanical Surprises

A large proportion of materials act in an expected way, for example, this is responsible for the definition of Poisson's ratio. As we have discussed, there are materials that have a negative Poisson's ratio, demonstrating the property of auxeticity. Materials with unexpected and unnatural properties are a fervent area of research and are labelled as metamaterials.[80] Mechanical metamaterials demonstrate truly amazing properties, including a negative effective bulk modulus[81] and fluid-like behaviour.[82] These interesting mechanical phenomena have enabled researchers to produce equally amazing devices, such as an 'unfeelability' cloak.[1] These mechanical metamaterials achieve these properties through a defined periodic lattice arrangement, using a specific combination of composition and geometric distribution. These materials are usually manufactured on a mesoscopic scale; however, there are a growing number of atomistic systems that also display similar metamaterial properties.[83] Often these atomistic systems are examples of coordination frameworks and MOFs. Notably, many of the features observed in mechanical metamaterials can also be found in these framework materials, including buckling building units and complex dynamic arrangements as illustrated in Figure 3.15. The response, and hence displacement of metamaterials is usually macroscopic; however, the response of these atomistic systems occurs on the microscopic scale.

The unique mechanical properties exhibited by these atomistic materials include similar mechanical properties to that observed in traditional mechanical metamaterials. However, a nanoscale atomistic structure provides access to behaviour difficult to produce on a macroscopic scale, such as negative thermal expansion, which is challenging to engineer in micron-sized architectures.[84]

This understanding and eventual production of bizarre mechanical properties by atomistic systems is an interesting research area, employing important concepts of mechanics and thermodynamics, where simulations are important to quantify and understand these phenomena.

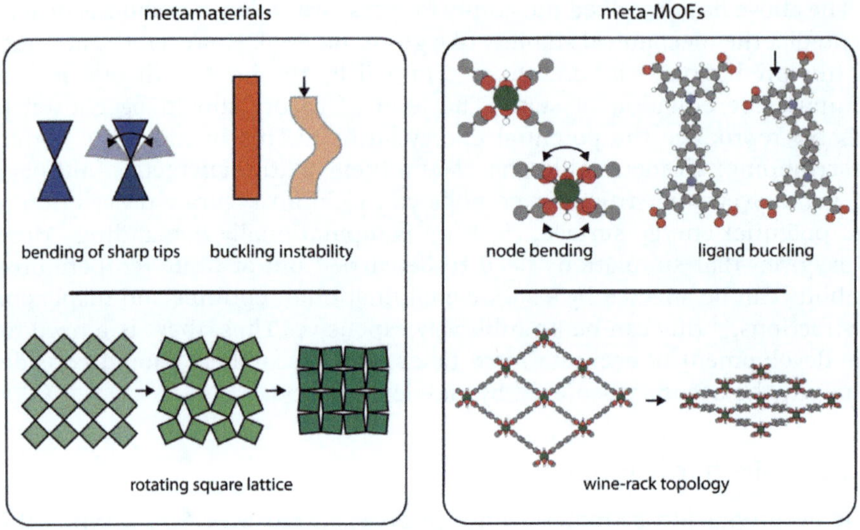

Figure 3.15 Local and structural features used in the construction of mechanical metamaterials and similar motifs that are observed in frameworks, identified as meta-MOFs.
Reproduced from ref. 83 with permission from Elsevier, Copyright 2019.

3.4.2.1 Auxeticity

In the linear elastic regime, many atomistic materials can display unusual mechanical behaviour. One important example is that of auxeticity, which is the occurrence of a negative Poisson's ratio, as we have defined in Section 3.2.4. Most systems have a positive Poisson's ratio, such as rubber that will get thinner in the opposite direction to the direction of stretch. Contrastingly, auxetic materials expand in one or two of the perpendicular directions. This auxeticity property is present in many porous foam macrostructures, but is relatively rare in atomistic, relatively dense systems. There are, however, a few atomistic framework structures that display this property including MIL-140,[85] ZIF-4[86] and some zeolites.[87]

These relatively few examples of structures that display auxeticity do so in a few directions, not all. Interestingly, the occurrence of a negative Poisson's ratio in all directions of space is extremely rare. This property results in an expansion in all perpendicular directions in the response to a tension applied in any direction. This is called complete auxeticity and can be designed in macroscopic structures; an illustrative example is the Hoberman sphere.[88] Complete auxeticity, however, is extremely rare in crystalline materials.

In proposing a classification for auxeticity, Siddorn *et al.* analysed the experimental elastic constants of almost 500 crystalline systems.[33] They demonstrated that some degree of auxeticity is observed in 37% of systems studied. However, average auxeticity, which is defined as average negative

Poisson's ratio for all directions, is only found in one compound, α-cristobalite. This material is able to produce this response by the two-dimensional rotation of rigid units, involving rotating rectangles. Notably this motif is commonly used to engineer auxeticity in macrostructures. No material in this experimental database was found to present complete auxeticity. In the same work, the authors tested 200 pure silica zeolites, using molecular simulation methods like those described in Section 3.3.1, and predicted the first known example of complete auxeticity for a crystalline system. This special system is the JST zeolite framework, which includes a characteristic structural motif of tetrahedra featuring four 'stars of David' that produces this rare property of complete auxeticity.

The ability for advanced methods to efficiently simulate the mechanical properties of crystalline materials has enabled researchers to screen large databases in the search for new examples of auxeticity. Dagdelen *et al.* investigated more than 67 000 materials in the Materials Project database for candidates exhibiting features correlating with auxetic behaviour.[89] They were able to significantly accelerate this discovery process by initially screening this large database with a set of criteria. These criteria were designed to reduce the number of the relatively expensive DFT elasticity calculations required to accurately identify auxetic properties (Figure 3.16). The first criterion

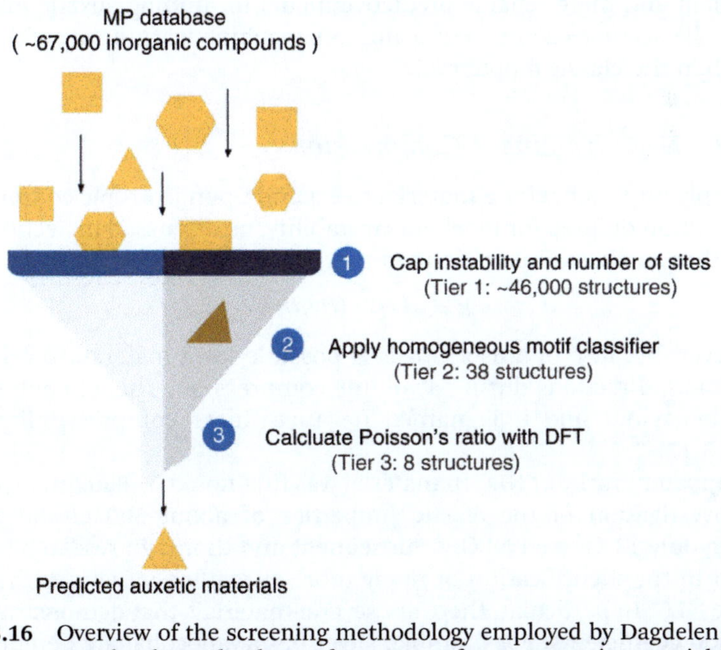

Figure 3.16 Overview of the screening methodology employed by Dagdelen *et al.* to search a large database of structures for new auxetic materials. Reproduced from ref. 89, https://doi.org/10.1038/s41467-017-00399-6, under the terms of the CC BY 4.0 license, http://creativecommons.org/licenses/by/4.0/.

narrowed the candidates using a stability benchmark, retaining candidate materials that have a formation energy less than 100 meV per atom above the convex hull, in an effort to remove materials that could prove challenging to experimentally produce. The second criterion then reduced the candidate pool to computationally tractable materials with fewer than 50 sites per unit cell. The application of these two criteria reduced the database to approximately 46 000 structures. The next screening step employed a homologous structure motif analysis. The goal of this step was to identify materials with structural features likely to exhibit auxetic behaviour, using α-cristobalite as an archetype. This analysis reduced the number of candidate materials that needed complete elasticity calculations to 38. As a result, the investigation predicted three previously unknown auxetic compounds, one of which was the relatively everyday solid, high-temperature AlPO4. The 38 investigated materials also included nine anepirretic materials, characterised by possessing a combination of both auxetic and non-auxetic directions to give an almost zero average Poisson's ratio. Clearly these cristobalite-like structures can produce surprising and important mechanical properties. Recently, Gaillac *et al.* analysed approximately 600 000 hypothetical zeolitic structures using a classical potential approach.[76] Results of almost 1000 examples were then compared to DFT calculations demonstrating clear limitations to predicting auxeticity using these classical potentials. The authors were able to leverage the DFT simulations to train a machine learning algorithm enabling an efficient and more reliable predictive model for finding auxetic materials. Indeed, the accuracies obtained using this machine learning algorithm were better than the classical potential.

3.4.2.2 Negative Linear Compression

It is simply impossible for a material to expand upon isotropic compression. The condition defined for mechanical stability, as discussed in Section 3.2.1, means that compressibility, defined as κ, must be positive.

$$\kappa = -(1/V)\ (\partial V/\partial P)_T$$

However, because of anisotropy it is possible for a material to expand in one or more directions during a volume compression. This phenomenon is a rare behaviour and it is named negative linear compressibility (NLC) (Figure 3.17).[90]

The apparent rarity of NLC in material was first noted by Baughman *et al.* in their investigation of the elastic properties of about 500 crystal phases, in which only 13 showed NLC.[91] Subsequent and thorough research has now resulted in the identification of many more crystalline structures that demonstrate NLC. In particular, there are several materials that demonstrate large expansion coefficients. For example, silver hexacyanocobaltate demonstrates outstanding mechanical properties[92] and was experimentally demonstrated to exhibit NLC of an extremely large magnitude, made possible by a wine-rack-like framework structure.[93] An even rarer phenomenon is the expansion

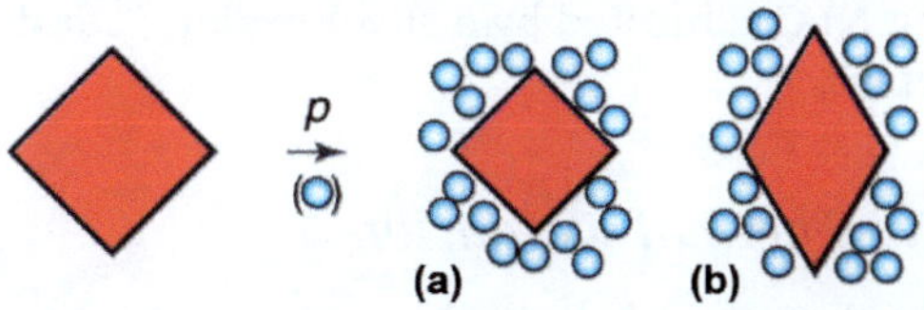

Figure 3.17 Examples of some compressibility characteristics for a material in response to an applied hydrostatic pressure: positive compressibility observed by contraction in all directions (a), negative linear compressibility observed by linear expansion in one direction (b). The system is represented by the red area and the blue particles represent the pressure transducing media, which transfers hydrostatic pressure to the system.
Reproduced from ref. 90 with permission from the Royal Society of Chemistry.

of a material in two different crystallographic directions when compressed. This behaviour has been named negative area compressibility.[94]

There are a variety of mechanisms that can result in NLC, including phase changes induced by ferroelasticity, tilting of networks and helical compression. The details of these mechanisms have been described by Cairns *et al.*[90] Framework-like structures formed by atomistic systems produce many examples of NLC, where the bonds throughout the system can act like a macroscopic trusses and hinges. The metal–organic chemistry, which is used to construct these frameworks, can produce significant flexibility from deformation of the metal coordination environment.[95] Particularly, the two most prevalent framework topologies that inherently display NLC are the 'wine rack' and 'honeycomb' networks. For example, a material that demonstrates large negative area compressibility comprises a hydrogen-bonded two-dimensional supramolecular coordination complex featuring puckered honeycomb-like layers.[96] In this material, the negative area compressibility is produced by the complex interactions of intra-layer coordination, hydrogen-bonding and inter-layer van der Waals interactions.

Atomistic simulations play an important role in identifying and understanding NLC. In particular, simulations can be employed to reveal the presence of soft phonon modes. This has been used to understand the structural motifs present in structures like zinc cyanide that produce NLC.[97] Moreover the simulations of the complete elastic tensor can be directly used to compute compressibility in all directions. Simulations of the complete elastic tensor of 121 zeolites by Coudert identified that a total of 16 zeolite frameworks exhibited NLC.[24] This study showed the largest NLC is observed for the GIS zeolite topology and importantly demonstrated that the NLC phenomenon is less rare than initially believed, as 13% of the studied zeolites showed NLC. Furthermore, simulations can be employed to explicitly optimise structural features and cell parameters for different applied external pressures to compute compressibility in a similar way to experiments. This approach, specifically using atomistic DFT simulations, was used to predict outstanding NLC exhibited by a silver oxalate[98] and can be applied to any system of interest.

3.4.2.3 Negative Thermal Expansion

A peculiar property that combines both mechanical understanding, as discussed in this chapter, and thermal effects is negative thermal expansion (NTE). This property is in stark contrast to the behaviour of most materials that expand upon heating. Thus, negative thermal expansion is described as an unusual physical property and is limited to certain types of structures.[99] Despite being considered unusual, NTE is exhibited by water, a very common material. Liquid water demonstrates an increase in density when heated between 273 K and 277 K and the hexagonal form of ice exhibits NTE at 45 K.[100] This behaviour also appears to occur more readily in solids of various types, including framework or network structures that encompass a large number of crystalline solids: zeolites, aluminium phosphates, metal oxides and metal–organic frameworks.[101]

Materials that exhibit NTE are highly sought after, particularly those materials that feature NTE with a large amplitude, in an isotropic manner, across a wide temperature range. The most astonishing example of an NTE material is zirconium tungstate, a cubic oxide, that demonstrates isotropic NTE from 0.3 to 1050 K.[102]

Negative thermal expansion is desirable for several specific applications. It is often sought after to be used in combination with a positive thermal expansion material to produce a composite material with tuneable thermal expansion characteristics. For example, these tuneable composites are a possible approach for producing zero thermal expansion materials. Examples of zero thermal expansion materials are extremely rare[103] but are desirable in the design of precise instruments and devices. The behaviour of negative thermal expansion appears to occur more frequently in atomistic framework structures, for example, it is observed in several zeolites.[103]

Interestingly, atomistic simulations were fundamental to predicting this special property of framework structures before experimental confirmation. Initial simulations employed lattice dynamics that can incorporate temperature by calculating the vibrational frequencies in a periodic structure. Specifically, a quasi-harmonic approximation is applied, where vibrational frequencies are assumed to be harmonic oscillators with frequencies that vary with cell volume. The atoms in a lattice dynamics simulation are fixed during evaluation of vibrational frequencies and as a result this method samples the energy at these specific lattice sites.[104] Lattice dynamics represents an efficient method for determining a wide range of thermodynamic properties, when anharmonic effects are not significant. These simulations were applied to rationalise NTE in zeolite frameworks by the occurrence of transverse vibrations of bridging oxygen atoms between two rigid polyhedrons, or the cooperative motions of linked quasi-rigid polyhedral units ('rigid-unit modes').[105]

This behaviour observed in zeolites and other framework materials is in direct contrast to simple dense materials, where positive thermal expansion arises from the anharmonicity of the pair interaction potentials (Morse-like). As a result, vibrations make bonds longer on average that in turn propagate in three dimensions to produce volume expansion. Contrastingly, the vibrational nature of molecular framework structures is different. The modes of lowest frequency (and higher amplitude) in molecular frameworks are associated with vibration of the molecule often perpendicular to node–molecular–node axis, which shortens the average node–node distance. This local shortening upon heating of the node distances can be propagated to a three-dimensional network. This explains why NTE is relatively common in zeolites and other framework materials, and how it can be straightforward to simulate for simple materials using a lattice dynamics approach.

Alternatively, atomistic simulation can be applied to more complex (all atom detail) structures and can explicitly test the changes in unit cell shape, and volume, with increasing temperature. Recently, the NTE properties of a selection of mesoporous metal–organic frameworks were investigated using classical molecular dynamics simulations (Figure 3.18).[106] The framework response to temperature was simulated using (N, P, $\sigma_a = 0$, T) ensemble molecular dynamics, specifically employing the Nosé–Hoover chain thermostat and Martyna–Tuckerman–Tobias–Klein barostat. This approach was used to simulate the heating of the systems of interest from 40 to 400 K, in 20 K intervals. To ensure equilibrium properties for each temperature an equilibration period of 0.5 ns simulation was followed by a production period of 1.0 ns, where the average volume and further dynamics were investigated. Notably, the complete mechanical properties using the fluctuation method can then be calculated from the trajectories at these finite temperatures. These structural simulations can be applied to most materials, such as other MOFs, to investigate negative thermal expansion coefficients over the full temperature ranges of stability.[107]

Lattice dynamics or all atom molecular dynamics can be used to provide unsurpassed insight into the process of NTE. This can be achieved by studying the phonon modes and their respective Grüneisen parameters, which characterise their contribution to thermal expansion. Interestingly, the Grüneisen parameter can be expressed as a dimensionless combination of properties (expansion coefficient, bulk modulus, density and specific heat) and can also be represented using elastic moduli and their pressure derivatives.[108] It provides a quantitative link between thermal and mechanical parameters.

Simulations have predicted NTE in many material classes. For example, Bouëssel du Bourg *et al.* predicted that zeolitic imidazolate framework (ZIF) materials can exhibit thermal expansion coefficients that depend solely on their topology.[60] Moreover, detailed simulations of cubic siliceous zeolites have suggested that pressure-induced softening, where a material becomes progressively more compressible under pressure, is a common feature of

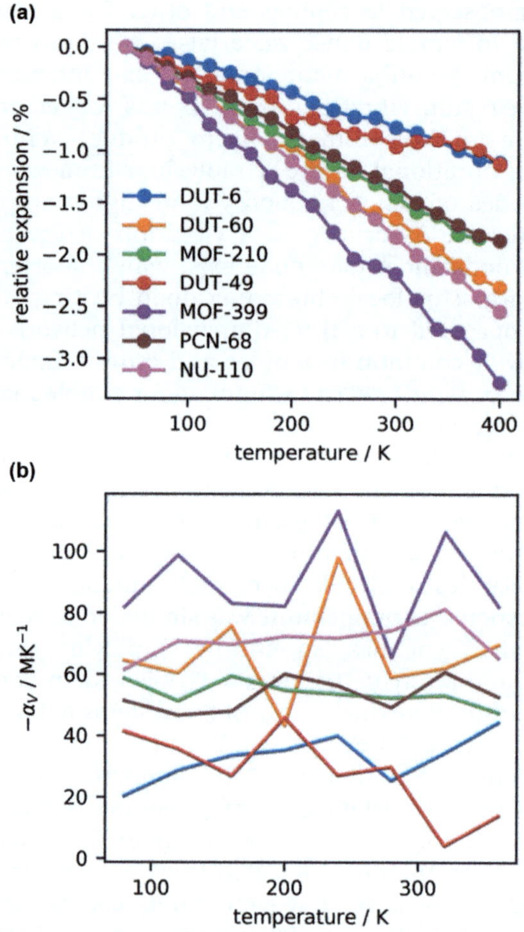

Figure 3.18 The relative expansion for a series of mesoporous MOFs simulated using (N, P, $\sigma_a = 0$, T) ensemble classical molecular dynamics (a). Volumetric thermal expansion coefficient, $\left(\alpha_V = \frac{1}{V}\left(\frac{\partial V}{\partial T}\right)_P\right)$, a metric that defines the change in volume with temperature, computed for intervals of 40 K (b).
Reproduced from ref. 106 with permission from the Royal Society of Chemistry.

framework materials that also show NTE.[109] The origin of pressure-induced softening is directly related to the dependence of the frequency of NTE phonon modes with strain. A simple relationship, where frequency has positive dependence with respect to pressure-induced strain, produces phonon modes that have negative Grüneisen parameters and also negative first and second derivatives with respect to pressure. This dependence would result in NTE in addition to pressure-induced softening of the system.

3.5 Summary

Atomistic simulations are an incredibly useful toolbox for understanding the large diversity of physical and chemical phenomena that are observed among materials. Throughout this chapter, I have demonstrated the importance in understanding the mechanical properties of systems using simulation. These properties are key to thoroughly elucidate the physical and chemical properties of a material. This is especially useful for considering the mechanical stability of materials at the outset of discovery and during the process of design. Moreover, they can shed light on potential applications in several areas, such as mechanical energy storage.[110]

To accurately simulate these properties, the effects of geometry, chemistry and electronics must be carefully considered, and this is achieved using atomistic simulations. A first-order approximation, for a linear response to external force, is completely described by the elastic constants, which describe the full response of a material to external mechanical constraints in any direction. These constants can be characterised with comparatively cheap computational cost, depending on the level of theory and computational approaches employed, and should be considered the initial simulations conducted on the path to fully characterising a system. From these constants, a whole host of different elastic moduli can be calculated to provide direct comparison with more macroscopic observations, for example, bulk modulus. However, this is a first-order approximation and only provides information for the mechanical behaviour within this elastic regime. To capture the mechanical behaviour of materials beyond this regime it is necessary to produce a pressure–volume equation of state, for a specific region of volume. This equation of state provides crucial insight into any mechanical transformations possible, including the existence of any metastable phases.

Highly accurate density functional theory or more approximate and, computationally cheaper, classical potentials can be used to describe the interactions between the atoms within the structure to eventually result in the simulation of mechanical properties. These methods are used to obtain an energy or force for a specific deformation. Without considering the effects of finite temperature, as a static approach, the complete elastic constants can be calculated by simply combining a series of deformations and geometry optimisations. Specifically, starting from an optimised or relaxed structure, a set of distorted volume (or strained) structures can be generated and optimised to solve the linear response to provide the complete elastic properties. Elastic properties can also be calculated by using the natural deformations that occur at finite temperature, representing a dynamic approach. This approach represents a pathway for investigating the mechanical properties at realistic conditions or specific temperature and external pressure. Moreover, molecular dynamics methods can be used to produce a pressure–volume equation of state and provide further insight to the changes in free energy with respect to changes in volume or applied force.

Finally, machine learning has emerged as a totally different way to simulate the mechanical properties of a material. Computational approaches based on machine learning provide an alternative and novel pathway to discover new materials with a specific mechanical response, with much lower computational cost than traditional static or dynamic methods.

This chapter has also described notable examples from materials that display surprising mechanical phenomena that can be fully characterised using the computational approaches described here. This includes properties such as negative compressibility, complete auxeticity and negative thermal expansion. Often the discovery process of these exotic and novel behaviours is a product of chance. The studies described here have shed light on the important microscopic roots of the phenomena using molecular simulation. Similarly, molecular simulation has provided insight for the modification and tuning of these exciting behaviours. The process of identification of novel mechanical behaviour is still wide open. Finding these needles in the haystack, which represent the endlessly growing number of known atomistic structures, is a grand challenge where prediction methods are playing an important role. Recently, Chibani and Courdert performed a large-scale exploration of the mechanical properties of 13 621 inorganic crystals, examining both average mechanical properties, such as bulk and shear moduli, in addition to the anisotropic elastic behaviour.[111] Importantly, they highlighted the importance of the anisotropy of elastic properties, which can only be considered using tensorial analysis of the complete elastic stiffness matrix. They also were able to identify crystalline materials with surprising elastic properties, such as negative linear compressibility and negative Poisson's ratio. This enabled, for the first time, a quantification of how frequently these properties were encountered: negative linear compressibility is found in 3% of inorganic materials and partial auxeticity (a negative Poisson's ratio in at least one direction) is found in 30%. Notably, total auxeticity is incredibly rare and observed in only 0.3% of crystals studied.

There remains a host of properties that are of growing interest where simulation is only beginning to be applied. For example, crystal size has a significant influence on flexibility[112] and large-scale, expensive, simulations are beginning to be applied in this area.[113,114] The presence of defects in the crystalline structure can also produce significant changes in elastic properties,[115] which produces further complication for simulation.[116] However, in the past few years, the state of the art in molecular simulation of materials has evolved considerably from simply simulating the elastic properties of highly symmetric materials to now large-scale screening studies, capturing complex effects such as inelastic behaviour. The computational characterisation of materials and their properties now covers a large range of length and time scales, including high-accuracy quantum chemistry calculations of complex structures, providing insight into electronic properties, to classical approaches that enable large-scale molecular dynamics sampling.

Atomistic simulation is a key element in collaborative studies of novel materials. Here, in combination with *in situ* and *in operando* experimental techniques, a very detailed picture can be obtained. Together with experiment, the simulations described here can produce unprecedented insight into the behaviour of extremely complex systems leading to new imaginative devices.

Abbreviations

MOF	metal–organic framework
COF	covalent organic framework
DFT	density functional theory
NTE	negative thermal expansion
AFM	atomic force microscopy
ZIF	zeolitic imidazolate framework

References

1. T. Bückmann, M. Thiel, M. Kadic, R. Schittny and M. Wegener, *Nat. Commun.*, 2014, **5**, 4130.
2. J. Zhao, N. Wei, Z. Fan, J.-W. Jiang and T. Rabczuk, *Nanotechnology*, 2013, **24**, 095702.
3. O. L. Blakslee, D. G. Proctor, E. J. Seldin, G. B. Spence and T. Weng, *J. Appl. Phys.*, 1970, **41**, 3373–3382.
4. H. J. McSkimin and P. Andreatch, *J. Appl. Phys.*, 1972, **43**, 2944–2948.
5. A. M. Ito, A. Takayama, Y. Oda and H. Nakamura, *J. Phys.: Conf. Ser.*, 2014, **518**, 012011.
6. F. P. Beer, E. R. Johnston, J. T. DeWolf and D. F. Mazurek, *Mechanics of Materials*, 2015.
7. W. Weaver and J. M. Gere, *Matrix Analysis of Framed Structures*, D. Van Nostrand Co., 1980.
8. F.-X. Coudert, *Chem. Mater.*, 2015, **27**, 1905–1916.
9. N. L. Rosi, J. Eckert, M. Eddaoudi, D. T. Vodak, J. Kim, M. O'Keeffe and O. M. Yaghi, *Science*, 2003, **300**, 1127–1129.
10. H. Furukawa, K. E. Cordova, M. O'Keeffe and O. M. Yaghi, *Science*, 2013, **341**, 1230444.
11. J. C. Tan and A. K. Cheetham, *Chem. Soc. Rev.*, 2011, **40**, 1059–1080.
12. R. Wyckoff, *Cryst. Struct.*, 1963, **1**, 85–237.
13. H. E. Robson, D. P. Shoemaker, R. A. Ogilvie and P. C. Manor, in *Molecular Sieves*, American Chemical Society, 1973, vol. 121, ch. 9, pp. 106–115.
14. J. An, S. J. Geib and N. L. Rosi, *J. Am. Chem. Soc.*, 2009, **131**, 8376–8377.
15. D. Tabor, *Polymer*, 1994, **35**, 2759–2763.
16. H. M. Ledbetter, *J. Appl. Phys.*, 1981, **52**, 1587–1589.
17. J. J. Neumeier, *J. Phys. Chem. Ref. Data*, 2018, **47**, 033101.
18. R. J. Angel, M. Bujak, J. Zhao, G. D. Gatta and S. D. Jacobsen, *J. Appl. Crystallogr.*, 2007, **40**, 26–32.
19. C. Kittel, *Introduction to Solid State Physics*, Wiley, 8th edn, 2004.
20. F. Mouhat and F.-X. Coudert, *Phys. Rev. B*, 2014, **90**, 224104.
21. M. Born, *Math. Proc. Cambridge Philos. Soc.*, 1940, **36**, 160–172.

22. D. C. Wallace, *Thermodynamics of Crystals*, Wiley, New York, 1972.
23. C. Baerlocher and L. B. McCuske, *Database of Zeolite Structures*, http://www.iza-structure.org/databases/, 2020.
24. F.-X. Coudert, *Phys. Chem. Chem. Phys.*, 2013, **15**, 16012–16018.
25. M. Vanlandingham, S. McKnight, G. Palmese, R. Eduljee, J. Gillespie and J. R. McCulough, *J. Mater. Sci. Lett.*, 1997, **16**, 117–119.
26. N. A. Burnham, R. J. Colton and H. M. Pollock, *Nanotechnology*, 1993, **4**, 64–80.
27. A. Marmier, Z. A. D. Lethbridge, R. I. Walton, C. W. Smith, S. C. Parker and K. E. Evans, *Comput. Phys. Commun.*, 2010, **181**, 2102–2115.
28. A. B. Cairns, J. Catafesta, C. Levelut, J. Rouquette, A. van der Lee, L. Peters, A. L. Thompson, V. Dmitriev, J. Haines and A. L. Goodwin, *Nat. Mater.*, 2013, **12**, 212–216.
29. M. Anderson, P. Mott and C. Roland, *Rubber Chem. Technol.*, 2004, **77**, 293–302.
30. M. A. Fortes and M. Teresa Nogueira, *Mater. Sci. Eng., A*, 1989, **122**, 227–232.
31. I. W. Gilmour, A. Trainor and R. N. Haward, *J. Appl. Polym. Sci.*, 1979, **23**, 3129–3138.
32. A. Makishima and J. D. Mackenzie, *J. Non-Cryst. Solids*, 1975, **17**, 147–157.
33. M. Siddorn, F. X. Coudert, K. E. Evans and A. Marmier, *Phys. Chem. Chem. Phys.*, 2015, **17**, 17927–17933.
34. T. Loiseau, C. Serre, C. Huguenard, G. Fink, F. Taulelle, M. Henry, T. Bataille and G. Férey, *Chem. – Eur. J.*, 2004, **10**, 1373–1382.
35. A. U. Ortiz, A. Boutin, A. H. Fuchs and F.-X. Coudert, *Phys. Rev. Lett.*, 2012, **109**, 195502.
36. R. Gaillac, P. Pullumbi and F.-X. Coudert, *J. Phys.: Condens. Matter*, 2016, **28**, 275201.
37. W. Voigt, *Lehrbuch der Kristallphysik*, Teuber, Leipzig, 1928.
38. A. Reuss, *ZAMM J. Appl. Math. Mech./Z. Angew. Math. Mech.*, 1929, **9**, 49–58.
39. R. Hill, *Proc. Phys. Soc., A*, 1952, **65**, 349–354.
40. D. H. Chung and W. R. Buessem, *J. Appl. Phys.*, 1967, **38**, 2010–2012.
41. S. I. Ranganathan and M. Ostoja-Starzewski, *Phys. Rev. Lett.*, 2008, **101**, 055504.
42. S. M. J. Rogge, S. Caroes, R. Demuynck, M. Waroquier, V. Van Speybroeck and A. Ghysels, *J. Chem. Theory Comput.*, 2018, **14**, 1186–1197.
43. J. Wieme, S. M. J. Rogge, P. G. Yot, L. Vanduyfhuys, S.-K. Lee, J.-S. Chang, M. Waroquier, G. Maurin and V. Van Speybroeck, *J. Mater. Chem. A*, 2019, **7**, 22663–22674.
44. F. Birch, *Phys. Rev.*, 1947, **71**, 809–824.
45. A. U. Ortiz, A. Boutin, A. H. Fuchs and F.-X. Coudert, *J. Phys. Chem. Lett.*, 2013, **4**, 1861–1865.
46. M. de Jong, W. Chen, T. Angsten, A. Jain, R. Notestine, A. Gamst, M. Sluiter, C. Krishna Ande, S. van der Zwaag, J. J. Plata, C. Toher, S. Curtarolo, G. Ceder, K. A. Persson and M. Asta, *Sci. Data*, 2015, **2**, 150009.

47. S. P. Ong, W. D. Richards, A. Jain, G. Hautier, M. Kocher, S. Cholia, D. Gunter, V. L. Chevrier, K. A. Persson and G. Ceder, *Comput. Mater. Sci.*, 2013, **68**, 314–319.
48. W. F. Perger, J. Criswell, B. Civalleri and R. Dovesi, *Comput. Phys. Commun.*, 2009, **180**, 1753–1759.
49. R. Dovesi, A. Erba, R. Orlando, C. M. Zicovich-Wilson, B. Civalleri, L. Maschio, M. Rérat, S. Casassa, J. Baima, S. Salustro and B. Kirtman, *WIREs Comput. Mol. Sci.*, 2018, **8**, e1360.
50. M. Parrinello and A. Rahman, *J. Chem. Phys.*, 1982, **76**, 2662–2666.
51. J. R. Ray and A. Rahman, *J. Chem. Phys.*, 1984, **80**, 4423–4428.
52. J. R. Ray, *Comput. Phys. Rep.*, 1988, **8**, 109–151.
53. S. M. J. Rogge, L. Vanduyfhuys, A. Ghysels, M. Waroquier, T. Verstraelen, G. Maurin and V. Van Speybroeck, *J. Chem. Theory Comput.*, 2015, **11**, 5583–5597.
54. H. J. C. Berendsen, J. P. M. Postma, W. F. van Gunsteren, A. DiNola and J. R. Haak, *J. Chem. Phys.*, 1984, **81**, 3684–3690.
55. G. J. Martyna, M. E. Tuckerman, D. J. Tobias and M. L. Klein, *Mol. Phys.*, 1996, **87**, 1117–1157.
56. M. A. Hopcroft, W. D. Nix and T. W. Kenny, *J. Microelectromech. Syst.*, 2010, **19**, 229–238.
57. S. Plimpton, *J. Comput. Phys.*, 1995, **117**, 1–19.
58. W. Shinoda, M. Shiga and M. Mikami, *Phys. Rev. B*, 2004, **69**, 134103.
59. B. W. Dodson, *Phys. Rev. B*, 1987, **35**, 2795–2798.
60. L. B. D. Bourg, A. U. Ortiz, A. Boutin and F.-X. Coudert, *APL Mater.*, 2014, **2**, 124110.
61. R. Demuynck, S. M. J. Rogge, L. Vanduyfhuys, J. Wieme, M. Waroquier and V. Van Speybroeck, *J. Chem. Theory Comput.*, 2017, **13**, 5861–5873.
62. S. Bureekaew, S. Amirjalayer, M. Tafipolsky, C. Spickermann, T. K. Roy and R. Schmid, *Phys. Status Solidi B*, 2013, **250**, 1128–1141.
63. N. Hansen and W. F. van Gunsteren, *J. Chem. Theory Comput.*, 2014, **10**, 2632–2647.
64. R. W. Zwanzig, *J. Chem. Phys.*, 1954, **22**, 1420–1426.
65. G. M. Torrie and J. P. Valleau, *J. Comput. Phys.*, 1977, **23**, 187–199.
66. L. Maragliano and E. Vanden-Eijnden, *Chem. Phys. Lett.*, 2006, **426**, 168–175.
67. J. Wieme, L. Vanduyfhuys, S. M. J. Rogge, M. Waroquier and V. Van Speybroeck, *J. Phys. Chem. C*, 2016, **120**, 14934–14947.
68. J. Kästner and W. Thiel, *J. Chem. Phys.*, 2005, **123**, 144104.
69. M. Souaille and B. T. Roux, *Comput. Phys. Commun.*, 2001, **135**, 40–57.
70. C. Jarzynski, *Phys. Rev. Lett.*, 1997, **78**, 2690–2693.
71. O. Valsson, P. Tiwary and M. Parrinello, *Annu. Rev. Phys. Chem.*, 2016, **67**, 159–184.
72. O. Valsson and M. Parrinello, *Phys. Rev. Lett.*, 2014, **113**, 090601.
73. J. Schmidt, M. R. G. Marques, S. Botti and M. A. L. Marques, *npj Comput. Mater.*, 2019, **5**, 83.

74. M. de Jong, W. Chen, R. Notestine, K. Persson, G. Ceder, A. Jain, M. Asta and A. Gamst, *Sci. Rep.*, 2016, **6**, 34256.
75. J. D. Evans and F.-X. Coudert, *Chem. Mater.*, 2017, **29**, 7833–7839.
76. R. Gaillac, S. Chibani and F.-X. Coudert, *Chem. Mater.*, 2020, **32**, 2653–2663.
77. P. Z. Moghadam, S. M. J. Rogge, A. Li, C.-M. Chow, J. Wieme, N. Moharrami, M. Aragones-Anglada, G. Conduit, D. A. Gomez-Gualdron, V. Van, Speybroeck and D. Fairen-Jimenez, *Matter*, 2019, **1**, 219–234.
78. S. M. J. Rogge, M. Waroquier and V. Van Speybroeck, *Acc. Chem. Res.*, 2018, **51**, 138–148.
79. J. Wieme, K. Lejaeghere, G. Kresse and V. Van Speybroeck, *Nat. Commun.*, 2018, **9**, 4899.
80. M. Kadic, G. W. Milton, M. van Hecke and M. Wegener, *Nat. Rev. Phys.*, 2019, **1**, 198–210.
81. Y. Ding, Z. Liu, C. Qiu and J. Shi, *Phys. Rev. Lett.*, 2007, **99**, 093904.
82. M. Kadic, T. Bückmann, N. Stenger, M. Thiel and M. Wegener, *Appl. Phys. Lett.*, 2012, **100**, 191901.
83. F.-X. Coudert and J. D. Evans, *Coord. Chem. Rev.*, 2019, **388**, 48–62.
84. J. Qu, M. Kadic, A. Naber and M. Wegener, *Sci. Rep.*, 2017, **7**, 40643.
85. M. R. Ryder, B. Civalleri and J.-C. Tan, *Phys. Chem. Chem. Phys.*, 2016, **18**, 9079–9087.
86. M. R. Ryder and J.-C. Tan, *Dalton Trans.*, 2016, **45**, 4154–4161.
87. J. N. Grima, R. Jackson, A. Alderson and K. E. Evans, *Adv. Mater.*, 2000, **12**, 1912–1918.
88. C. Hoberman, Radial expansion/retraction truss structures, US5024031A, *US Pat.*, 1988.
89. J. Dagdelen, J. Montoya, M. de Jong and K. Persson, *Nat. Commun.*, 2017, **8**, 323.
90. A. B. Cairns and A. L. Goodwin, *Phys. Chem. Chem. Phys.*, 2015, **17**, 20449–20465.
91. R. H. Baughman, S. Stafström, C. Cui and S. O. Dantas, *Science*, 1998, **279**, 1522–1524.
92. A. L. Goodwin, M. Calleja, M. J. Conterio, M. T. Dove, J. S. O. Evans, D. A. Keen, L. Peters and M. G. Tucker, *Science*, 2008, **319**, 794–797.
93. A. L. Goodwin, D. A. Keen and M. G. Tucker, *Proc. Natl. Acad. Sci.*, 2008, **105**, 18708–18713.
94. W. Cai, A. Gładysiak, M. Anioła, V. J. Smith, L. J. Barbour and A. Katrusiak, *J. Am. Chem. Soc.*, 2015, **137**, 9296–9301.
95. J. M. Ogborn, I. E. Collings, S. A. Moggach, A. L. Thompson and A. L. Goodwin, *Chem. Sci.*, 2012, **3**, 3011–3017.
96. G. Feng, W.-X. Zhang, L. Dong, W. Li, W. Cai, W. Wei, L. Ji, Z. Lin and P. Lu, *Chem. Sci.*, 2019, **10**, 1309–1315.
97. M. K. Gupta, B. Singh, R. Mittal, M. Zbiri, A. B. Cairns, A. L. Goodwin, H. Schober and S. L. Chaplot, *Phys. Rev. B*, 2017, **96**, 214303.
98. F. Colmenero, *Adv. Theory Simul.*, 2019, **2**, 1900040.
99. J. Grima, V. Zammit and R. Gatt, *Xjenza*, 2006, **11**, 17–29.

100. K. Rottger, A. Endriss, J. Ihringer, S. Doyle and W. F. Kuhs, *Acta Crystallogr., Sect. B*, 1994, **50**, 644–648.
101. W. Miller, C. W. Smith, D. S. Mackenzie and K. E. Evans, *J. Mater. Sci.*, 2009, **44**, 5441–5451.
102. T. A. Mary, J. S. O. Evans, T. Vogt and A. W. Sleight, *Science*, 1996, **272**, 90–92.
103. S. Margadonna, K. Prassides and A. N. Fitch, *J. Am. Chem. Soc.*, 2004, **126**, 15390–15391.
104. M. T. Dove, *Introduction to Lattice Dynamics*, Cambridge University Press, 1993.
105. P. Tschaufeser and S. C. Parker, *J. Phys. Chem.*, 1995, **99**, 10609–10615.
106. J. D. Evans, J. P. Dürholt, S. Kaskel and R. Schmid, *J. Mater. Chem. A*, 2019, **7**, 24019–24026.
107. D. Dubbeldam, K. S. Walton, D. E. Ellis and R. Q. Snurr, *Angew. Chem., Int. Ed.*, 2007, **46**, 4496–4499.
108. E. Grüneisen, *Ann. Phys.*, 1912, **344**, 257–306.
109. H. Fang and M. T. Dove, *Phys. Rev. B*, 2013, **87**, 214109.
110. P. G. Yot, Q. Ma, J. Haines, Q. Yang, A. Ghoufi, T. Devic, C. Serre, V. Dmitriev, G. Férey, C. Zhong and G. Maurin, *Chem. Sci.*, 2012, **3**, 1100–1104.
111. S. Chibani and F.-X. Coudert, *Chem. Sci.*, 2019, **10**, 8589–8599.
112. Y. Sakata, S. Furukawa, M. Kondo, K. Hirai, N. Horike, Y. Takashima, H. Uehara, N. Louvain, M. Meilikhov, T. Tsuruoka, S. Isoda, W. Kosaka, O. Sakata and S. Kitagawa, *Science*, 2013, **339**, 193–196.
113. S. M. J. Rogge, M. Waroquier and V. Van Speybroeck, *Nat. Commun.*, 2019, **10**, 4842.
114. J. Keupp and R. Schmid, *Adv. Theory Simul.*, 2019, **2**, 1900117.
115. A. K. Cheetham, T. D. Bennett, F.-X. Coudert and A. L. Goodwin, *Dalton Trans.*, 2016, **45**, 4113–4126.
116. S. M. J. Rogge, J. Wieme, L. Vanduyfhuys, S. Vandenbrande, G. Maurin, T. Verstraelen, M. Waroquier and V. Van Speybroeck, *Chem. Mater.*, 2016, **28**, 5721–5732.

CHAPTER 4

Modelling Sorption and Diffusion Behaviour in Porous Solids

JOSEPH GLOVER AND ELENA BESLEY*

School of Chemistry, University Park, University of Nottingham, Nottingham NG7 2RD, UK
*Email: Elena.Besley@nottingham.ac.uk

4.1 Introduction

Since the birth of synthetic zeolites in the 1950s, porous solids have received much attention for their practical applications in gas storage, separation, purification, sensing and catalysis.[1] Today, over 200 known zeolites have been synthesised with an additional 2 600 000+ hypothetical structures predicted from theoretical work.[2,3] Over the past 30 years, several new and exciting classes of porous solids have emerged such as porous polymer networks (PPNs), metal–organic frameworks (MOFs), zeolitic imidazolate frameworks (ZIFs) and covalent organic frameworks (COFs). The main advantage of these materials is that researchers are able to develop structures with distinct pore sizes, shapes and functionalities by choosing an appropriate combination of molecular building blocks. As such, the highly tuneable nature of these materials has motivated research into their role as zeolite replacements in several industrial applications. Due to the limitless number of potential structures that can be generated from different building block combinations, it is not surprising that only a small fraction have been realised due to the high cost and time-consuming nature of synthesis.

Inorganic Materials Series No. 8
Computer Simulation of Porous Materials: Current Approaches and Future Opportunities
Edited by Kim Jelfs

Published by the Royal Society of Chemistry, www.rsc.org

As such, the adsorption and transport properties of many novel materials remain largely unexplored. In these scenarios, molecular modelling is a valuable tool that can improve our understanding of adsorption and transport processes at the microscopic level. In addition, models of new porous solids may be generated and used to predict their adsorption and transport properties. The emergence of high-throughput screening techniques has played a vital role in accelerating the discovery of promising materials and understanding structure–performance relationships which govern an application of interest. These allow design principles to be generated which focus experimental efforts towards the generation of new materials.

In this chapter, we provide an overview of the computational approaches that are commonly employed to characterise and study the adsorption and transport properties of nanoporous materials. We review examples of their application in the literature, discussing the opportunities and limitations presented by these different methods and their use in high-throughput screening. Finally, we provide a summary of the current challenges and outlook for future studies in the field.

4.2 Molecular Simulations of Adsorption Behaviour

4.2.1 Basics of the Grand Canonical Monte Carlo Method

When investigating the adsorption equilibria of gases and their mixtures in porous materials, it is common to employ Monte Carlo simulations in the grand canonical ensemble (GCMC). In a GCMC simulation, the temperature (T), chemical potential (μ) and volume (V) of the system remain fixed, allowing the number of molecules within the system (N) to fluctuate. This mimics an adsorption experiment where a solid interface or porous material is exposed to a bulk gas reservoir until an equilibrium is established between the adsorbed and bulk phases. An adsorption isotherm is determined experimentally by carrying out multiple adsorption measurements at a given T and at various bulk pressures. In a GCMC simulation, the bulk fugacity is used as an input to account for non-ideal gas behaviour at higher pressures. This is generally converted to the relevant value of μ using an equation of state (EOS), such as the Peng–Robinson EOS.[4] To represent the adsorbent, an atomistic model can be generated using experimental techniques such as X-ray diffraction (XRD). Alternatively, adsorbent models can be generated computationally using structural optimisations at the classical or quantum mechanical (QM) level of theory.

During each GCMC simulation, random perturbations are performed which explore different configurations of the system. For single-component studies, these perturbations may include random particle translations, rotations, insertions and deletions. For multi-component simulations, an identity swap move is generally included which attempts to replace a particle of one component with a particle of another, randomly selected component. After each perturbation, the new configuration of the system is accepted or rejected depending on a Boltzmann-type weighting. The acceptance probabilities for translation (P_{trans}), rotation (P_{rot}), insertion (P_{ins}) and deletion (P_{del}) perturbations are:

$$P_{trans} = \min\left(1, e^{-\beta \Delta U}\right)$$

$$P_{rot} = \min\left(1, e^{-\beta \Delta U}\right)$$

$$P_{ins} = \min\left(1, \frac{\beta f V}{(N+1)} e^{-\beta \Delta U}\right)$$

$$P_{del} = \min\left(1, \frac{N}{\beta f V} e^{-\beta \Delta U}\right)$$

where β is $1/k_B T$ (k_B is the Boltzmann constant), f is the gas-phase fugacity and ΔU is the change in the potential energy as a result of the perturbation. As more perturbations are applied, U and N begin to converge and fluctuate around mean values and the remaining steps of the simulation are used to generate statistical averages of N. In the context of a GCMC simulation, N is the absolute loading inside the pores (n_{abs}) and must be converted to an excess loading (n_{exc}) for comparison with experiment. n_{exc} can be quantified as the difference between the amount of gas adsorbed and the amount that would be present in the same pore volume (V_{pore}) under bulk conditions:[5]

$$n_{exc} = n_{abs} - V_{pore} \cdot \rho_{gas}$$

where ρ_{gas} is the fluid density in the bulk phase.

4.2.2 Brief Overview of Classical Force-fields

To compute ΔU in a fast and efficient manner, a force-field is typically employed. This is an analytical expression that describes U as a function of atomic coordinates, $\boldsymbol{r}$. In general, the force-field is grouped into a series of bonded and non-bonded terms; bonded terms describe intramolecular interactions such as bond stretching, bending and torsions, while non-bonded terms describe intermolecular interactions such as dispersive and electrostatic interactions. In most classical simulations that study the adsorption behaviour of zeolites, MOFs, ZIFs and COFs, the host structure is modelled as a rigid entity, with the host atoms fixed in their crystallographic or geometry optimised positions. The advantage of this is that bonded interactions can be neglected, and $U(\boldsymbol{r})$ is calculated from the non-bonded interactions. Furthermore, as these atoms are fixed, the potential energy surface (PES) induced by the framework can be pre-tabulated onto a 3D grid. $U(\boldsymbol{r})$ can then be obtained through interpolation algorithms during a GCMC simulation, increasing the computational efficiency of each Monte Carlo step. We note that a number of force-fields that describe flexibility in these adsorbents have been developed and are discussed in more detail in Section 5.1.2.

To accurately model the adsorption behaviour of porous materials, a set of parameters for the host–guest and guest–guest interactions is required by the force-field. For simulations involving zeolites, the host–guest interactions are commonly modelled using a Kiselev-type model.[6] These employ a 12-6 Lennard-Jones (LJ) function to describe dispersive interactions:[7]

$$U_{ij}(\boldsymbol{r}) = \sum_{ij} 4\varepsilon_{ij}\left[\left(\frac{\sigma_{ij}}{r_{ij}}\right)^{12} - \left(\frac{\sigma_{ij}}{r_{ij}}\right)^{6}\right]$$

where ε_{ij} is the interaction well depth, r_{ij} is the internuclear separation and σ_{ij} represents the collision diameter for atoms i and j. Within the Kiselev model, the zeolite is assumed to have a rigid structure and the silicon atoms are generally not considered as LJ interaction sites due to the effective shielding of surrounding oxygen atoms. The parameters describing host–guest interactions are typically fitted to reproduce experimental data such as isotherm inflections or isosteric heats of adsorption (Q_{st}).[8–13] In recent years, Bai *et al.* developed the TraPPE-zeo force-field for modelling all-silica zeolites.[14] While this force-field was also parameterised to reproduce experimental isotherms, it differs by including LJ interaction sites and partial charges on silicon atoms, leading to a better balance of the dispersive and electrostatic interactions. This was found to improve transferability across a wide range of polar and non-polar compounds.

For the study of MOFs, COFs and ZIFs, dedicated force-fields that describe host–guest interactions are scarce due to the wide chemical diversity of these materials. Instead, many studies employ general force-fields such as the universal force-field (UFF),[15] DREIDING,[16] or combinations of both. These force-fields contain parameters that were estimated from general rules based on the element, hybridisation, and connectivity for most elements in the periodic table. Another popular choice is the all-atom, optimised potentials for liquid simulations (OPLS-aa) force-field.[17] Unlike DREIDING or UFF, OPLS-aa was developed to study organic liquids and was parameterised to reproduce their bulk thermodynamic and structural properties. As such, each element possesses multiple sets of parameters which are able to describe different chemical environments of organic liquids. In all of these force-fields, pairwise dispersive interactions are modelled using a 12-6 LJ function and each force-field provides the homonuclear pair parameters, σ_{ii} and ε_{ii}. To derive the cross terms, a series of mixing rules are employed, such as the Lorentz–Berthelot mixing rules.[18,19] When electrostatic interactions are important, an additional Coulomb's law term is introduced for a series of point charges. For periodic structures, techniques such as the Ewald summation can provide an accurate solution to the electrostatic interaction energy to a specified degree of precision.[20]

Guest–guest interactions are typically described by force-fields which have been parameterised to reproduce the thermophysical and structural properties of adsorbates in the bulk phase. One of the most popular examples of these is the TraPPE family developed by Siepmann and co-workers.[21,22] These are a transferable and highly accurate set of force-fields used in the simulations of hydrocarbons, small gases and organic molecules. The parameters in these force-fields are typically found through empirical fitting to vapour–liquid coexistence curves (VLCC), liquid densities, critical temperatures, and vapour pressures across multiple temperatures and/or compositions. Small molecules such as CO_2, N_2, O_2, He and NH_3 are modelled with rigid bonds, however larger adsorbates with intrinsic flexibility are modelled with additional bonded terms. For some molecules (*e.g.* CH_4, CH_3, CH_2 and CH groups), a united-atom (UA) approach is employed to represent multiple atoms by single interaction sites or pseudo-atoms.[22] This improves the computational efficiency of the force-field with a small sacrifice in accuracy. In addition to the TraPPE family, several alternative force-fields exist within the literature for describing various guest interactions, including the SPC[23] and TIP[24,25] H_2O potentials, the EPM2 CO_2 potential[26] and the Buch[27] and Darkrim–Levesque[28] H_2 models.

4.2.3 Atomic Partial Charge Calculation

In order to compute electrostatic interactions between species, knowledge of the atomic partial charge (APC) distribution is required. It is important to realise that APCs are not physical observables, meaning that they cannot be directly measured. In order to estimate their values, a number of different schemes can be applied with varying degrees of accuracy. The computational cost of these methods differs considerably, and the choice is mainly governed by the size of the system or number of individual structures to be studied. For many adsorbates, the APCs assigned to each atom are already determined as part of the force-field parameterisation process to ensure they reproduce equilibrium properties (*i.e.* VLCCs). For porous adsorbents, the most reliable methods of determining the APCs are based on QM electronic structure calculations which calculate their values based on population analysis (*e.g.* Mulliken,[29] Löwdin charges[30] and natural population analysis[31]) or by partitioning the electron density distribution into atomic contributions (*e.g.* Bader,[32] Hirshfeld[33] and iterative Hirshfeld schemes[34]).

Another popular approach is to use the electrostatic potential (ESP) of the adsorbent to derive APCs that fit the QM ESP on selected grid points. The CHELPG,[35] RESP[36] and Merz–Kollman[37] schemes are examples of algorithms which employ this methodology and mainly differ from one another by the choice of grid points used to calculate the ESP. One of the limitations of these schemes is that they poorly describe the charges on buried atoms, which belong to the internal regions of a chemical system and, hence, are located far from the ESP surface. Buried atoms can possess a range of physical or unphysical APCs which reproduce the ESP with similar accuracy. In addition, these methods are applied to clusters which are assumed to provide an accurate representation of the ESP in the periodic structure. As such, the size of the cluster and how cleaved bonds are terminated can have a non-negligible impact on the predicted charge distribution. To address these issues, Campañá *et al.* developed the REPEAT scheme,[38] which employs a modified error functional into the fitting procedure that acts on the relative differences between the two ESPs, instead of their absolute values. Within this error functional, a penalty function is added to address unphysical charges assigned to buried atoms, similar to the RESP scheme of Bayly and co-workers.[36] These schemes may be applied to periodic structures to circumvent the issues of using finite clusters to represent the periodic material. In recent years, Manz and Sholl developed the density derived electrostatic and chemical (DDEC) method, which has become one of the most popular choices for characterising MOF APCs from QM calculations.[39] This algorithm determines the APCs from partitioning of the electron density but incorporates spherical averaging and reference ion densities to generate a set of chemically meaningful charges that accurately reproduce the QM ESP. These APCs exhibit good transferability between chemically related systems and do not require any fine-tuned constraints to treat buried atoms, as employed in the REPEAT and RESP schemes.

One of the natural drawbacks of QM calculations is the high cost of each electronic structure calculation. For many porous adsorbents, both cluster and periodic models may require several hours of computation time to obtain the APCs. For identifying the adsorption properties in a large number of structures, this is extremely demanding and has motivated the development of faster approximative methods. One example of these computationally cheaper methods is the semi-empirical charge equilibration (Qeq) scheme of Rappé and Goddard.[40] This popular algorithm is based on Sanderson's concept of electronegativity equalisation and assumes that certain properties of isolated atoms are similar to those bonded within a molecule. Therefore, the APCs of a system in a specific geometry may be rapidly obtained from the ionisation potentials, electron affinities and atomic radii of its isolated atoms using a truncated second-order Taylor series expansion of U as a function of the APCs. These properties may be extracted from experimental data or QM calculations. Over the years, a number of variants of the Qeq algorithm have been proposed which modify the reference state of the Taylor expansion and extend the algorithm to periodic systems (*e.g.* the extended Qeq (EQeq),[41] split-charge Qeq (SC-Qeq),[42] formal-charge Qeq (FC-Qeq)[43] and periodic Qeq (PQeq)).[44] The accuracy and transferability of different Qeq variants was recently reviewed by Ongari *et al.* in a diverse range of 2388 MOFs.[45] For each framework, the APCs derived from each Qeq variant were directly compared with DDEC charges. While no single Qeq variant was found to be the best, a number of recommendations were made when employing these methods. For example, they noted that the choice of a neutral atom reference state for the second-order Taylor expansion is not a good choice for estimating the APCs of alkali metals. Furthermore, ionisation energies calculated at the coupled-cluster level of theory are fairly accurate for most elements but suffer from a basis set dependence. Therefore, the use of experimental data was suggested for methods that employ higher-order reference states such as the EQeq and FC-Qeq algorithms. Approximate charges based on Qeq algorithms are commonly used for high-throughput screenings in gas storage and separation applications. While these charges can be used to give a good initial assessment of a target application, it has been suggested that charges derived from QM-based methods should be employed for comparison and for the ultimate identification of top performing materials.[46]

In another study, Xu and Zhong developed the CBAC method for rapidly estimating framework charges in MOFs and COFs.[47] This method assumes that atoms with the same bonding connectivity have identical charges in different frameworks. Using a training set of 30 MOFs, the CHELPG scheme was employed to determine the partial charges of different atoms. The transferability of these charges was then tested in 13 MOFs with similar connectivity. Simulated adsorption isotherms of CO_2, CO and N_2 were found to give nearly identical results to those calculated using charges from QM calculations and were also in excellent agreement with experimental isotherms. More recently, a similar methodology based on the assumption that MOF molecular building blocks present similar charges was proposed by Argueta *et al.*[48] In this method, the CHELPG scheme was applied separately to ligands and metallic nodes in order to determine the APCs of different molecular building blocks. In their analysis, the authors found the APCs were comparable to charges from DFT calculations and produce similar adsorption property predictions such as Henry coefficients (K) and gas uptakes at different pressures.

4.2.4 Enhanced Sampling Monte Carlo Techniques

While conventional GCMC simulations are suitable for the study of gases containing small molecules, they become much more inefficient for larger adsorbates or dense systems. In these scenarios, insertion and deletion perturbations suffer from low acceptance rates which leads to inefficient sampling of phase space. These issues have motivated the development of several techniques that employ biased Monte Carlo algorithms and code parallelisation to improve the efficiency of sampling.

4.2.4.1 Cavity and Energy Bias Methods

Larger adsorbates occupy more space and often fit very tightly within the pore volume. This leads to a low probability of a successful insertion due to the potential overlap with neighbouring adsorbates or adsorbent atoms. To address this issue, a number of biasing techniques have been proposed such as cavity biasing, energy biasing and combinations of both. In cavity-bias GCMC,[49] particle insertions are no longer attempted at random and are biased to cavities or regions that can accommodate the adsorbate. In this method, a number of trial points are randomly selected across the simulation volume and tested to see whether they are inside a cavity of suitable radius, R_c (*i.e.* a space large enough to accommodate a newly inserted adsorbate). One of these points is randomly selected and trial insertion is attempted with the following acceptance probability:

$$P_{\mathrm{ins}}^{\mathrm{cav}} = \min\left(1, P_{\mathrm{c}}(N)\frac{\beta f V}{(N+1)}\mathrm{e}^{-\beta \Delta U}\right)$$

where $P_{\mathrm{c}}(N)$ is the probability of finding a cavity with a radius of R_{c} or larger in a system of N molecules. In order to preserve detailed balance, a deletion is performed without bias by randomly removing a molecule from the system, with an acceptance probability:

$$P_{\mathrm{del}}^{\mathrm{cav}} = \min\left(1, \frac{1}{P_{\mathrm{c}}(N-1)}\frac{N}{\beta f V}\mathrm{e}^{-\beta \Delta U}\right)$$

For simulations of a LJ fluid, Mezei[49] showed that cavity-bias insertions have up to an 8-fold increase in efficiency over random insertions. In later work, Snurr *et al.* built on the cavity-bias methodology to develop an energy biasing scheme when studying aromatic hydrocarbon adsorption in an MFI zeolite.[50] This technique divides an adsorbent's unit cell into a 3D grid, assigning a weight to each voxel. The weight in voxel i (ρ_i) is determined by inserting a probe molecule at the centre of the voxel and evaluating the potential energy felt by the probe (U_i). This is normalised over the total sum of voxel weights to give:

$$\rho_i = \frac{\mathrm{e}^{-U_i \beta}}{\sum_j \mathrm{e}^{-U_j \beta}}$$

An energy biased insertion is accepted with the following probability ($P_{\mathrm{ins}}^{\mathrm{en}}$):

$$P_{\mathrm{ins}}^{\mathrm{en}} = \min\left(1, \frac{1}{\rho_i}\frac{V_{\mathrm{voxel}}}{V}\frac{\beta f V}{(N+1)}\mathrm{e}^{-\beta \Delta U}\right)$$

where V_{voxel} and V are the volumes of the voxel and simulation box, respectively. To preserve detailed balance, the deletion move is performed without bias, using a modified acceptance probability:

$$P_{\mathrm{del}}^{\mathrm{en}} = \min\left(1, \rho_i \frac{V}{V_{\mathrm{voxel}}}\frac{N}{\beta f V}\mathrm{e}^{-\beta \Delta U}\right)$$

Snurr *et al.* investigated the efficiency of energy biasing GCMC by simulating Ar adsorption on silicalite.[50] In their analysis, they observed that both cavity-biasing and energy-biasing lead to a substantial improvement of the insertion and deletion acceptance ratios. This could be increased further by applying both methods in the same simulation. However, the authors stated that the ultimate goal of this methodology is to improve the relative efficiency of the GCMC simulation. Therefore, it is important to balance the added computational cost with the gains in efficiency.

4.2.4.2 Configurational-bias Method

For flexible or long-chain molecules, insertion and deletion moves can be even more challenging than for small molecules. Flexible molecules and chains possess a large configurational space that can take a long time to sample with conventional Monte Carlo techniques. In these cases, configurational-bias Monte Carlo (CBMC) simulations can help to improve the sampling efficiency. Instead of inserting a long chain molecule as a whole, the CBMC scheme grows these molecules segment by segment.[7,51] Similarly, the configuration of an existing molecule can be altered by re-growing part of the chain. For the growth of each segment, i, a set of trial orientations, k, is generated according to the bonded intramolecular interactions of this segment with the other inserted segments. A trial position, j, is selected based on a probability (P_i) proportional to its Boltzmann factor:

$$P_i = \frac{e^{-U_i^{ext}(j)\beta}}{\sum_{n=1}^{k} e^{-U_i^{ext}(n)\beta}} = \frac{e^{-U_i^{ext}(j)\beta}}{w(i)}$$

Here, $U_i^{ext}(j)$ and $U_i^{ext}(n)$ are the external components of the potential energy from insertions at trial positions j and n, which account for non-bonded interactions between segment i and all other segments and molecules in the system. This step is repeated for each segment until the entire chain is grown. The Rosenbluth factor for this new configuration, W^{new}, can be determined as:

$$W^{new} = \prod_{i=1}^{l} w(i)$$

where l is the total number of segments in the chain, and the probability of accepting this new configuration, P_{accept}, is determined by:

$$P_{accept} = \min\left(1, \frac{W^{new}}{W^{old}}\right)$$

W^{old} is the Rosenbluth factor calculated by retracing over the old configuration. Over the years, CBMC has been expanded to treat a large range of linear, branched and cyclic hydrocarbons.[52–55]

4.2.4.3 Continuous Fractional-component Monte Carlo Method

For simulations involving dense systems, Shi and Maginn developed the continuous fractional-component Monte Carlo (CFCMC) algorithm to improve insertion and deletion acceptance ratios.[56] In a CFCMC simulation, a 'fractional' molecule is placed within the system. While this molecule is allowed to interact with surrounding adsorbates, the interaction strength is scaled using a pseudo-continuous coupling parameter, λ. In addition to standard translation, rotation, insertion and deletion moves, CFCMC simulations include perturbations that allow the value of λ to change between 0 (no interactions with surroundings) and 1 (full interactions with surroundings). When λ becomes greater than 1, the fractional molecule is converted into a real one and a new fractional molecule is randomly placed in the simulation cell with a λ value equal to $\lambda - 1$. While the newly inserted fractional molecule may overlap with other molecules in the system, its small value of λ means that it will interact relatively weakly with its surroundings, leading to a higher acceptance probability. Alternatively, if λ becomes smaller than 0, the fractional molecule is removed from the system and another molecule is randomly chosen to be the new fractional molecule with a λ value equal to $\lambda + 1$. To ensure that the system does not become 'stuck' around a small range of λ values, an adaptive bias factor can be introduced using Wang–Landau (WL) sampling to ensure that the probability distribution of λ is relatively flat.[57]

The main advantage of this method over conventional GCMC is that new molecules are gradually inserted or deleted, allowing the system to slowly adapt to these changes. In contrast, conventional GCMC insertion moves require the spontaneous formation of void spaces that are large enough to accommodate the adsorbates, or the deletion of an existing adsorbate which is energetically unfavourable. During the testing of their algorithm, Shi and Maginn compared the efficiency of CFCMC and GCMC at estimating H_2O densities. While both methods yielded identical densities, less than 0.01% of H_2O insertion and deletion moves were accepted using standard GCMC, corresponding to a change in composition once every 10 000 attempts. Using CFCMC, 2.4% of moves changing the λ value led to a successful insertion or deletion of a water molecule. On average, this meant that the composition of the system changed for every 42 attempts that were made to alter the value of λ, providing a much higher sampling efficiency. Zhang and Snurr recently employed CFCMC to study H_2O adsorption in **ZIF-8**.[58] For simulations at the saturation pressure, equilibration using conventional GCMC took over 30 days of simulation time due to the system becoming trapped in several meta-stable states. By combining energy-biased insertions with CFCMC, the authors were able to accelerate their simulations by a factor of 6.7, corresponding to an equilibration period of 4.5 days. Furthermore, the CFCMC algorithm was found to increase the number of accepted insertions from 0.0052 per second (using conventional GCMC) to 0.285 per second.

4.3 Computational Approaches for Characterising the Structural Properties of Porous Solids

Many interesting adsorption and transport phenomena can be understood from knowledge of the structural properties of an adsorbent, such as its surface area, pore size, pore volume and pore network topology. Therefore, we need methods that are able to accurately characterise these properties in fast and efficient ways. This is especially important for computational studies which attempt to understand different adsorption and transport behaviours of materials, and whether these can be explained through relationships with the structural properties of the adsorbents. In this section, we discuss various computational approaches for characterising the structural properties and pore topology of porous materials, and how these properties relate to experimental methods.

4.3.1 Surface Area

The specific surface area (SSA) is one of the most important physical properties of an adsorbent for gas storage and separation applications. It is routinely estimated by applying the theory of Brunauer, Emmett and Teller (BET theory) to gas adsorption isotherms.[59] Developed in 1938, BET theory extends on Langmuir's work[60] by explaining the physical adsorption of gases onto surfaces through multi-layer adsorption. To estimate the SSA of an adsorbent, BET theory is employed to calculate the amount of gas forming an adsorbed monolayer on the surfaces of the adsorbent. This quantity, referred to as the monolayer capacity, is typically extracted from N_2 adsorption experiments at 77 K. However, other gases such as Ar and CO_2 are also commonly used. From the adsorption isotherm, the monolayer capacity can be calculated using the linearised BET equation:

$$\frac{p}{V(p_0 - p)} = \frac{1}{V_m C} + \left(\frac{C-1}{V_m C}\right)\frac{p}{p_0}$$

where p_0 is the saturated pressure of the adsorbate gas, p is the partial pressure of the adsorbate gas at equilibrium conditions, V is the volume of gas adsorbed at the equilibrium pressure p, V_m is the volume of gas adsorbed to produce a monolayer of the, and C is a dimensionless constant related to the heat of adsorption of the monolayer.

Using the BET equation, V_m can be calculated from regression analysis over a restricted linear region of the BET plot. In the original publication, Brunauer, Emmett and Teller studied several catalysts, supports and adsorbents which possessed type II N_2 adsorption isotherms due to their mesoporosity.[61] In these materials, they found that the BET plots typically became linear between $0.05 \leq p/p_0 \leq 0.35$. While this range may be suitable for mesoporous adsorbents, there are many zeolites, MOFs, PPNs and COFs which possess microporosity. In microporous materials, surface coverage and pore filling occur at much lower pressures, typically below a relative pressure of 0.05. Furthermore, the BET plots of adsorbents that combine micropores with mesopores may contain multiple linear ranges, making it more difficult to identify the most appropriate region. The location and extent of these regions will depend on both the adsorbent and the adsorbate, making it important to have a consistent method for reporting BET surface areas. To address these issues, Rouquerol *et al.*[62] devised a series of consistency criteria for selecting an appropriate linear range:

(1) Only a range where $V(p_0 - p)$ increases monotonically with p/p_0 should be selected.
(2) The value of constant C resulting from linear regression should be positive.
(3) V_m should correspond to a relative pressure that falls within the selected linear region.
(4) The relative pressure corresponding to the monolayer loading calculated from BET theory $(1/\sqrt{C}+1)$ should be equal to the relative pressure determined in criterion (3) with a tolerance of 10–20%.

After calculating V_m, the SSA (A_{BET}) is determined using the following formula:

$$A_{BET} = \frac{\sigma N_A V_m}{m_s}$$

σ is the adsorption cross-sectional area of the adsorbate, N_A is Avogadro's constant and m_s is the mass of the adsorbent sample. Brunauer and Emmett[63] suggested that σ could be calculated from the liquid density in the bulk liquid state, as:

$$\sigma = f(M/\rho N_A)^{2/3}$$

where f is a packing factor, ρ is the liquid adsorbate density and M is the mass of the adsorbate. It is commonly assumed that the arrangement of the adsorbed monolayer is hexagonal close-packing, making $\sigma = 0.162$ nm^2 for a N_2 molecule at 77 K.[64]

GCMC simulations can be employed to simulate adsorption isotherms for BET analysis. GCMC simulations normally use perfect crystal structures that are free of defects or collapsed pores, which may exist in the experimental sample. As such, the computational BET analysis gives an indication of the theoretical upper limit to SSA. This is especially useful for materials such as MOFs, where the upper limit can be used to determine the effectiveness of the activation procedure.

4.3.1.1 Applicability of the BET Method

While BET theory provides a method for ranking surface area across different materials, the underlying assumptions of the theory tend to break down for several porous materials. For example, BET theory assumes that gases physically adsorb onto surfaces forming multiple layers, of which an infinite number exist at the saturation pressure. By applying Langmuir theory to each layer, it assumes all sites are energetically homogenous and that adsorbate–adsorbate interactions and co-operative adsorption events do not impact the adsorption process.[65] It is clear that these assumptions oversimplify the adsorption process and are more appropriate for describing an ideal gas adsorbing onto a solid surface. In reality, many adsorbents possess finite-sized pores which contain a great deal of heterogeneity with respect to the adsorption energy. For many of these materials, it has also been demonstrated that gas adsorption is more appropriately described by a pore-filling mechanism rather than layer-by-layer formation.[66] Consequently, this can influence the arrangement of the adsorbed gas, resulting in a range of σ values that provide different estimations of the SSA.[64] This has led to many questions about the physical meaning of the BET surface area and whether the method is valid for systems such as microporous adsorbents. Walton and Snurr used molecular simulations to calculate BET surface areas in six microporous IRMOFs.[67] Although simulation snapshots indicated a pore-filling mechanism in these materials, the authors reported very good agreement between the BET surface areas and the accessible surface area (ASA) determined using geometric methods (see Section 4.3.1.2). This agreement was further improved upon application of the consistency criteria proposed by Rouquerol *et al.* and discussed in Section 4.3.1, thus indicating the importance of identifying the correct linear region of the BET plot. In a later study, Bae and co-workers[68] extended Walton and Snurr's research[67] by expanding the range of porous materials to zeolites and MOFs containing a combination of ultra-micropores (diameter (d) less than 7 Å) and super-micropores (d in the region of 7–20 Å). They reported consistent findings that showed good agreement between the BET and ASAs when applying the consistency criteria. Similarly, simulation snapshots indicated that gas adsorption did not form distinct layers, further demonstrating the weakness of this assumption in BET theory.

While the use of the consistency criteria has proven to be very important for microporous adsorbents, their application and interpretation can become a non-trivial matter for larger pore systems. Recently, Wang and co-workers employed isoreticular expansion[69] to develop a series of water stable, zirconium-oxo/hydroxo cluster-based MOFs with **ftw** topology.[70] The **NU-110X** MOF series possess a series of tetratopic linkers that were predetermined to afford ultra-high surface area frameworks from *in silico* computational work.[71] The **ftw** topology of these four MOFs provides two types of pore inside a cubic unit cell; one central large pore (17.2 Å$\leq d \leq$24.2 Å) surrounded by eight smaller pores (9.5 Å$\leq d \leq$13.5 Å), which centre at the vertices of the unit cell. During the calculation of the BET SSA, the authors noted difficulty in fulfilling the third and fourth consistency criteria. To satisfy these criteria, a tolerance between 10–20% is required across all experimental and simulated isotherms. In the case of **NU-1103**, this led to a large discrepancy between the simulated BET surface area and geometric ASA (6553 *vs.* 5646 $m^2 g^{-1}$). Further insights were provided using GCMC simulation snapshots across several points on the adsorption isotherm. In all materials, snapshots revealed that varying amounts of pore filling occurs before monolayer coverage is even achieved, leading to an overestimation of the SSA. By calculating the distance between each adsorbate and the framework, an estimate of the amount of pore-filling contamination (PFC) that affects the monolayer capacity can be achieved. These results showed that PFC was highest in MOFs containing the largest pores (**NU-1103** and **NU-1104**), resulting in a larger overestimation of the BET SSA.

PFC was further investigated by Gómez-Gualdrón *et al.* who employed molecular simulations to calculate the BET SSA for a diverse set of MOFs.[72] Many of these MOFs contained different pore-size distributions (PSD), with varying degrees of micro- and mesoporosity, allowing an in-depth analysis into the application of the consistency criteria in these materials. In their assessment, the authors concluded that even in the case where all four consistency criteria are fulfilled, the monolayer loading can be overestimated as a result of PFC. In MOFs containing multiple pores with an appreciable size disparity, an overlap of pore-filling and surface adsorption can occur between both pores. Trends across 25 MOFs indicated that this effect was worse in frameworks that combine large micropores with mesopores, leading to high levels of PFC and BET surface areas that were considerably larger than their ASAs. For a more accurate representation of the monolayer, snapshots of simulations at the saturation pressure were analysed. By removing any adsorbates that were more than 1 Å away from the surface, the remaining monolayer molecules gave a much-improved agreement with the ASA. Using this method, Glover and Besley demonstrated how PFC can influence experimental conclusions, such as the effect of linker substitution on MOF surface areas.[73] One example of this is the substitution of phenyl rings for alkyne subunits, which has previously been proposed as an efficient way of increasing the SSA.[74,75] However, when accounting for PFC, this strategy is not always as effective as predicted from experimental analysis, particularly for **rht** MOFs which combine large micropores and mesopores. As an example, **NU-111** possesses an experimental BET surface area of 5020 $m^2 g^{-1}$.[74] This agrees reasonably with the simulated value of 5249 $m^2 g^{-1}$ predicted over the same p/p_0 range from applying the four consistency criteria. However, analysis of simulation snapshots showed high degrees of PFC which overestimate the surface area by ~700 $m^2 g^{-1}$. By substituting each alkyne subunit for a phenyl ring, the isoreticular **NOTT-119** MOF is obtained. This MOF possesses an experimental BET surface area of 4118 $m^2 g^{-1}$, which is approximately 20% smaller than **NU-111**. However, Glover and Besley showed that the difference in surface area in these materials is negligible if PFC is taken into account. This demonstrates the challenges of comparing experimental BET surface areas for materials that are susceptible to PFC, and how molecular simulations can act as a powerful tool to clarify experimental observations regarding the surface areas of porous materials.

Overall, molecular simulations have provided a great deal of insight into the applicability of the BET method across many different porous systems. Despite its flaws, the BET method is still considered the experimental standard for characterisation of surface area due to its simplicity. The application of the consistency criteria is of high importance for deriving SSAs across all materials, and extra attention should be given to ensure the fulfilment of as many criteria as possible with the smallest amount of error. Additional calculations to account for PFC are very important for providing accurate estimations of surface areas and meaningful comparisons of structural properties between different materials.

4.3.1.2 Surface Areas Derived from Geometric Algorithms

For high-throughput studies which contain thousands of materials, the BET method is too expensive to estimate the SSA of each material in a time-efficient manner. Therefore, it is more beneficial to employ faster methods that can calculate the structural properties geometrically. In this context, a geometric surface area can be determined mathematically from the atomic coordinates of the porous material alone. Depending on the application of the material, a number of different surface areas may be defined independently. For porous materials, there are two popular definitions of the geometric surface area, as depicted by Düren *et al.* (Figure 4.1).[76]

The ASA is calculated from the total area that is available to a probe molecule travelling across the surface of a material. Lee and Richards were the first to describe an algorithm for estimating the ASA of ribonuclease-S, lysozyme and myoglobin.[77] Their program computes the ASA by drawing an arc, centred around each protein atom, with a radius equal to the sum of the protein atom and solvent molecule radii. Regions of the arc which do not overlap with any other protein atoms are deemed accessible and the total ASA is therefore proportional to the summed length of accessible regions. This algorithm has been made more efficient over the years. Shrake and Rupley introduced a series of uniformly distributed grid test points around each atom to speed up the calculation.[78] In another study, Gelb and Gubbins used a Monte Carlo approach to estimate the ASA in porous glasses.[79] A test probe is randomly placed at locations surrounding the surface of each atom and tested for overlap with any other atoms in the porous glass. The fraction of insertions that do not overlap with other atoms are then used to calculate the ASA.

An alternative to the ASA is the re-entrant surface area described by Richards.[80] The re-entrant surface area is more commonly known as the Connolly surface area (CSA), named after Michael Connolly who developed a number of numerical and analytical algorithms for computing the CSA.[81] Like the ASA, the CSA is generated by rolling a probe molecule across a surface. It consists of the contact surface between the probe and the surface atoms plus the re-entrant surface where the probe comes into contact with more than one surface atom. With regards to sorption-based applications, Düren *et al.* demonstrated that the ASA is more appropriate for characterising porous solids than the CSA.[76] Using a 2D square box, they explain that the CSA is similar both inside and outside the box. In contrast, the ASA is considerably smaller on the inside than it is on the outside. Consequently, if the adsorption capacity of a material is estimated by dividing the total

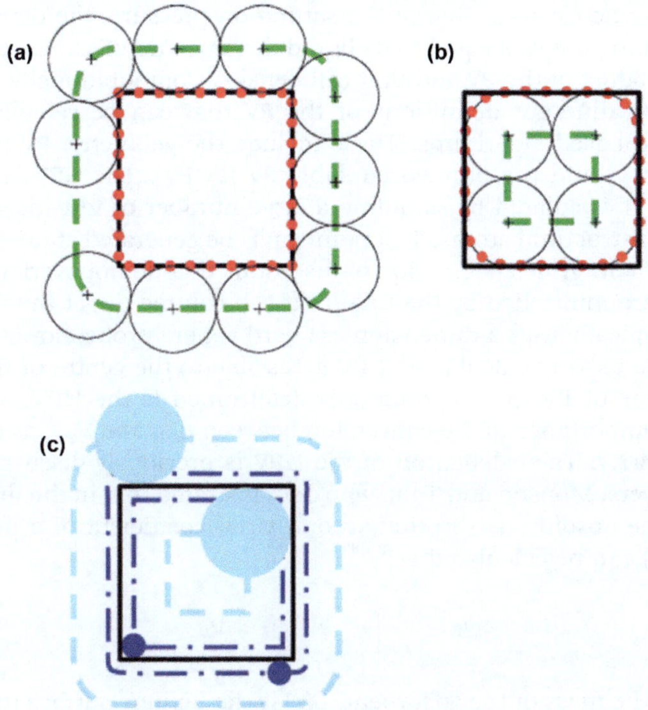

Figure 4.1 (a) 2D schematic illustration of molecules adsorbed on the outside of a box and the corresponding surface areas (Connolly surface: red, dotted line; accessible surface: green, dashed line). (b) 2D schematic illustration of molecules adsorbed inside the same box. (c) 2D schematic illustration of the accessible surface area measured on the inside and outside of the box with a small (dark blue, dash-dotted line) and large (light blue, dashed line) probe molecule.
Reproduced from ref. 76 with permission from American Chemical Society, Copyright 2007.

surface area by the cross-section of the adsorbate molecules, then the CSA would estimate the same adsorption capacity both inside and outside the box. As shown in Figure 4.1, this is clearly not the case and the ASA is better suited for estimating the adsorption capacity. The ASA calculations[76] across 11 MOFs gave excellent agreement with the corresponding BET surface areas using the consistency criteria, further highlighting the suitability of this method for evaluating surface areas both efficiently and accurately.

4.3.2 Pore Volume and Size Distribution

For many gas storage applications, the specific pore volume (PV) and void fraction (VF) are also important structural properties that govern the available volume of the unit cell which adsorbates can occupy. Experimentally, the PV can be derived from N_2 adsorption isotherms using the Gurvich rule.[61] This rule assumes that at the saturation pressure, the density of N_2 adsorbed in the pores is equal to its liquid density. The VF can be calculated from the product of the PV and unit cell density. Computationally, there are a number of different definitions of the PV that can be calculated from Monte Carlo based algorithms. These include the geometric PV (GPV), helium PV (HPV) and the probe-occupiable PV (POPV). The GPV is fairly intuitive and is calculated by sampling a large number of test insertions for overlap with structural atoms. Test points may be generated at random or be situated on a 3D grid. The fraction of insertions that do not overlap with the adsorbent are multiplied by the total unit cell volume to get the GPV. This approach typically uses a dimensionless hard sphere probe; however, larger radii may be used to calculate the PV accessible to the centre of the probe. The next type of PV that is commonly determined is the HPV, which has significant importance in the conversion between n_{ads} and n_{exc}, as described in Section 4.2.1. The calculation of the HPV is previously discussed in the works of Myers, Monson and Talu, who demonstrated that in the limit of low pressure, the absolute adsorption second virial coefficient of a porous adsorbent (B^a) can be calculated as:[82,83]

$$B^{a} = \frac{1}{m}\int e^{-\beta U(\boldsymbol{r})}\,d\boldsymbol{r} - V_{g}$$

where m is the mass of the adsorbent, $U(\boldsymbol{r})$ is the helium–surface interaction energy for a single helium atom, and V_g is the specific HPV. At room temperature, the adsorption of helium is considered negligible, therefore:

$$V_{g} = \frac{1}{m}\int e^{-\beta U(\boldsymbol{r})}\,d\boldsymbol{r}$$

The above integral can be estimated using the Widom insertion technique,[7] which samples the helium–solid interaction energy across the pore space. Consequently, the HPV will be sensitive to the choice of force-field and the temperature used in the simulation. While a standard temperature of 298 K is typically used, direct comparison with experimental work will be achieved using the same specified temperature. The final PV to be mentioned is the POPV that was recently described by Ongari and co-workers.[84] The POPV is defined by the accessible and non-accessible void space to a probe of radius r. While this definition appears very similar to the GPV, the POPV accounts for the entire volume that can be occupied by the probe (not just the centre). Ongari *et al.* compared multiple methods of calculating the PV in a large number of MOFs and COFs. The authors observed that the POPV from a N_2 probe gave the closest agreement with experimentally derived PVs and thus concluded that it is the best method of identifying discrepancies such as low crystallinity, presence of adsorbed solvent molecules, or pore shrinking in the experimental sample.

Another important property when characterising porous solids is the PSD. A number of experimental methods exist for analysing the PSD of a porous material such as mercury porosimetry,[85] NMR cryoporometry[86] and small-angled X-ray/neutron scattering (SAXS and SANS).[87,88] The PSD can also be interpreted from gas adsorption isotherms by applying assumptions to the shape of the pores. Work described by Barrett, Joyner and Halenda,[89] Broekhoff and de Boer,[90] Horvath and Kawazoe,[91] Saito and Foley,[92] and Cheng and Yang[93] are some of the many methods that may be employed for micro- and mesoporous materials. Over the past 30 years, computational techniques involving non-local density functional theory (NLDFT) have become popular for determining PSDs.[94–97] In this method, the local fluid density of an adsorbed species $\rho(\boldsymbol{r})$ is calculated by minimisation of the grand potential. This knowledge can be used to generate a series of adsorption isotherms (a kernel) across a set of model systems with varying pore sizes. PSDs can then be extracted by solving the generalised adsorption isotherm (GAI) equation:

$$N(P/P_0) = \int_{W_{\min}}^{W_{\max}} N(P/P_0, W)\, f(W)\, \mathrm{d}\boldsymbol{W}$$

where $N(P/P_0)$ is the experimental isotherm data, $N(P/P_0, W)$ is the isotherm data of a model pore system with pore width W, and $f(W)$ is the PSD function. The GAI equation assumes that an experimental isotherm can be represented by multiple model pore isotherms, each multiplied by their relative distribution.[95] While NLDFT has been validated in multiple materials with varying pore size ranges, it struggles to characterise systems with energetic heterogeneity or surface roughness, such as carbonaceous materials. One solution to this is a similar approach known as quenched solid density functional theory (QSDFT), which characterises the surface heterogeneity using a roughness parameter.[96]

While the NLDFT and QSDFT methods are valuable, Gelb and Gubbins developed a cheap and exact way of determining the PSD geometrically.[98] Their algorithm uses a Monte Carlo simulation to generate a series of random test points within the pore space of a material. At each test point, the largest sphere that contains the test point and does not overlap with atoms of the structure is recorded. By dividing the total radius range into a series of small bins, a cumulative pore volume function can be determined over a large number of insertions. This is a monotonically decreasing function that corresponds to the volume of void space that is coverable by spheres of a given radius or smaller. The PSD is then determined from the derivative of this function with respect to the sphere radius.

4.3.3 Pore Connectivity and Analysis of Topological Features

The structural properties defined so far give little indication of the topology of a porous network. Depending on the class and composition of a porous material, the pore network can possess varying shapes and connectivity.

Therefore, its characterisation is essential to gain an understating of how certain topological features impact adsorption and transport properties. To establish the pore connectivity within a given material, a number of approaches[99,100] have been proposed that employ a grid-based analysis using the Hoshen–Kopelman algorithm.[101] In these methods, the unit cell of a material is uniformly discretised into a number voxels and a probe particle of a given size is inserted at the centre of each one. If the inserted probe does not overlap with any framework atoms, then the voxel is considered accessible. After each voxel has been sampled, the Hoshen–Kopelman labelling algorithm groups accessible voxels together into clusters. If a cluster connects two opposite points across the entire length of a unit cell, it is referred to as a spanning cluster, which under periodic boundary conditions, represents a diffusion path of infinite length. This algorithm can efficiently identify pore channels and cages that exist within a structure and whether they are interconnected with one another. In addition, this algorithm provides an efficient way of evaluating the pore-limiting diameter (PLD) and largest cavity diameter (LCD) of a material. The PLD corresponds to the largest probe diameter that can freely diffuse through a material (*i.e.* the narrowest part of a spanning cluster). The LCD corresponds to the largest probe diameter that can be placed inside the pore system of a material without overlapping with framework atoms. Both of these dimensions are highly important for understanding diffusion in gas separation processes as they define what kinds of adsorbate are accessible to the pores and how well mixtures can pass one another. The results from this method are dependent on the 3D grid spacing used in the calculation; as the size of the spacing is reduced, the level of uncertainty in the predictions decreases.

An alternative approach is to use Voronoi decomposition, which provides a graph representation of the void space in a porous material.[102] In this method, the space closest to each of the framework atoms is divided into a series of polyhedral cells. The faces, edges and vertices of these cells correspond to planes, lines and nodes that are equidistant between neighbouring atoms. The Voronoi network mapped by the edges and vertices of each cell corresponds to the pore system in the material and can be analysed to extract geometrical descriptors such as the PLD, LCD, ASA, PV and VF. This methodology is implemented into the Zeo++ software package developed by the Harancyzk group and is commonly employed for high-throughput characterisation of large structural databases.[102]

4.4 Classical Molecular Simulations for Adsorption-based Applications

We provide a few examples of early studies that have applied classical simulations to study gas storage and separations, before moving on to more recent high-throughput discoveries. We also discuss the limitations of general forcefields and how they can be improved upon using QM methods.

4.4.1 H_2 Gas Storage

Large rises in anthropogenic greenhouse gas emissions since the beginning of the 20th century have prompted the scientific community to search for cleaner alternatives to fossil fuels. H_2 is considered to be one of the most promising candidates, as it possesses a higher energy density than gasoline (123 MJ kg^{-1} *vs.* 47.2 MJ kg^{-1}) and only emits H_2O as a product of combustion.[103] In order to use H_2 as fuel in vehicular transport, it must be stored under high pressures or cryogenic temperatures, which is expensive and inherently increases the risk of safety issues. An alternative approach is to adsorb H_2 into a porous solid with large storage capacities, but at more practical conditions. In order for the material to be viable for gas storage applications, the US Department of Energy (DoE) have designed a series of uptake targets for onboard H_2 storage systems. These are typically re-evaluated every few years and have ranged from 4.5 wt% and 28 g cm^{-3} in 2009 to 4.5 wt% and 30 g cm^{-3} in 2020, with an ultimate goal of 6.5 wt% and 50 g cm^{-3}.[104]

In comparison with studies on many zeolites[105,106] and carbon nanotube (CNT) based materials,[107–110] most attention has been received by MOFs and their derivatives due to their highly tuneable structures that can be easily optimised for H_2 storage. To establish the H_2 storage performance of these materials, molecular simulations are commonly employed to provide initial estimates and insights into the relationships that influence the storage process. In a recent review, Basdogan and Keskin provided a comparison of simulated and experimental H_2 uptakes in various MOFs, COFs and ZIFs at 77 K and 1 bar.[111] As seen in Figure 4.2, molecular simulations using the general UFF are in good agreement with experimental values even when neglecting quantum effects that are important for H_2 at low temperatures or high densities. Due to the generality of the force-field, it is not surprising that its accuracy is highly dependent on the material and its composition. In most simulations, MOFs are treated as infinite, rigid, defect-free crystal structures that have been fully activated which provide estimates of the upper boundaries of gas uptakes. For the practical applications of MOFs, the materials must be compacted from powders into pellets or monoliths, which can lead to reductions in the PV and ASA,[112] so the upper boundary for gas uptake needs to be exceptionally high. If the force-field lacks a fundamental description of certain interactions, the gas uptake can to be underpredicted and one of the current computational challenges is to develop transferable force-fields that accurately describe host–guest interactions in a wide range of gas storage systems (see Section 4.4.5).

In order to develop design principles for future materials, molecular simulations can predict the structure–performance relationships that govern H_2 storage. Frost *et al.* employed GCMC simulations to study the factors that influence H_2 storage using 10 IRMOFs.[113] These materials possess the same topology but have varying linker lengths leading to a diverse range of ASAs and PVs. By plotting the H_2 uptake as a function of different properties, the authors found that the adsorption capacity was strongly correlated with the

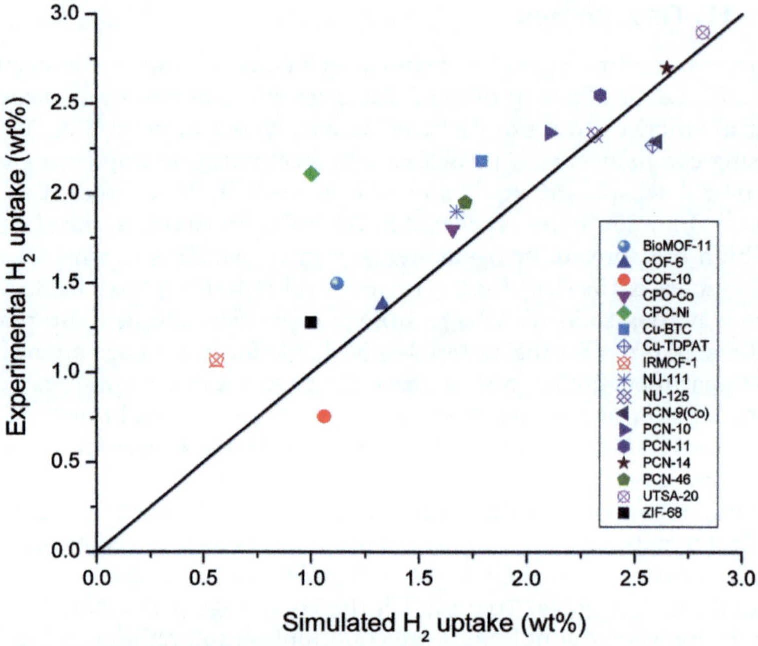

Figure 4.2 Comparison of simulation results with experimental data for hydrogen uptake of widely studied MOFs at 77 K and 1 bar.
Reproduced from ref. 111 with permission from the Royal Society of Chemistry.

Q_{st}, SSA and PV at low, medium and high pressures, respectively. In addition, a number of these materials were able to surpass the 2015 DoE storage targets, highlighting their potential as H_2 adsorbents. Adsorption isotherms were predicted at 77 K using the H_2 Darkrim–Levesque and DREIDING potentials. In a follow-up study, Frost and Snurr validated these correlations with simulations at 298 K.[114] They also ran multiple simulations with adjusted host–guest interaction strengths and suggested that values of Q_{st} ranging between 10 kJ mol^{-1} and 15 kJ mol^{-1} and PV between 1.6 cm^3 g^{-1} and 2.4 cm^3 g^{-1} are optimal for surpassing DoE targets of 6 wt% and 45 g cm^{-3}. Similar findings were obtained by Bhatia and Myers who employed GCMC simulations to determine the optimal conditions for gas storage.[110] In some studies, framework catenation has been suggested as a reasonable method to improve H_2 Q_{st} values.[115] However, from investigations using **IRMOF-9** and **IRMOF-10**, Frost and Snurr reported that this was generally not enough to offset the loss of PV leading to poorer uptakes in the high-pressure region. Therefore, alternative methods may be more promising for achieving the DoE goals, such as co-ordinately unsaturated sites (CUS) or charged frameworks (*i.e.* lithium doping).[116,117] These findings were also in agreement with other studies presented by Jung *et al.* and later by Ryan *et al.*[118,119]

From the correlations presented above, Farha *et al.* used computational tools to design and predict the gas storage properties of a novel MOF.[120] **NU-100** contains Cu(II) sites that are linked together by large hexacarboxylate ligands, forming a (3,24)-paddlewheel connected network. This material combines ultra-high surface areas (6143 $m^2 g^{-1}$) and PVs (2.82 $cm^3 g^{-1}$) with strong adsorption sites that are expected to raise the H_2 Q_{st} values. The net of this material cannot undergo catenation which avoids losses in the surface area and PV of the material. Molecular simulations of **NU-100** predicted a H_2 capacity of about 18 $mg g^{-1}$ at 1 bar and about 100 $mg g^{-1}$ at 56 bar, thus prompting experimental efforts to synthesis the new MOF. These simulated values were later confirmed from the measured H_2 isotherms, showing one of the highest H_2 capacities to date. Such studies demonstrate that molecular simulations play a vital role in the determination of design principles and the predictive analysis of gas uptakes that can lead to the discovery of exceptional materials.

4.4.2 Natural Gas Storage

Natural gas (NG) is another alternative fuel source that has the potential to replace petroleum and diesel in vehicular transportation applications. Although NG is not as clean as H_2, it still provides less polluting emissions than with petrol- and diesel-based vehicles, as it is mainly composed of CH_4. In order to use NG in onboard storage systems, it must either be compressed to very high pressures (>250 bar) or liquified at cryogenic temperatures.[121] Due to the inefficiency and high cost of these methods, research efforts have focused on employing porous materials to adsorb and release CH_4 at more reasonable conditions. In 2012, the DoE set a series of CH_4 uptake targets for new materials, corresponding to 0.5 g (CH_4) g^{-1} (adsorbent) and 263 cm^3 (CH_4) cm^{-3} (adsorbent) at 298 K.[122] Assuming a 25% packing loss during processing, this would require a volumetric uptake of approximately 330 $cm^3 cm^{-3}$. Some of the first materials to be studied for adsorbed natural gas (ANG) technologies were activated carbons and zeolites. However, in the same respect as H_2 storage, these materials suffer from limited tuneable properties which restrict the overall diversity of structural properties that can be attained by these materials. Furthermore, the hydrophilicity of many zeolites can result in reduced adsorption capacities due to competitive adsorption in the presence of moisture or polar adsorbates.[123] As such, attention has mainly focused towards MOFs and COFs, with molecular simulations being employed to predict their storage capacities and optimal design characteristics.

In 2004, Düren *et al.* employed GCMC simulations to study the storage capabilities of several classes of porous material, including IRMOFs, zeolites, molecular squares, mesoporous silicas and CNTs.[124] In particular, they were interested in rationalising the large CH_4 capacity of **IRMOF-6** by comparing its structural properties and gas uptakes with other porous materials.[69] Using general force-fields, the authors reported excellent quantitative agreement between the simulated and experimental isotherms of **IRMOF-1** and **IRMOF-6**, suggesting the strong uptake in these materials is a result of reversible physisorption. By comparing different materials, the authors suggested that an ideal material for NG storage should possess large ASA, large PV, low framework density, and strong host–guest interactions. Many of these properties were found to exist in several IRMOFs (including **IRMOF-6**), which could explain their high performance as CH_4 adsorbents. With these conclusions in mind, the authors designed a number of new MOFs using 1,4-tetrabromobenzenedicarboxylate and 9,10-anthracenedicarboxylate as linker molecules, showing that volumetric CH_4 capacities could be increased above the DoE target of 180 $cm^3\,cm^{-3}$ at 35 bar and 298 K (Figure 4.3). Using this information, Ma *et al.* were able to synthesise a new MOF, **PCN-14**, which possesses a similar anthracene ligand derivative.[125] This material was found to exhibit a CH_4 storage capacity of 230 $cm^3\,cm^{-3}$, which was ~28% higher than the DOE target. This highlights another classic example of where computational design and prediction can provide useful insights for experimental design of new materials.

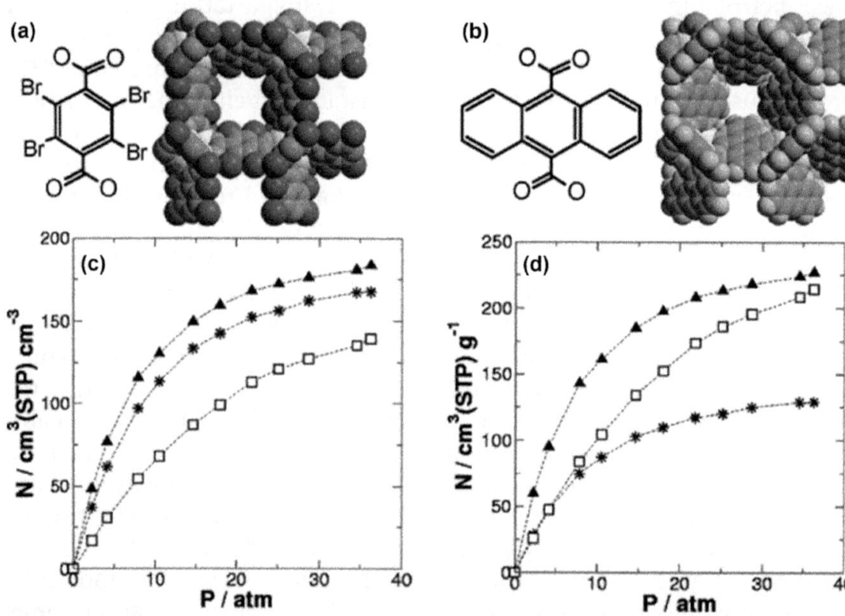

Figure 4.3 Molecularly designed IRMOF materials and the predicted adsorption isotherms: (a) **IRMOF-992**, which uses 1,4-tetrabromobenzenedicarboxylate as the linker molecule; (b) **IRMOF-993**, which uses 9,10-anthracenedicarboxylate; (c) adsorption isotherms as amount adsorbed per volume; (d) adsorption isotherms as amount adsorbed per mass adsorbent (✳, **IRMOF-992**; ▲, **IRMOF-993**; □, **IRMOF-6**).
Reproduced from ref. 124 with permission from American Chemical Society, Copyright 2004.

The correlations discovered by Düren *et al.* have been similarly observed in later studies. Wang *et al.* employed a systematic approach to compare the CH_4 storage performance in various MOFs with different topologies, including several IRMOFs and **CuBTC**.[126] After a slight adjustment to some of the OPLS-aa force-field parameters, they were able to obtain excellent agreement between simulated and experimental isotherms for several materials. They found that the ASA and PV play lead roles in influencing the CH_4 capacity at 35 bar, and that adsorbents with high Q_{st} values and low framework density would be ideal for CH_4 storage. Bae *et al.*[127] employed GCMC simulations to determine whether DoE targets could be obtained through the optimisation of Q_{st} by varying the LJ parameters that describe methane–metal interactions. In many studies, it is common to report the storage capacity rather than the working capacity (ΔN), which determines the usable amount of gas between a loading and discharge pressure. Therefore, high uptakes below the discharge pressure can be detrimental to ΔN_{CH_4}, as a large proportion of the stored gas is unusable. The authors discovered that an optimal Q_{st} value of about 20 kJ mol^{-1} for PV less than 1.3 cm^3 g^{-1} and between 20 kJ mol^{-1} and 40 kJ mol^{-1} for PV greater than 1.8 cm^3 g^{-1} is sufficient to achieve the maximum ΔN_{CH_4} in 8 different MOFs. Garberoglio studied the adsorption performance of several light gases, including CH_4 in **COF-102**, **COF-103**, **COF-105** and **COF-108**.[128] Due to the low framework densities and high volume of strong adsorption sites, **COF-102** and **COF-103** were found to possess volumetric CH_4 capacities in excess of the DoE target of 180 cm^3 cm^{-3} at 35 bar. On the other hand, the lighter mass of **COF-108** and **COF-105** was found to be advantageous for gravimetric capacity.

4.4.3 Gas Separation

Gas separation plays a significant role in several environmental and industrial applications.[129] For example, increases in atmospheric CO_2 emissions have urged the development of materials that can selectively remove CO_2 from flue gas mixtures. In the petroleum industry, olefin, paraffin, aliphatic and xylene separations are of high importance for generating chemical feedstocks. The separation of air provides O_2 and N_2 products that are used in many medical applications and chemical reactions such as the Haber process. Many noble gases are important in lighting and biomedical applications, but occur naturally in mixtures with other gases.[129–131] As an alternative to expensive and energy intensive distillation methods, research into the application of porous materials has led to the development of many adsorption-based technologies such as pressure swing adsorption, temperature swing adsorption and membrane-based separations (see Section 4.5.3). Due to the difficulties with measuring mixed-gas adsorption equilibria, many researchers rely on ideal adsorbed solution theory (IAST) to predict multi-component adsorption data using single-component adsorption isotherms.[132] Alternatively, molecular simulations are routinely employed to study many classes of porous solids for various gas separations.

In this section we present some examples from the literature where molecular simulations were employed to predict a materials performance in CO_2 based separations, determine the factors which influence its performance and how they can be used to design and test new materials.

García-Pérez *et al.* combined experimental studies with GCMC calculations to study the adsorption of CO_2, N_2 and CH_4 in MFI, DDR, ITE, MOR, CHA and ISV zeolites.[133] For all gases, the computed single-component isotherms were in excellent agreement with several sets of experiment data, validating the choice of force-fields. To study the separative properties of these zeolites, binary and ternary mixture calculations were performed at various compositions, temperatures and pressures. In all cases, the authors observed that CO_2 is the most strongly adsorbed component in each material, with the CO_2 adsorption selectivity (S_{ads,CO_2}) depending highly on the shape and the size of the pores (where S_{ads,CO_2} defines a materials propensity to selectively adsorb CO_2 over other gases). In strong confinements, CO_2 adsorbs more strongly than the other gases due to its large size and strong coulombic interactions resulting in the highest host–guest interaction strength of each gas. Therefore, S_{ads,CO_2} is largest in zeolites with small cages and narrow windows (*e.g.* CHA) and weakest in those which possess interconnected pore networks (*e.g.* MFI). Huang *et al.* investigated the effect of temperature, pressure and pore size on the adsorption behaviour of CO_2/CH_4 mixtures in a range of CNT materials.[134] Their simulation results showed that CO_2 is preferentially adsorbed over CH_4 with CO_2 loadings increasing as a function of the CNT's diameter. These values range from 4 $mol\,kg^{-1}$ in (6,6) CNTs to 9 $mol\,kg^{-1}$ in (10,10) CNTs. In addition, CNT materials demonstrated S_{ads,CO_2} greater than 8 across all studied temperatures (283–343 K) and pressures (10–300 bar). At 343 K and 10 bar, the S_{ads,CO_2} reaches a peak of 11.2 in the (6,6) CNT, which is higher than previous reports in some activated carbons, zeolites and MOFs.[135–137]

Liu and Smit performed a systematic study to compare CO_2/N_2 and CH_4/N_2 separations in various zeolites and MOFs.[138] These included 3 zeolites (MFI, LTA and DDR) and 7 MOFs (**CuBTC**, **MIL-47(V)**, **IRMOF-1**, **IRMOF-12**, **IRMOF-14**, **IRMOF-11** and **IRMOF-13**) covering a range of chemical compositions, pore shapes and sizes. Analysis of K_{Henry}, Q_{st} and adsorbate loadings indicated that MOFs were generally better for gas storage than zeolites; however, their separation performances were comparable with one another. **CuBTC** possessed one of the best separation performances ($S_{ads,CO_2/CH_4} \sim 40$; $S_{ads,CH_4/N_2} \sim 5$) due to the large number of exposed metal sites. The authors investigated the role of electrostatics by switching on and off the CO_2 and N_2 host–guest and guest–guest contributions. The results indicated that if the selective component has a larger quadrupolar moment than its partner, then S_{ads} is enhanced by electrostatic interactions. Likewise, if the selective component has a smaller quadrupolar moment, then S_{ads} is reduced. Similar results were also obtained by Liu *et al.* who performed GCMC simulations to evaluate the separation performance of COFs and MOFs for $CH_4/CO_2/H_2$ mixtures.[139] Many of the COFs and MOFs possessed similar selectivities and were in good agreement with previous calculations by Liu and Smit.[138] While the electrostatic contributions in the COFs were found to be smaller than in MOFs, their impact still needs to be considered to accurately predict adsorption isotherms and mixture behaviour. In both of these studies, the results from mixture GCMC simulations were also found to be in good agreement with predictions from IAST, highlighting the potential to use single-component data to characterise the mixture behaviour. Yang *et al.* explored the impact of ligand functionalisation on the separation of CO_2/CH_4 mixtures by functionalising the BDC ligands of **UiO-66**[140] with –Br, $-NH_2$, $-NO_2$, $-(CF_3)_2$, $-(OH)_2$, $-SO_3H$ and $-CO_2H$ groups.[141] Polar functional groups were found to enhance the MOFs' affinity for CO_2, leading to improvement of its separation performance at 1 and 10 bar (Figure 4.4). Moreover, bulky substituents such as $-NO_2$ and $-CF_3$ were found to reduce the ASA and PV of the materials, leading to enhanced interactions through the confinement of the material. The materials functionalised with the $-SO_3H$ and $-CO_2H$ groups show the highest $S_{ads,CO_2/CH_4}$ of 16.8 and 19.2 at 1 bar, and 22.6 and 23.2 at 10 bar, respectively. In a later study, Babarao *et al.* systematically investigated the effects of exposed metals, catenation and the inclusion of extra framework ions on the separation of CO_2/CH_4 mixtures using several MOFs.[142] In all materials, CO_2 was preferentially adsorbed over CH_4 due to its strong affinity with the framework through a combination of dispersion and electrostatic interactions. MOFs with exposed metal sites such as **CuBTC** and **PCN-6** were found to enhance the $S_{ads,CO_2/CH_4}$ due to stronger electrostatic interactions with the framework. Framework catenation forms additional adsorption sites at the cost of PV, which boosts low pressure uptake and reduces high pressure uptake. As such, materials with catenated frameworks were found

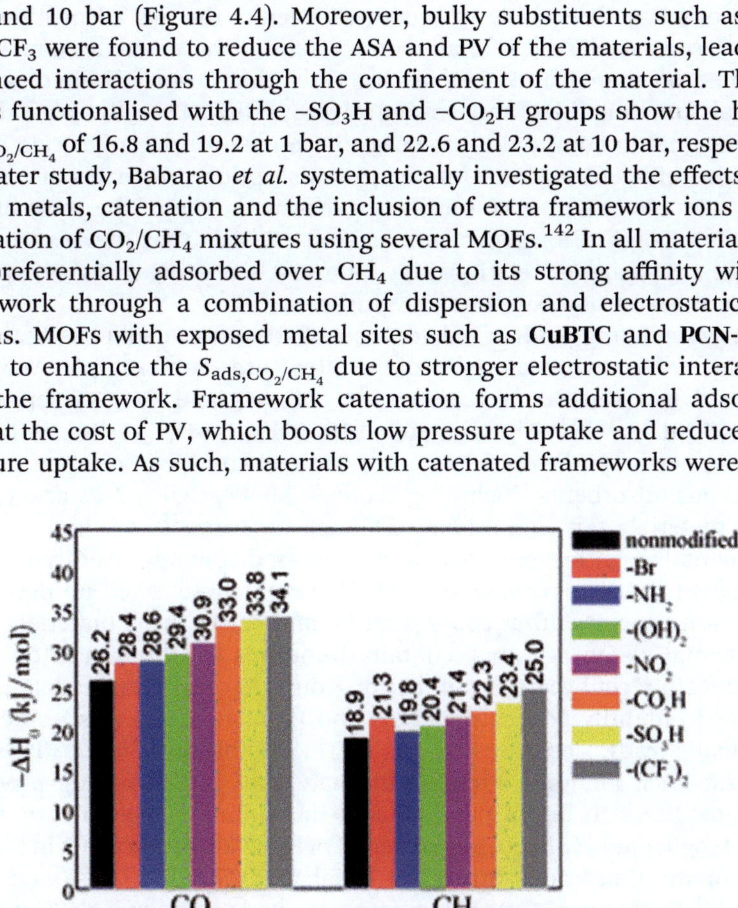

Figure 4.4 Calculated low-coverage adsorption enthalpies of pure CO_2 and CH_4 gases in the non-modified **UiO-66(Zr)** and its functionalised derivatives at 303 K.
Reproduced from ref. 141 with permission from the Royal Society of Chemistry.

to possess higher $S_{ads,CO_2/CH_4}$ than non-catenated frameworks. The presence of extra framework ions such as NO_3^- in **soc-MOF** can add additional adsorption sites which preferentially adsorb CO_2. In this MOF, the $S_{ads,CO_2/CH_4}$ was found to range between 22 and 36 at 0–50 bar pressure. These values for selectivity are some of the highest reported amongst MOFs, indicating that charged MOFs may be promising candidates for CO_2/CH_4 separations.

4.4.4 High-throughput Screening Studies

For many adsorption-based applications, finding a top performing candidate is a challenging task due to the large number of individual porous structures. Although this chapter has presented examples where classical simulations have been able to accurately predict the gas adsorption properties of several classes of nanoporous solids, these studies are usually performed on a small number of structures that cover a limited range of chemical functionalities and structural properties. Over the past decade, improvements in computational power have meant that large-scale screenings which efficiently determine the gas adsorption properties of a diverse set of materials can now be routinely performed. One of the most exciting prospects of high-throughput simulations (HTS) is that they can uncover structure–performance relationships on a vast scale that cannot be practically determined from experimental synthesis.

In early screening studies, a diverse set of structures could be obtained by selecting well-known materials in the literature. For example, Wu *et al.* performed two separate screenings for CH_4/H_2 and CO_2/N_2 separations using 105 MOFs.[143,144] These included IRMOFs, **Cu-BTC**, PCNs, MILs, ZIFs and some newly synthesised structures. Dubbeldam *et al.* screened over 100 nanoporous adsorbents, including zeolites, MOFs, COFs, ZIFs and carbonaceous materials for separations of alkane mixtures.[145] As the number of experimentally synthesised structures increased, the opportunity to screen much larger datasets presented itself. However, these relied on developing automation processes that could identify and select these materials out of experimental databases that contain hundreds of thousands of crystal structures. In recent years, a number of groups have performed data-mining exercises to identify porous materials and populate them in specific 'computationally-ready' databases for use in HTS (Table 4.1).[2,146–162] These materials are then analysed with the ultimate goal of discovering promising candidates that can be further validated in laboratory experiments. For example, Moghadam *et al.* recently screened over 82 000 compounds in the MOF subset of the Cambridge Structural Database (CSD) for O_2 storage applications.[163] In this work, room-temperature single-component GCMC simulations were performed at 5 and 140 bar, corresponding to the desorption and adsorption pressures. From the analysis of the volumetric and gravimetric O_2 capacities, the authors identified **UMCM-152**,[164] a MOF which delivers 22.5% more O_2 than the record holding material (**NU-125**).[165] The MOF was then synthesised and the measured O_2 adsorption isotherms were in excellent

Table 4.1 Examples of structural databases for different real and hypothetical porous materials.

Type	Number of structures[a]	Notes/alternate name	Reference
Zeolites	2000	Hypothetical zeolites	146, 149
	600 000		148
	2 700 000		147
	2 700 000		2
MOFs	5109	CoRE-MOF 2014	150
	137 953	Northwestern hypothetical	151
	116	MOF-5 analogues	152
	22 700	UM MOFs	153
	18 383	Mg MOFs	154
	13 512	ToBaCCo MOFs	155, 156
	69 666	CSD MOF subset	157
	14 142	CoRE-MOF 2019	158
COFs	4147	3D porous COFs	159
	69 840	2D and 3D COFs	160
	319	CoRE COFs	161
PPNs	17 846	**dia** topology	162

[a]Number of structures in the database upon publication.

agreement with those predicted from GCMC simulations. In addition to **UMCM-152**, many of the most promising MOFs were found to possess similar structural properties with LCDs between 7.5 Å and 8 Å, VF greater than 0.8 and ASAs greater than 2600 $m^2\,g^{-1}$.

Many of the databases shown in Table 4.1 have been generated by employing simulated annealing or *in silico* design principles to generate libraries of hypothetical zeolites, MOFs, COFs and PPNs. These hypothetical materials can improve the overall understanding of structure–performance relationships and identify potential candidates that do not exist in the literature. Kim *et al.* screened 30 000 hypothetical zeolites from the database of Pophale *et al.* for C_2H_6/C_2H_4 separations.[166] These structures were randomly chosen from approximately 130 000 candidates that possess a PLD greater than 3.25 Å and GCMC simulations were employed to record single-component adsorption isotherms at 1 bar and 300 K. IAST was then employed to determine the multi-component adsorption isotherm for an equimolar C_2H_6/C_2H_4 mixture. Many of these materials were identified to be unsuitable due to the low $S_{ads,C_2H_6/C_2H_4}$ and $\Delta N_{C_2H_6}$. Of the top performing hypothetical zeolites, 50 structures were found to possess a score greater than 3, which was determined from the product of S_{ads,C_2H_6} and $\Delta N_{C_2H_6}$. In a later study, the same authors screened 73 liquid solvents and over 87 000 zeolites for CH_4 capture in dilute and medium-concentrated mixtures using the IZA and hypothetical databases.[167] From calculations of the CH_4, N_2 and CO_2 solubilities in each of the liquid solvents, it was clear that K_{CH_4} and

S_{ads,CH_4} were too small for these liquids to be technologically promising. Each zeolite was evaluated for the selective removal of CH_4 from a 30/70 CH_4/CO_2 mixture at 70 bar. The results instantly revealed an IZA zeolite, SBN, that showed excellent potential with a high CH_4 uptake of 2.75 mol kg^{-1}, due to the highly localised positions of CH_4 adsorption sites that were separated by an average distance of 4.33 Å (closely aligned with the minimum energy separation for two CH_4 molecules). For the separation of CH_4 under dilute conditions (1/1/98 mixture of $CH_4/CO_2/N_2$), uptakes were found to correlate strongly with Henry coefficients. Two IZA zeolites, ZON and FER, were identified as performing candidates due to their large K_{CH_4} values of 1.29 mol kg^{-1} bar^{-1} and 1.12 mol kg^{-1} bar^{-1}, respectively. Additionally, the screening identified numerous hypothetical candidates with $S_{ads,CO_2/CH_4}$ greater than 2 that were able to outperform several zeolites in the IZA database. Many of these structures consist of 1D channels which create strong binding sites for CH_4 due to the significant number of oxygen atoms that are optimally distant from the adsorbed CH_4.

In 2012, Wilmer *et al.* developed a database containing 137 953 hypothetical MOF structures from a diverse library of ligand and metal building blocks that were combined according to applied geometric rules.[151] To validate their methodology, the authors generated crystal structures that resembled **CuBTC**, **IRMOF-1**, **PCN-14** and **MIL-47** and used them to simulate CH_4 adsorption isotherms. In comparison with simulations that used experimental crystal structures, excellent agreement was obtained, which motivated the authors to perform a HTS of the entire database for CH_4 storage applications. GCMC simulations were employed to calculate CH_4 storage capacity at 35 bar and 298 K over three stages: (1) all 137 953 were screened in a short simulation (500 cycles) to identify the top 7000 MOFs, (2) these 7000 MOFs were screened in a medium simulation (2500 cycles) to identify the top 350 MOFs, (3) the remaining 350 MOFs were then screened in a long simulation (12 500 cycles). In the third stage of calculations (Figure 4.5), the top 300 MOFs were found to possess volumetric uptakes that were greater than the world-record holder, **PCN-14**. Furthermore, 75% of MOFs with a volumetric uptake greater than 205 cm^3 cm^{-3} were found to contain methyl, ethyl, or propyl functional groups. In the final stages of their study, the authors identified that one of top performing candidates was an isostructural derivative of **PCN-14** with methyl functionalised linkers that had recently been reported as MOF **NOTT-107** by Lin and co-workers.[168] Wilmer *et al.* synthesised **NOTT-107** and their measured CH_4 adsorption isotherms were in good agreement with their simulations, validating the predictions of their screening.

Having confirmed that HTS can provide accurate predictions of top-performing materials, the database was screened for various storage and separation applications. Later that year, the same authors evaluated the structure–performance relationships that lead to exceptional CO_2 capture and separation in MOFs under three different conditions.[169] They found that with such a large number of data points, clear trends begin to emerge that

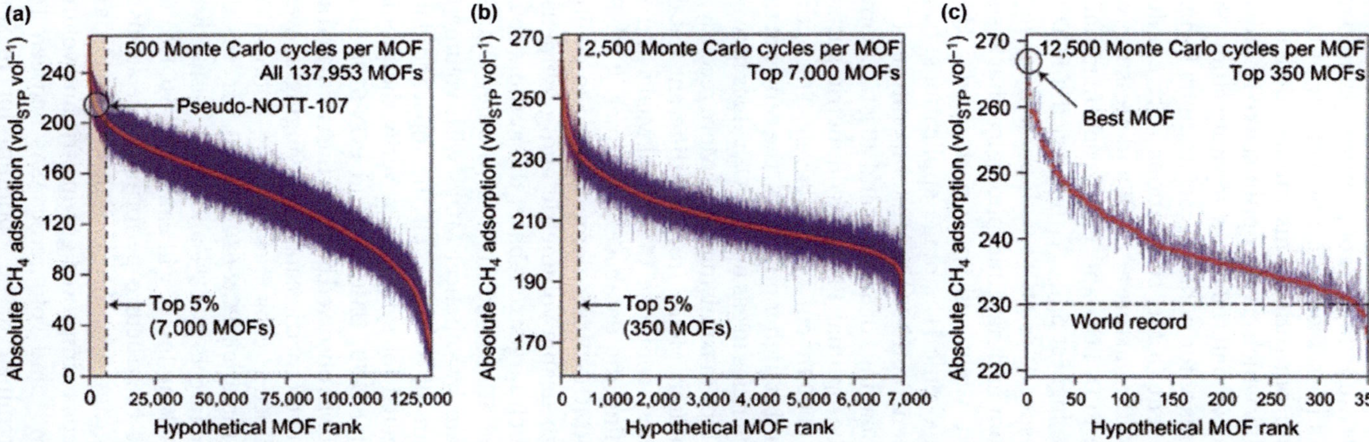

Figure 4.5 (a) In the initial stage, ∼137 000 MOFs were screened for methane storage at 35 bar through short simulations. (b,c) The top 5% of the first stage (b) and then the top 5% of the second stage (c), highlighted in orange, were recalculated using successively longer simulations, which significantly reduced the statistical error each time. In a–c, MOFs are rank-ordered for their methane adsorption abilities (best to worst runs from left to right), with statistical error indicated by purple bars. Adapted from ref. 151 with permission from Springer Nature, Copyright 2012.

do not exist when considering a much smaller dataset. They also discovered that MOFs with –F and –Cl functional groups appeared frequently among the top performers, while $-NH_3$ functional groups were less impactful than expected. The top candidates possessed a narrow window of Q_{st} values, but a broader range of ASAs and porosities. This implies that these properties possess a degree of flexibility which can be optimised to improve the stability or reduce the cost of the new materials. In a separate study, Sikora and coworkers screened the database for noble gas separations focusing on Xe/Kr mixtures.[170] Multi-component GCMC simulations were employed to study the adsorption of a 20/80 Xe/Kr mixture at 273 K and pressures of 1.0, 5.0 and 10 bar. By computing the PLD, LCD and PV of each material, the authors found that MOFs with the highest $S_{ads,Xe/Kr}$ possessed a LCD/PLD ratio between 2 and 6. This corresponds to MOFs with pores that are just large enough to fit a single Xe atom and have morphologies that resemble narrow uniform channels. Bobbitt, Chen and Snurr performed a HTS to identify suitable candidates for cryogenic H_2 storage applications.[171] In this work, ΔN_{H_2} was computed between 2 and 100 bar at a temperature of 77 K. A total of 29 807 MOFs exceeded H_2 loadings of 5.5 wt% and 40 $g\,L^{-1}$, corresponding to the 2015 DOE targets. The optimal VF and volumetric surface area were determined to be 0.9 and 2000 $m^2\,cm^{-3}$, respectively. The optimal pore size ranges from 12 Å to 15 Å, and at higher pore sizes, the centre of the pore is less saturated leading to inefficiencies in gas uptake. Q_{st} values between 2 $kJ\,mol^{-1}$ and 4 $kJ\,mol^{-1}$ are optimal for maximising ΔN_{H_2}. More recently, Ahmed *et al.* screened nearly half a million MOFs from 11 different databases for H_2 storage applications.[172] These comprise 15 235 real MOFs and 478 205 hypothetical structures, including the database of Wilmer *et al.* Based on the authors' previous work, **IRMOF-20** was demonstrated to exhibit one of the best combinations of high usable volumetric and gravimetric H_2 capacities.[173] By screening a much larger database, the authors found 102 compounds that could surpass **IRMOF-20** in both volumetric and gravimetric capacity. Many of these are hypothetical MOFs with volumetric capacities reaching a maximum value of 40 $g\,L^{-1}$. Of the top candidates that were identified, three were targeted for synthesis to verify the predictions of the computational screening (**SNU-70**, **UMCM-9** and **PCN-610/NU-100**).[120,174,175] In general, good agreement was observed between the simulated and measured adsorption isotherms with volumetric capacities ranging between 33.4 $g\,L^{-1}$ and 35.5 $g\,L^{-1}$ and gravimetric capacities ranging between 5.7 wt% and 10.1 wt%. While the top material, **PCN-610/NU-100**, exceeds the 2020 DoE volumetric and gravimetric targets by 18% and 124%, respectively, it is likely that considering crystal-packing losses will cause this MOF to fall below the volumetric target. Analysis of structure–performance relationships between different structural properties and ΔN_{H_2} indicates that maximising PV, gravimetric ASA and VF leads to improved gravimetric and volumetric capacities. Densities of 0.6 $g\,cm^{-3}$ and volumetric ASAs between 1600 $m^2\,cm^{-3}$ 2200 $m^2\,cm^{-3}$ were determined to be optimum for maximising ΔN_{H_2}, correlating well with the results of Bobbitt *et al.*[171]

In addition to zeolites and MOFs, a number of databases have been developed for other classes of nanoporous material. Martin *et al.* performed *in silico* design of 17 846 PPN materials with **dia**-type nets using a large library of tetrahedral and linear building blocks.[162] To illustrate the use of this database, each material was screened for CH_4 storage applications at room temperature across two different pressure regions (5–65 bar and 1–35 bar). A total of three candidates possessed ΔN_{CH_4} that exceeded the DoE targets of 180 $cm^3\,cm^{-3}$. Many of the PPNs possessed favourable separation distance between CH_4 adsorption sites (approximately 4.2 Å apart), which was identified as an important criterion for improving ΔN_{CH_4}. The same group also used similar principles to design 4147 hypothetical 3D COF materials based on tetrahedral building blocks.[159] Like the PPN database, these materials were screened for CH_4 storage applications, but predicted few exceptional candidates. The top two COF structures with ΔN_{CH_4} of 181 and 182 $cm^{-3}\,cm^{-3}$ were found to marginally exceed the DOE targets of 180 $cm^{-3}\,cm^{-3}$. As these predictions are based on perfect structures and conditions, it is likely that the uptakes are infeasible to achieve experimentally. More recently, Mercado *et al.* presented a much larger COF database of 69 840 structures.[160] These include 2D and 3D hypothetical structures that were assembled from a library of 666 organic linkers following four different synthetic routes. With a more diverse set of materials, the authors were able to identify 304 structures with a ΔN_{CH_4} greater than 190 $cm^3\,cm^{-3}$ at 65 bar and 34 with ΔN_{CH_4} greater than 200 $cm^3\,cm^{-3}$, many of which are based on carbon–carbon bonded 3D networks. Analysis of the structure–performance relationships indicated that the materials with high CH_4 storage performances were generally found to possess densities greater than 400 $kg\,m^{-3}$, LCDs of approximately 9 Å and ASAs between 4000 $m^2\,g^{-1}$ and 6000 $m^2\,g^{-1}$. However, the authors noted that the real challenge in using this information is to determine the exact combination of linkers, bonds and topologies that will result in a material with the predicted properties. Finally, in order to evaluate several classes of porous materials, Haranczyk *et al.* screened the nanoporous materials genome containing databases of *in silico* designed MOFs, COFs, ZIFs, zeolites and PPNs for Xe/Kr separations.[176] This database contains over 670 000 materials in total making it one of the most diverse collections of porous materials. Due to the large amount of structures being screened, brute force calculations using GCMC simulations are impractical. Therefore, the authors trained a machine learning algorithm using 15 000 structures to predict the separation performance from structural descriptors such as the ASA, VF, PLD, LCD and crystal density. The results suggested a number of tested materials that were able to compete with a porous organic cage, **CC3**, highlighted in the literature as one of the best performing materials for Xe/Kr separations.[177] These included **JAVTAC**, an aluminophosphate zeolite analogue, and **KAXQIL**, a calcium-based coordination network, both of which possess a $S_{ads,Xe/Kr}$ of 82. Both of these materials appear in the CoRE MOF database, consist of 1D channels, and have been previously synthesised. However, their Xe/Kr separation performance has yet to be validated through experimental methods.

4.4.5 Challenges and Limitations of Using General Force-fields

Despite their widespread use in the literature, general force-fields do not always provide an accurate description of the underlying interactions in a chemical system. Pérez-Pellitero *et al.* demonstrated that both the UFF and DREIDING overestimate the CH_4 uptake in **ZIF-8** at pressures above 1 bar.[178] To improve the agreement with experimental isotherms, the authors adjusted the UFF parameters by rescaling σ_{UFF} and ε_{UFF} to 95% and 69% of their original value, respectively. These scaling factors and new parameters were also found to be transferable to CO_2 and N_2 adsorption in **ZIF-76** and **ZIF-69**. A similar approach was applied by Fairen-Jimenez *et al.*, who studied the conformational changes of **ZIF-8** upon adsorption of CH_4.[179] Vandenbrande investigated the effects of different general force-fields in predicting CH_4 uptakes in Zr-based MOFs.[180] Their simulations show that the UFF largely overestimates the loading in the **UiO-66** MOF at both low and high pressures, relative to experiment. This agreement is even worse for **UiO-67** and **DUT-52**, a series of isoreticular MOFs which differ from **UiO-66** by the length of their linker molecules. Lennox *et al.* further tested the suitability of general force-fields for use in predicting CH_4 adsorption in MOFs.[181] While the UFF, DREIDING and OPLS-aa force-fields are found to give good qualitative agreement, both the UFF and OPLS-aa are likely to overestimate the amount of adsorbed CH_4, as well as the values for Q_{st} and K_{CH_4} obtained from the low-pressure region of the isotherm. When comparing to experiment, Garberoglio found that the UFF and DREIDING overestimate the amount of adsorbed Ar in **COF-102** and **COF-103** by almost 25%.[128] While it is possible that the experimental sample suffered from poor activation, poor agreement was observed across all pressure ranges indicating a poor description of the host–guest interactions.

In addition to these examples, it is now well understood that general force-fields fail to describe interactions between adsorbates and coordinately unsaturated sites (CUS). CUS appear frequently in MOFs where solvent molecules compete with labile linkers for coordination sites around metal centres. Upon activation, these solvents are removed, leaving a vacant site or CUS such as those seen in the widely studied **CuBTC** MOF[182] (Figure 4.6).[183] MOFs containing CUS have beneficial value in gas storage, separation and catalytic applications, where the CUS can act as potential binding sites for adsorbing guest species. Peterson *et al.* demonstrated this using *in situ* neutron powder diffraction (NPD) to study the binding sites of H_2 in **CuBTC** from D_2 adsorption.[184] A total of six sites were identified, which are primarily located above the CUS with a Cu–D_2 distance of 2.39 Å, suggesting significant interaction between D_2 and the Cu(II)

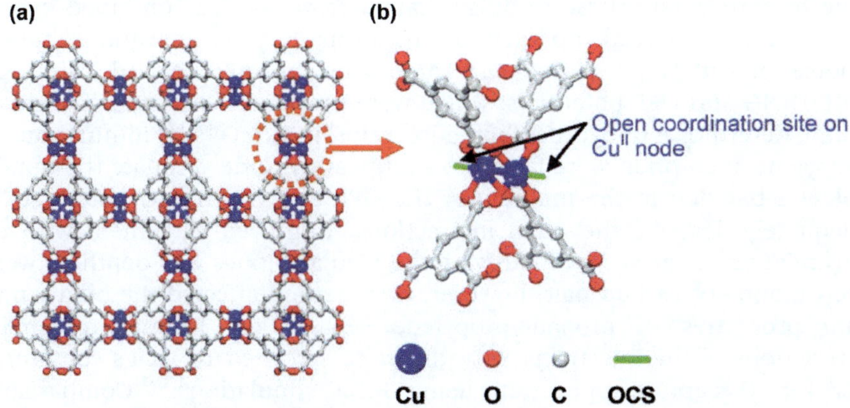

Figure 4.6 Illustrations of (a) the **CuBTC** structure in a 2D view along the (100) direction and (b) a **CuBTC** node with the Cu-paddlewheel environment. Reproduced from ref. 183 with permission from American Chemical Society, Copyright 2012.

cations. Wu *et al.* explored the role of CUS on the adsorption of CH_4 by comparing the M-MOF-74 family (M = Mg, Mn, Co, Ni and Zn) and selected PCNs with the fully coordinated structure of **IRMOF-1**.[185] For all M-MOF-74 members, values of Q_{st} were recorded between 18.3 kJ mol^{-1} and 20.2 kJ mol^{-1}, almost 50% larger than that of **IRMOF-1**. NPD experiments on **Mg-MOF-74** at a loading of 0.73 CD_4 per Mg reveal that the primary site occurs on the CUS, directly above the square-pyramidal MgO_5 cluster. Caskey *et al.* noted that **Mg-MOF-74** can achieve an affinity for CO_2 capture that is competitive with several liquid amines, thus highlighting the impact which CUS can have on the physisorption properties of an adsorbent.[186] In another study, Bloch *et al.* investigated olefin/paraffin separations in **Fe-MOF-74**.[187] From the obtained isotherms, **Fe-MOF-74** displays a strong affinity for all C_1–C_3 unsaturated hydrocarbons. Stoichiometric uptakes of 1 guest molecule per Fe(II) centre recorded at 1 bar infers guest binding to each of the CUS surrounding the Fe centres. This was confirmed through NPD experiments, which reveal a single binding site where each unsaturated hydrocarbon resides side-on above the CUS with an Fe–C distance ranging between 2.42 Å and 2.46 Å.

While experimental analysis of gas-CUS interactions has been successful in improving our understanding of their role in adsorption applications, many attempts to model these interactions *via* classical simulations have failed. Yang and Zhong designed a computational study to assess the separation of gas mixtures containing CO_2, CH_4 and C_2H_6 in **CuBTC**.[188] Good agreement between experimental and simulated isotherms was only achieved by adjusting the energy parameters of the oxygen and carbon atoms in the carboxyl groups in **CuBTC**. While experimental evidence from NPD has determined the CUS to be the primary binding site for CO_2 in **CuBTC**,[189] it is not clear whether snapshots from the authors' CO_2 simulations were able to corroborate these findings. Lamia *et al.* used a combined experimental and theoretical approach to study propane, propylene and isobutane uptake in **CuBTC**.[190] Host–guest interactions were described using the DREIDING and UFF potentials, which were previously shown to give an accurate description of hydrocarbon adsorption in IRMOFs.[124] Simulations of propylene adsorption were found to significantly underestimate the uptake below 1 bar due to the inability of the UFF and DREIDING potentials to adequately describe the Cu–π interactions. Improved agreement with experiment was achieved by modifying the Lennard-Jones well depth between the Cu and sp^2 carbon pair; however, this method affected the binary mixture properties of propane/propylene separations, leading to underestimations of the selectivity. Yazyadin *et al.* screened 14 MOFs containing CUS for CO_2 capture applications using GCMC simulations.[191] Comparisons between experimental and simulated CO_2 uptakes reveal poor agreement for MOFs containing a high density of CUS, such as those in the M-MOF-74 family. Hamon *et al.* noted large discrepancies between experimental and simulated CO_2 adsorption in the mesoporous **MIL-100(Cr)**, which they attributed to the inadequate description of the interactions between the quadrupolar molecules and the CUS.[192] Finally, Fischer *et al.* studied H_2 adsorption in seven MOFs employing the UFF and DREIDING potentials to only describe dispersion interactions.[193] For **CuBTC** in particular, density maps of the H_2 adsorption sites indicated low density surrounding the CUS suggesting the Cu–H_2 interaction is not dominated by dispersive forces. Similar cases were also observed for other CUS containing MOFs, including the Cu_2 paddlewheel containing **MOF-505**.

4.4.5.1 *Improving the Performance of General Force-fields*

As mentioned previously, a simple approach to improving agreement between simulations using general force-fields and experimental data is to apply a scaling factor to the force-field parameters. While this is a simple approach, the method does not have a physical basis and can lead to the wrong potential energy profile.[194] Over the past 20 years, the scientific community has invested considerable efforts in developing new interatomic potentials that can improve on the performance of general force-fields. In particular, QM calculations have become powerful tools for parameterising highly accurate and reliable force-fields for predicting gas adsorption properties in porous materials. In this approach, a number of host–guest configurations are sampled to determine the PES along a specific interaction coordinate. These reference data are then used to parameterise an appropriate classical potential to reproduce the QM PES. To obtain high-quality reference data, QM methods that capture dispersion should be employed, such as coupled-cluster theory, Møller–Plesset second-order perturbation theory (MP2), symmetry-adapted perturbation theory (SAPT) and dispersion corrected density functional theory (DFT-D). For many of these methods, the large computational cost makes it infeasible to study complex periodic solids. As a compromise, an appropriately sized finite cluster is used, which captures the most important structural characteristics and electronic environment of the host periodic solid. Maurin *et al.* developed a set of force-fields for studying CH_4 and CO_2 adsorption on siliceous faujasites using *ab initio* cluster calculations.[195,196] Simulations using the new force-fields show good agreement with microcalorimetry experiments. Furthermore, the transferability of the parameters was later demonstrated in three different faujasites – DAY, NaY and NaLSX. In a later study, these parameters were extended to include both Na^+ and Li^+ cations to study CO_2 adsorption in NaX, NaY and LiY.[197,198] Simulated isotherms from GCMC simulations were found in excellent agreement with experimental work, which allow a microscopic insight into the preferential adsorption sites in NaY and NaX structures. Reasonable accuracy was also obtained for the LiY zeolite.

Liu *et al.* utilised 160 *ab initio* calculations at the resolution of the identity (RI)-MP2/TZVPP level of theory to parameterise a new force-field that describes NH_3 interactions in the MFI zeolite.[199] The new force-field was determined by adjusting the Lennard-Jones parameters of NH_3 from the OPLS force-field to reproduce the *ab initio* data. While a satisfactory fit was achieved, the number of adjustable parameters in the Lennard-Jones potential limits the overall flexibility of the parameterisation process. Despite this, the new force-field parameters provided a predictive insight into the NH_3 temperature desorption behaviour of several MFI zeolites. Han *et al.* employed the RI-MP2 method to develop a force-field that accurately predicts the interaction of H_2 with cubic MOFs based on $M_4O(CO_2)_6$ building blocks (M = Zn, Be, or Mg).[200] Three sets of PES were determined for the interactions between H_2 and benzene, H_2 and naphthalene and H_2 and $M_4O(CO_2)_6$ structural units. A Morse potential was fitted to reproduce these energy profiles in order to study the adsorption thermodynamics of MOFs. By combining the accurate force-field with GCMC simulations, the authors learned that **Mg-MOF-C30** possesses one of the largest H_2 uptakes reported to date (8.08 wt% storage at 77 K and 0.33 wt% at 300 K, 20 bar). In later work, this methodology was applied to study CH_4 and H_2 interactions with various porous materials, including COFs, ZIFs and lithium-doped materials.[116,201–203] McDaniel *et al.* developed a methodology employing SAPT to construct force-fields that describe CO_2 interactions in ZIF materials.[204] One of the main advantages of SAPT is that it naturally decomposes the total interaction energy into electrostatic, exchange, induction and dispersion energy contributions. As these components can be fitted separately, the resulting force-field provides an improved interpretation of the weightings of each individual contribution to the overall interactions. To generate reference data, approximately 700 single-point density fitting DFT-SAPT (DF-DFT-SAPT) calculations were employed to calculate the interactions of various CO_2 configurations with two imidazolate-type cluster models that represented **ZIF-8** and **ZIF-71**. The individual components of the decomposed interaction energy were fitted to generate a new force-field. Simulated adsorption isotherms using the new force-fields parameters were in good agreement with experimental results, demonstrating how

QM-derived force-fields can predict the correct shape and gas uptake loading of an adsorption isotherm without resorting to arbitrary scaling of force-field parameters. This method was later extended to develop a series of transferable force-fields for a variety of ZIFs and small molecules.[205,206]

In CUS containing MOFs, QM-derived force-fields can provide a more accurate description of the interactions between adsorbates and the CUS than general force-fields. Dzubak *et al.* used MP2 theory to describe interactions of CO_2 and N_2 with the CUS containing **Mg-MOF-74.**[207] The MP2 interaction energy between the adsorbate and each of the cluster atoms was decomposed using the non-empirical modelling approach (NEMO).[208] NEMO decomposition splits the interaction energy into electrostatic, repulsive and attractive contributions (including polarisation and dispersion). In this work, the repulsive and attractive energies were modelled using a modified Buckingham potential that contained an additional r^5 term in the attractive part. The newly parameterised force-field is found to reproduce the MP2 energies with excellent agreement for both CO_2 and N_2. As seen in

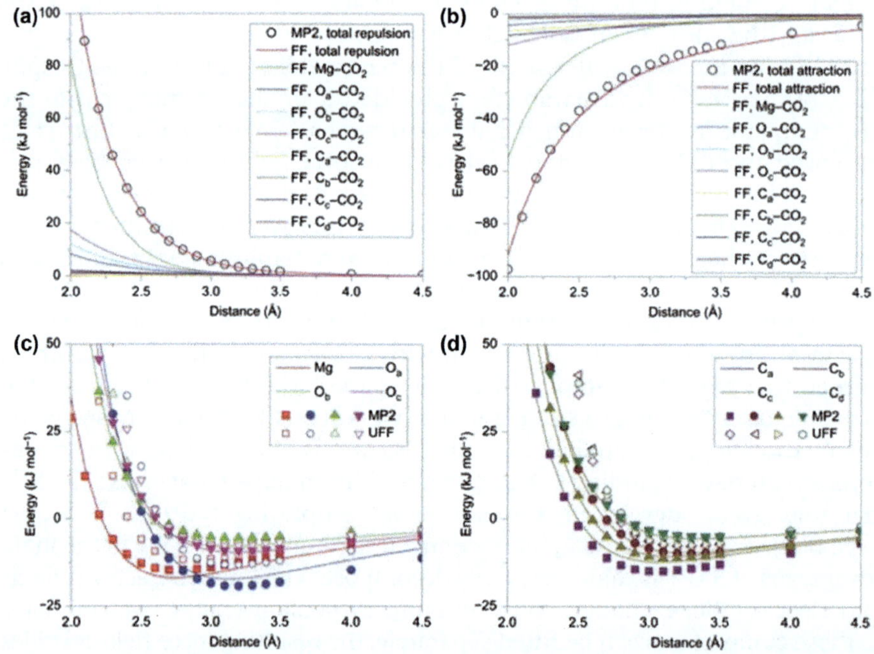

Figure 4.7 (a, b) NEMO decomposition of the MP2 energies on the Mg path into repulsive (a) and attractive (b) interactions. The black circles are the MP2 results and the solid lines are the fitted force-fields for the various atoms. The red line gives the contribution of Mg. (c, d) Comparison of the MP2 repulsive and attractive energies (filled symbols) for the eight different paths with the force-field results (lines). Mg and O paths (c) and C paths (d) are compared with the predictions from the UFF (open symbols). Reproduced from ref. 207 with permission from Springer Nature, Copyright 2012.

Figure 4.7c, the UFF does not allow for CO_2 to approach the Mg CUS as closely as predicted by the MP2 method. This conclusion further highlights the inability of UFF to correctly capture the short-range nature of the CUS interaction. The newly derived force-field was also used to predict Q_{st} and single-component adsorption isotherms for CO_2 and N_2. An inflection in the CO_2 Q_{st} profile is observed at a loading of precisely 1.0 CO_2 molecule per Mg site, quantitatively reproducing the experimental result. Single-component CO_2 and N_2 adsorption isotherms also show improved agreement over UFF. Differences between the experimental and simulated isotherms were rationalised by the fact that simulations use a perfect crystal with all CUS being accessible.

One of the highlights of this approach is that the force-field parameters are found to be fully transferable to the similar MOF, **Zn-MOF-74**. By re-parameterising the interaction with the metal, the remaining parameters are taken from the **Mg-MOF-74** set, yielding CO_2 adsorption isotherms in good agreement with experimental data. This motivated other studies to explore and analyse the gas capture properties of different metal centres in **MOF-74**. Borycz *et al.* employed restricted open-shell MP2 to parameterise CO_2 interactions with **Fe-MOF-74**.[209] MOFs containing Fe centres are challenging to model computationally as high-spin and low-spin configurations may share similar energies, making it difficult to predict the correct ground-state configuration. Previous studies on the **MOF-74** family suggest that these MOFs typically maintain a high-spin (quintet state) configuration when binding adsorbates such as H_2 or hydrocarbons.[187,210] This state was further confirmed using complete-active space self-consistent field (CASSCF) calculations. The ROMP2 interaction energy between CO_2 and a small **Fe-MOF-74** cluster was decomposed using the NEMO approach and the CO_2–metal interactions fitted to a Buckingham potential, whilst non-metal interactions were taken from the UFF. This combination has fewer parameters in total and still yields Q_{st} values and single-component adsorption isotherms that are in good agreement with experiment. A similar approach was employed by Haldoupis and co-workers who expanded the range of metal centres to Mn, Co, Ni and Cu.[211] Gibbs ensemble Monte Carlo simulations were used to simulate single-component CO_2 adsorption isotherms for the different M-MOF-74 members. Unlike UFF, the derived force-fields agree with experimental results, which indicates that **Ni-MOF-74** possesses the highest affinity toward CO_2 at low pressures.

In another study, Fischer *et al.* used all-electron DFT calculations using the PBE exchange-correlation functional to describe H_2–CUS interactions in **CuBTC**.[212] To generate the PES, the paddlewheel node was represented by a finite cluster consisting of two Cu(II) ions coordinated by four benzene-monocarboxylates. Based on a side-on coordination, the DFT PES predicts an energy minimum at 2.4 Å in excellent agreement with experimental studies. Fitting a Morse function to the PEC is found to give remarkable agreement across all regions of the PEC and was used in GCMC simulations to obtain adsorption isotherms and values of Q_{st}. In comparison to UFF, which greatly underestimates the H_2 loading, the DFT derived force-field provides much better agreement in the low-pressure region with the experimental isotherms of both **CuBTC** and **PCN-12**. This agreement is further improved by accounting for the rotational degrees of freedom of H_2. As an end-on binding mode has negligible impact on the adsorption of H_2, the well depth of the interaction was multiplied by 2/3. This orientation-averaged potential shows almost excellent low-pressure agreement but overestimates the high-pressure uptake.

The applicability of the method presented by Fischer *et al.* has been demonstrated for other adsorbates,[213] as it provides significant improvements over general force-fields in the description of the single-component adsorption isotherms. Furthermore, Fischer's force-field parameters are found to exhibit reasonable transferability when tested in other copper-containing MOFs such as $(Cu_4Cl)_3$(benzene-1,3,5-tristetrazolate)$_8$. However, it is important to note that this approach does not directly account for dispersion dominated interactions. It is well known that one of the main problems with DFT is its failure to capture long-range electronic correlation effects, such as dispersion. While dispersion corrections for DFT methods exist,[214,215] Fischer *et al.* noted that the parameterisation process would become more difficult as the contributions from non-metal atoms would have to be fully removed. In a subsequent study, the authors proposed a way to include dispersion contributions by parameterising olefin interactions in **CuBTC**.[194] The authors assume that the DFT interaction energy, $U_{\mathrm{DFT}}(\boldsymbol{r})$, can be split into two contributions:

$$U_{\mathrm{DFT}}(\boldsymbol{r}) = U_{\mathrm{rep}}(\boldsymbol{r}) + U_{\mathrm{Cu-}\pi}(\boldsymbol{r})$$

$U_{\mathrm{rep}}(\boldsymbol{r})$ represents the energy from repulsion interactions and $U_{\mathrm{Cu-}\pi}(\boldsymbol{r})$ represents the specific interactions between Cu atoms and the π orbitals in the unsaturated bonds of the olefins. The dispersion contribution to the adsorption energy is taken to be zero and any nonspecific electrostatic interactions are implicitly included in these contributions. By assuming that general force-fields such as the DREIDING or UFF correctly account for the dispersion and repulsion contributions of the host–guest interactions, the Weeks–Chandler–Andersen (WCA) approach[216] can be employed to extract the repulsive contribution:

$$U_{\mathrm{rep}}(r) = \begin{cases} U_{\mathrm{LJ}}(r) - U_{\mathrm{LJ}}(r_{\mathrm{min}}) & r < r_{\mathrm{min}} \\ 0 & r \geq r_{\mathrm{min}} \end{cases}$$

By isolating the repulsive part of the standard force-fields, an estimation of the attractive Cu–π interaction can be extracted and modelled with an appropriate function. GCMC simulations can then employ standard force-fields describing the host–guest interactions, in addition to the specific function derived for the CUS interactions. The accuracy of this methodology was demonstrated by comparing the PEC of different approaches used to model the Cu–π interaction (Figure 4.8).

The dashed blue curve represents the Cu–π interaction reported by Lamia *et al.* who adjusted the UFF parameters to match simulated C_3H_6 isotherms

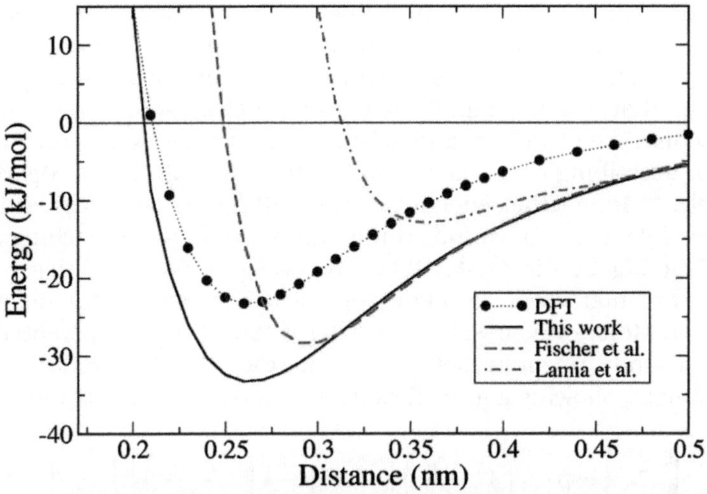

Figure 4.8 Potential energy curves obtained in classical Monte Carlo simulations using different approaches for including the Cu–π interactions. The circles represent the DFT energy profile obtained in this work. The blue dotted-dashed line was obtained from the LJ potential with the adjustments applied by Lamia *et al.*[190] The red dashed line was obtained by adding a Morse potential, fitted to reproduce the DFT curve, to the LJ potential, following the approach of Fischer *et al.*[212] The full black line was obtained by using the WCA approach.
Reproduced from ref. 194 with permission from American Chemical Society, Copyright 2012.

with experimental data.[190] This approach yields an incorrect description of the Cu–π interaction as the potential well is too shallow and appears at the incorrect separation. The red curve corresponds to the PES obtained by adding the fitted Morse function proposed by Fischer *et al.* to the UFF LJ potential.[212] While this PES yields a minimum separation in good agreement with the DFT calculations, a steep repulsive region is observed which begins at higher separation. Assuming that the DFT energy is composed of repulsive and Cu–π interactions, this implies that the Morse potential fitted by Fischer *et al.* contains both contributions and that its sum with the UFF LJ potential leads to the adsorbates experiencing twice the amount of repulsion. Finally, the black curve corresponds to the new approach of Fischer *et al.*, which identifies the appropriate attractive Cu–π interactions from the DFT energy using the WCA approach. As the repulsive contribution is fully removed from the fitting, the PES obtained from combining the fitted Morse-power law and UFF LJ potentials yields a more accurate description of the repulsive region. In later studies, the force-field parameters for the unsaturated olefin bonds were found to be transferable between ethylene and propylene,[217] and to other Cu-containing MOFs.[218] More recently, this approach has been generalised to polar adsorbates where electrostatic interactions play an important role in the adsorption mechanism.[219]

Many simple interatomic potentials such as the LJ, Buckingham and Morse potentials are commonly employed to fit force-fields. However, the small number of adjustable parameters allows little flexibility for describing complex interactions that change rapidly with internuclear separation. This was previously discussed by Chen *et al.* who developed a series of semi-empirical models for describing CH_4 and CO_2 interactions in **CuBTC** and **Mg-MOF-74**, respectively.[220] Interaction energies at the B2PLYP-D2/DEF2-TZVPP level of theory were calculated for various configurations of CO_2 approaching a finite cluster of the Mg CUS in **Mg-MOF-74.** To develop a force-field that correctly describes the interaction strength dependence at the CUS, the authors employed a Morse–Morse–Spline–van der Waals (MMSV) potential. This combines separate functions that best describe the CO_2–CUS interaction across different ranges, allowing a more flexible description of the CUS interaction:

$$E_{\mathrm{vdW}}(r_{ij}) = \begin{cases} D_{\mathrm{eq}}\left\{\exp\left[\alpha_1\left(1-\frac{r_{ij}}{R_{\mathrm{eq}}}\right)\right] - 2\exp\left[\frac{\alpha_1}{2}\left(1-\frac{r_{ij}}{R_{\mathrm{eq}}}\right)\right]\right\} & 0 \le r_{ij} \le R_{\mathrm{eq}} \\ D_{\mathrm{eq}}\left\{\exp\left[\alpha_2\left(1-\frac{r_{ij}}{R_{\mathrm{eq}}}\right)\right] - 2\exp\left[\frac{\alpha_2}{2}\left(1-\frac{r_{ij}}{R_{\mathrm{eq}}}\right)\right]\right\} & R_{\mathrm{eq}} \le r_{ij} \le R_1 \\ b_1 + (r_{ij}-R_1)\{b_2 + (r_{ij}-R_2)[b_3 + (r_{ij}-R_1)b_4]\} & R_1 \le r_{ij} \le R_2 \\ -C_6 r_{ij}^{-6} & R_2 \le r_{ij} \le \infty \end{cases}$$

Here, α_1 and α_2 are stiffness parameters, R_1 is an inflection point of the spline function corresponding the max force on the Morse functions, R_2 is a spline parameter equal to 1.4, R_{eq}, b_1–b_4 are spline parameters that are obtained algebraically outside the fitting, and C_6 is a dispersion constant fitted during the parametrisation process. For non-CUS interactions, a modified Buckingham, Carra–Konowalow (CK) potential[221] was employed:

$$E_{\mathrm{vdW}}(r_{ij}) = D_{\mathrm{eq}}\left(\frac{\beta+6}{\beta}\right)\left(\frac{R_{\mathrm{eq}}}{r_{ij}}\right)^6\left\{\left(\frac{6}{\beta+6}\right)\exp\left[\beta\left(1-\frac{r_{ij}}{R_{\mathrm{eq}}}\right)\right] - 1\right\}$$

In this equation, D_{eq} is the potential well depth, R_{eq} is the separation at the minimum potential, r_{ij} is the internuclear separation between atoms i and j, and β is a parameter that controls the steepness of the exponential repulsion. To fit such complex functions, the authors used a multi-objective genetic algorithm. The same methodology was used to describe the CH_4 interaction with Cu(II) in **CuBTC** using a PES taken from previous work employing the DFT/CC method (discussed in Section 4.4.5.2).[222] Using the new force-fields, the predicted adsorption isotherms for **Mg-MOF-74** and **CuBTC** are in excellent agreement with experimental results, demonstrating the importance of choosing an appropriate potential in the fitting procedure. Moreover, by employing the MMSV + CK potential, the correct isotherm shape and adsorption mechanism are predicted for the adsorption of CH_4 in **CuBTC**; results that could not be obtained from using a Morse or Mie function.

4.4.5.2 Periodic Models

While MP2 and SAPT methods scale too poorly to study periodic solids, the cost-efficiency of DFT-D methods permits their accurate study in an accessible timeframe. Fang *et al.* tested several DFT-D methods to study CO_2 adsorption in the siliceous Na-FER zeolite.[223] For each method, interaction energies between CO_2 and Na^+ cations present in these zeolites were recorded and compared to highly accurate DFT/CC calculations. The PBE-D2 method was able to reproduce DFT/CC interaction energies with deviations between 3 kJ mol^{-1} and 6 kJ mol^{-1} for single cation sites and between 6 kJ mol^{-1} and 12 kJ mol^{-1} for dual cation sites. Using this information, periodic PBE-D2 calculations were employed to calculate the interaction energy of 400 random CO_2 configurations inside the pores of the Si-CHA zeolite. This reference data was then used to parameterise a new force-field composed of a pairwise van der Waals term based on Grimme's empirical expression for dispersion, and a coulombic term:

$$U_{ij}(r_{ij}) = s_{12}\frac{C_{12}^{ij}}{r_{ij}^{12}} - s_6\frac{C_6^{ij}}{r_{ij}^{6}} + \frac{q_i q_j}{r_{ij}}$$

Here, r_{ij} is the internuclear separation between atoms i and j, C_6^{ij} and C_{12}^{ij} are the attractive and repulsive coefficients, q_i and q_j represent the APCs on atoms i and j, respectively, and s_6 and s_{12} are global scaling factors that depend on the DFT method. The C_{12}^{ij} term replaces a damping function in Grimme's original expression to represent repulsive interactions, and is defined as:

$$\frac{C_{12}^{ij}}{C_6^{ij}} = \frac{(r_0^i + r_0^j)^6}{2}$$

where r_0^i and r_0^j are the van der Waals radii of atoms i and j. C_{12}^{ij} parameters were determined using the C_6 and r_0 values available in Grimme's work.[214] The s_6 and s_{12} parameters were optimised to obtain the closest least-squares fit between the DFT reference data and the new force-field (identified hereafter as D2FF). The D2FF interaction energies for CO_2 adsorption in Si-CHA are in excellent agreement with those calculated using the PBE-D2 method and also outperform the Clay force-field (CLAYFF),[224] commonly used for clay or pillared materials. To test the predicative power of the force-field, adsorption isotherms and Q_{st} values were calculated for Si-CHA. Adsorption isotherms are in excellent agreement with experimental work over all pressure ranges and temperatures, and Q_{st} values agree across all loadings, including an acceptable 1 kJ mol^{-1} deviation in the limit of zero-coverage. The transferability of the force-field parameters is demonstrated for MFI and DDR zeolites, with GCMC simulations yielding adsorption isotherms in good agreement with several sets of experimental data.

Chen *et al.* employed a hybrid methodology combining periodic DFT calculations with CCSD(T) (DFT/CC) to study CH_4 adsorption in **CuBTC**.[222] The DFT/CC correction scheme was previously described by Bludský as a way of describing weakly interacting systems using DFT.[225] It is based on the pairwise representability of the DFT error, ΔE, which is defined as the difference between the interaction energies calculated at the CCSD(T) and DFT level:

$$\Delta E = E_{\mathrm{CCSD(T)}} - E_{\mathrm{DFT}} = \sum_{ij} \varepsilon_{ij}(R_{ij})$$

where ε_{ij} are a series of pairwise atom–atom correction functions and R_{ij} is the interatomic separation between atoms i and j. In essence, a series of precise reference calculations are used to improve the DFT description of systems that are too large to study at the CCSD(T) level. In their work, Chen *et al.* derived a number of correction functions describing interactions between H_2 and H_2, H_2 and C_6H_6, CH_4 and H_2, CH_4 and C_6H_6, H_2 and CO_2, CH_4 and CO_2, and CH_4 and $Cu(HCOO)_2$. Then, the DFT/CC method was employed to tabulate the **CuBTC**–CH_4 interaction energy across the periodic structure for direct use in GCMC simulations. The predicted CH_4 adsorption isotherms obtained using the DFT/CC scheme are found to be in excellent agreement with recently obtained experimental results, capturing the correct shape, steps and gas loading at 77 K and 87 K. In contrast, the UFF and PES derived from dispersion corrected DFT were not accurate enough to correctly predict the CH_4 adsorption isotherm in **CuBTC**, highlighting the value of using the DFT/CC method (Figure 4.9).

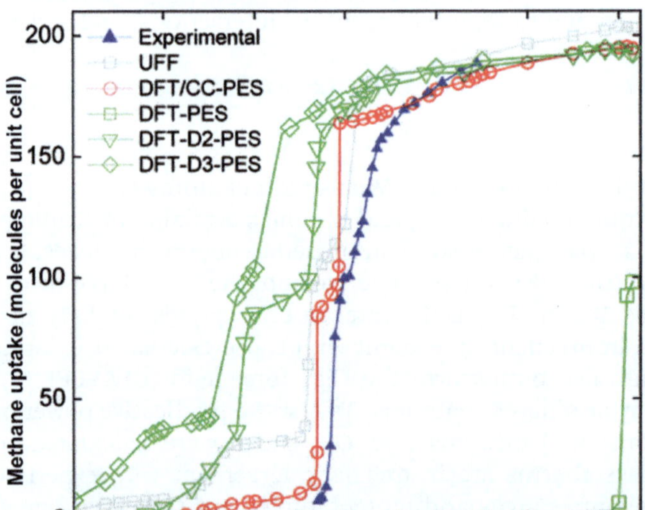

Figure 4.9 Comparison of methane adsorption isotherms at 77 K calculated using potential energy surfaces derived from different *ab initio* methods with experimental data and UFF calculations.
Reproduced from ref. 222 with permission from American Chemical Society, Copyright 2011.

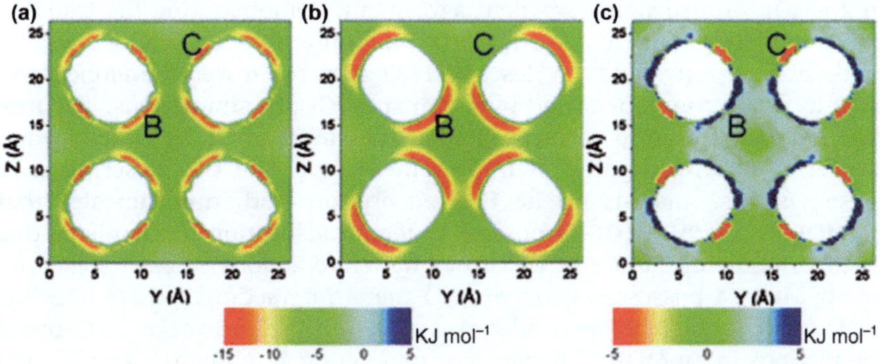

Figure 4.10 Contour plots of the potential energy between a methane molecule and the **CuBTC** framework on the (200) plane for (a) DFT/CC-PES, (b) UFF and (c) the difference, E(DFT/CC-PES) – E(UFF). Adsorption sites B and C (at the CUS) are labelled. The white space corresponds to the **CuBTC** framework.
Reproduced from ref. 222 with permission from American Chemical Society, Copyright 2011.

To analyse the differences between the adsorption isotherms predicted from the UFF and DFT/CC PES, a series of contour plots of the PES along the (200) plane were generated (Figure 4.10). The adsorption isotherm predicted from UFF shows a step at 40 molecules per unit cell, corresponding to the filling of the smallest pore in **CuBTC**. It should be noted that UFF fails to capture the interaction at the CUS, as made evident by the contour plot in Figure 4.10b (site C). Furthermore, the site closest to the corner of two neighbouring ligands in the largest pore (site B) appears to be overestimated by the UFF.

Lin *et al.* derived force-field parameters for **Mg-MOF-74** and **Zn-MOF-74** to study the separation of flue gas using periodic DFT calculations with the VdW-DF2 exchange correlation functional and a planewave basis. The total interaction energy between the adsorbates and several atom types was separated into electrostatic, repulsive and dispersive contributions, with the repulsive and dispersive contributions used to fit a Buckingham potential *via* a self-consistent optimisation loop. This force-field was used to predict CO_2 adsorption isotherms in **Mg-MOF-74**, which show excellent agreement with experimental data and previous simulations employing the MP2 NEMO derived force-field of Dzubak *et al.*[207] The periodic description of **Mg-MOF-74** is found to improve the force-field accuracy compared to when a finite cluster description is used, such as the MP2 NEMO derived force-field of Dzubak *et al.*[207] Furthermore, the MP2 force-field is found to systemically overestimate the CO_2 adsorption energy near the centre of the pore and incorrectly predicts the CO_2 binding angle with the Mg CUS site. For **Zn-MOF-74**, the authors calculated the Zn–CO_2 potential energy profile and the parameters for other atoms were taken directly from the **Mg-MOF-74** set. A comparison of the simulated and experimental CO_2 adsorption isotherms in **Zn-MOF-74** also yields excellent agreement, demonstrating the transferability of these parameters to other **MOF-74** family members. Using the same methodology, force-fields to describe H_2O adsorption were developed and used in single-component and binary mixture GCMC simulations. The new force-fields are able to correctly predict the shape of the H_2O adsorption isotherm, showing significant improvements over the UFF description of these systems. Analysis of the H_2O adsorption isotherms indicates that **Zn-MOF-74** and **Mg-MOF-74** possess different adsorption mechanisms due to the underlying nature of the host–guest interactions in each structure. As **Mg-MOF-74** possesses stronger H_2O–metal interactions, all Mg sites become fully occupied before condensation. In contrast, the weaker H_2O–metal interactions in **Zn-MOF-74** means that upon guest binding to a Zn site, it is energetically more favourable for the next H_2O to bind in the vicinity near the first H_2O, due to stronger guest–guest interactions. As a result of the weaker H_2O–Zn interaction, binary CO_2/H_2O simulations reveal that **Zn-MOF-74** can endure much higher H_2O concentrations than **Mg-MOF-74**, without sacrificing its ability to adsorb CO_2. This work was later extended by Mercardo *et al.* who employed the same methodology to study CH_4, CO_2 and H_2O adsorption in M-MOF-74 (M = Mg, Mn, Fe, Co, Ni and Zn) and isoreticular derivatives.

4.4.5.3 Polarisable Force-fields

In recent years, a number of force-fields that explicitly include polarisation effects have been developed for describing gas adsorption mechanisms in MOFs. In most simulations, standard force-fields describe electrostatic interactions between a series of static APCs, neglecting any effects of polarisation. To describe the effects of polarisation, a number of methods have been developed such as the fluctuating charge method (fQ), shell model and the induced point dipole (IPD) method.[226,227] In the fQ method, APCs are treated as dynamic properties, allowing them to fluctuate in response to an electric field. For a single neutral atom, the energy required to create a charge q can be expressed in terms of its electronegativity, χ^0, and atomic hardness, J^0, *via* a truncated second-order Taylor series:

$$U(q) = U^0 + \chi^0 q + \frac{1}{2} J^0 q^2$$

For a system of N atoms, the expression is expanded to account for the pairwise coulombic interactions between atoms:

$$U(q_1, q_2 \ldots q_n) = \sum_{i=1}^{N} \left(E_i^0 + \chi_i^0 q_i + \frac{1}{2} J_{ii} q_i^2 \right) + \sum_{i=1}^{N} \sum_{j>i}^{N-1} J_{ij}(r_{ij}) q_i q_j$$

During each timestep or Monte Carlo move, the APCs are obtained by minimising the above equation with respect to the total charge distribution, yielding a set of coupled linear equations that must be solved self-consistently through matrix inversion or extended Lagrangian methods.

In the shell model, polarisable sites are described by pairs of particles connected by harmonic springs with a force constant, k. Each pair consists of a heavy core particle that is usually positioned on the nuclei, and a light or massless shell particle. During a simulation, the charges on these sites are kept fixed, meaning polarisation is obtained from the relative displacement of the core and shell particles with respect to one another. Interactions that occur between pairs of particles in the same polarisable site are neglected. For each core–shell pair, the polarisation energy U_{pol} created by a shell particle that is displaced by a distance, d_i, away from the core particle is expressed as:

$$U_{\text{pol}} = \frac{1}{2} k_i d_i^2$$

Given that the dipole moment is expressed in terms of the shell charge and shell displacement, the polarisability of an isotropic shell model atom, α_i, can be expressed as:

$$\alpha_i = \frac{q_i^2}{k_i}$$

As the number of charge sites per atom is doubled in the shell model, the computational cost for evaluating the electrostatic interactions is much larger than other methods. Moreover, a small mass is required by the shell particle in order to observe fast changes of its position as a result of the electric field, prohibiting the use of large timesteps in molecular dynamics simulations.

Finally, in the IDP method, polarisation is modelled by assigning point dipoles to each interaction site. As a result of the electrostatic field, the induced dipole moment, μ_i, for a single, isolated point dipole, can be calculated assuming a linear response:

$$\mu_i = \alpha_i \cdot E_i(\boldsymbol{r}_i)$$

where α_i is a 3×3 polarisability tensor and $E_i(\boldsymbol{r})$ is the electric field strength at position $\boldsymbol{r}$. For a system of N inducible dipoles, $E_i(\boldsymbol{r})$ is a function of the permanent electric field caused by the static partial charges (E_i^0) and induced dipoles across all N sites. Therefore:

$$\mu_i = \alpha_i \cdot E_i(\boldsymbol{r}_i) = \alpha_i \cdot \left[E_i^0(\boldsymbol{r}_i) - \sum_{j \neq i}^{N} T_{ij} \cdot \mu_j \right]$$

where the dipole field tensor, T_{ij} is expressed as:

$$T_{ij} = \frac{1}{\boldsymbol{r}_{ij}^3} \boldsymbol{I} - \frac{3}{\boldsymbol{r}_{ij}^5} \begin{pmatrix} x^2 & xy & xz \\ yx & y^2 & yz \\ zx & zy & z^2 \end{pmatrix}$$

Here, $\boldsymbol{I}$ is an identity matrix, and x, y and z are the Cartesian coordinates of the vector $\boldsymbol{r}_{ij}$ connecting the interacting dipoles i and j. As the dipoles coupled through T_{ij} are interdependent of each other, self-consistent solutions with respect to μ_i must be employed. A common approach is to use matrix inversion; however, it scales at $O(N^3)$ making it infeasible for large systems. Alternative solutions involve extended Lagrangian methods or iterative processes. The total interaction energy of the IDP method (U_{IDP}) is composed of the polarisation energy, U_{pol}, describing the energy needed to create the induced dipoles, the dipole–dipole energy, $U_{\mu\text{-}\mu}$, describing the interactions among mutual dipoles, and the dipole–static energy, $U_{\mu\text{-stat}}$, describing the interaction between induced dipoles and the static charge distribution:[227]

$$U_{\text{IDP}} = U_{\text{pol}} + U_{\mu-\mu} + U_{\mu-\text{stat}} = -\frac{1}{2}\sum_{i=1}^{N} \mu_i \cdot E^0(\boldsymbol{r}_i)$$

Of these three approaches, the IDP method has been frequently employed to model the polarisation effects of adsorbing gases in zeolites and MOFs.[228,229] Polarisable force fields are particularly valuable in describing gas interactions in CUS-containing MOFs.[229–234] In many of these studies, the simulated adsorption isotherms derived from polarisable force-fields show improved agreement with experimental isotherms over standard force-fields such as the UFF, demonstrating the importance of polarisation effects. While the development of new polarisable force-fields is an important progression in generating more accurate classical potentials, their main drawback is the larger computational cost and overhead required to calculate polarisation effects. This is because many of the methods that account for polarisation require updates every timestep or Monte Carlo move.

4.5 Transport Properties of Gas and Flexibility of Porous Structures

In many applications of porous materials, the rates of guest transport can have significant impact on the overall performance and efficiency of a process. For example, gas separations are governed by differences in transport rates between diffusing adsorbates, which may define the selectivity of a material. Transport rates can also be a limiting parameter; the rate of diffusion around the active site of a catalyst often has a large impact on the efficiency of the chemical process. In the confinement of a porous material, the transport behaviour of a guest can deviate strongly from its bulk behaviour. This can become even more complicated for materials that exhibit large degrees of structural flexibility or undergo phase changes in response to guest adsorption. Therefore, it is crucial to develop a comprehensive understanding of these processes in order to exploit their potential in different applications.

To understand and predict the macroscopic transport properties in nanoporous media, molecular modelling tools can provide an excellent framework for obtaining detailed atomistic insights in the cases where experimental measurements are challenging. In this section, we will discuss approaches to modelling transport properties of pure gases and mixtures, rare events, and permeation across porous membranes and discuss the impact of the dynamic behaviour of porous solids on the transport properties of guest gases.

4.5.1 Modelling Guest Diffusion Using Molecular Dynamics

Molecular dynamics (MD) is a computational method used to study the time-dependant behaviour of a chemical system. MD simulations begin with the assignment of positions and velocities to all interacting particles in the system. From knowledge of the force acting on each particle, the subsequent positions and velocities (microscopic states) as well as accelerations can be derived by integrating Newton's equations of motion. The forces are determined from the gradient of the potential energy, which is usually described classically. Properties of interest are then sampled over the entire trajectory, corresponding to the time-averaged behaviour of a large number of microscopic states that are sequential in time.

When describing guest diffusion in a porous material, a number of different diffusion coefficients can be defined. The self-diffusion coefficient, D^{self}, describes the Brownian motion of individual particles and is calculated from the slope of the mean-squared displacement (MSD) using Einstein's relation:[7]

$$D^{\text{self}}(c) = \frac{1}{2\text{d}Nt} \lim_{t\to\infty} \frac{\text{d}}{\text{d}t} \left\langle \sum_{i=1}^{N} (\boldsymbol{r}_i(t) - \boldsymbol{r}_i(0))^2 \right\rangle$$

where d is the dimensionality of the system, t is the observation time, $\boldsymbol{r}_i(t)$ is the position of adsorbate i at time t, c is the concentration of the adsorbed species and N is the number of adsorbates in the system. The Einstein relation is only valid in the long-time limit where the MSD scales linearly with time. This can be tested by plotting a log–log plot of the MSD against the observation time and measuring the slope which is equal to 1 in the diffusive region. It is important to separate the long-time diffusive behaviour from non-diffusive behaviour, which commonly occurs during equilibration. In some cases, normal diffusive behaviour may not be observed over the timescale of the simulation. This is referred to as an anomalous diffusion process and is characterised by a non-linear relationship between the MSD and observation time. An extreme example of this occurs when the dimensionality and pore size of a porous material are decreased to the point where the mutual passage of molecules is no longer possible. This is referred to as single file diffusion (SFD) and mainly occurs in non-intersecting, 1D channel structures where the width of the pores is comparable to the size of the adsorbates.[235] During SFD, the ordering of the molecules in the channel remains fixed and the movements of each adsorbate become strongly correlated with the movement of surrounding molecules. The concept of SFD has been studied using various theoretical methods including random walker models. These describe diffusion on an infinite 1D chain where particles hop between sites that are separated by a distance λ. The MSD of a random walker can be expressed as:[236,237]

$$\langle r^2 \rangle = \lambda^2 \frac{1-\theta}{\theta} \sqrt{\frac{2}{\pi}} \sqrt{\frac{t}{\tau}}$$

where τ is the mean time between successive jump attempts and θ denotes the site occupancy. Similarly to Einstein's relation, this can be expressed as:[238]

$$\langle r^2 \rangle = 2Ft^{0.5}$$

where the single-file mobility factor, F, is expressed as:

$$F = \lambda^2 \frac{1-\theta}{\theta} \sqrt{\frac{1}{2\pi\tau}}$$

These expressions show the MSD of a particle experiencing SFD increases with the square root of time. This behaviour has also been studied experimentally[239–241] and using MD simulations,[242–245] indicating that single file behaviour is a complex issue and that some systems may exhibit single file and normal diffusive behaviour over different timescales.[246] Alternatively, D^{self} can be calculated using the velocity autocorrelation function (VACF) using the Kubo–Green relation.[7]

In the context of macroscopic diffusion, it is more convenient to define the transport diffusivity, D^{t}, which is the proportionality constant linking the macroscopic flux, N, to the concentration gradient, ∇c, described by Fick's Law:

$$N = -D^{\text{t}}(c)\nabla c$$

D^{t} is often described in terms of the corrected diffusivity, D^{c}, and a thermodynamic correction factor, Γ, associated with the curvature of the adsorption isotherm:[247]

$$D^{\text{t}}(c) = D^{\text{c}}(c)\frac{\partial \ln(f)}{\partial \ln(c)} = D^{\text{c}}(c)\Gamma$$

where f is the bulk phase fugacity. D^{c} can be calculated from MD simulations using the following expression:[248]

$$D^{\text{c}}(c) = \frac{1}{2\text{d}Nt} \lim_{t\to\infty} \frac{\text{d}}{\text{d}t} \left\langle \left(\sum_{i=1}^{N} (\boldsymbol{r}_i(t) - \boldsymbol{r}_i(0)) \right)^2 \right\rangle$$

Unlike D^{self}, which describes motion of individual particles, D^{c} is related to the collective centre-of-mass motion of all diffusing particles and must be calculated over a large number of independent trajectories. D^c is often referred to as the single-component diffusivity used in the Maxwell–Stefan (MS) formalism of mass transport (see Section 4.5.1.1).[249–251] It is worth noting that D^{self}, D^t and D^c are all functions of the adsorbate concentration and are only equal in the limit of zero loading.[249]

Due to the higher cost required to compute D^t, many early computational studies focused on the calculation of D^{self} for various gases. Demontis *et al.* employed MD simulations to study CH_4 diffusion in silicalite.[252] Simulations at a loading of 12 CH_4 molecules per unit cell predicted a D^{self} value of $(6.58 \pm 0.01) \times 10^{-9}$ m^2 s^{-1} in excellent agreement with the experimental value of $(6.58 \pm 1.00) \times 10^{-9}$ m^2 s^{-1}. Leherte *et al.* studied H_2O diffusion in ferrierite using MD.[253] A predicted D^{self} value of 5.0×10^{-6} cm^2 s^{-1} is found to be in excellent agreement with experimental measurements, and the confinement of the zeolite was found to reduce D^{self} by a factor of four, relative to that of liquid H_2O. June *et al.* reported the dynamic, structural and thermodynamic properties of Xe and CH_4 in silicalite.[254] Comparison of the predicted D^{self} values with NMR experiments showed good agreement for both gases. At low temperatures near 200 K, the D^{self} values for both gases exhibit a small maximum at low gas loadings before decaying at higher gas loadings, which is attributed to the trapping of adsorbate molecules in corresponding potential wells. By increasing the gas loading, the number of collisions increases leading to a larger number of molecules possessing enough energy to escape the potential well. However, as the gas loading is increased further, steric hinderance becomes an issue which leads to the decay in D^{self} values. Sastre *et al.* investigated the diffusion behaviour of benzene and propylene in the **MCM-22** zeolite at 650 K.[255] Simulations revealed that the rate of propylene diffusion was 2–3 orders of magnitude larger than that of benzene. This was attributed to the large size of benzene molecules, which are unable to freely diffuse through the pores due to the presence of large energy barriers. These results provide a better understanding of the transport properties of these gases as reactants in the synthesis of cumene.

Calculations of D^t are important for many industrial applications. Maginn *et al.* were some of the first to calculate D^t values from non-equilibrium MD methods. These included the gradient relaxation MD method, which monitors the relaxation of a concentration gradient as a function of time, and 'colour-field' MD in which an external force is applied and D^c is calculated using linear response theory. In subsequent years, improvements to computational resources made the calculation of D^t from equilibrium MD simulations more feasible. Skoulidas *et al.* reported D^{self} and D^t values for H_2 and CH_4 diffusion in CNTs, ZSM-5 and silicalite.[256] Interestingly, in the limit of zero loading, CNTs were found to possess D^{self} values that were 3–4 orders of magnitude higher than in either zeolite. Similar behaviour was also observed for D^t across 0–100 bar range of pressure. These results are attributed to the inherent smoothness of the PES inside the CNT, which results in a very small energy cost – 0.054 and 0.066 kJ mol^{-1} for CH_4 and H_2, respectively – to move molecules along the pore. Skoulidas and Sholl employed GCMC and MD simulations to study the temperature and concentration dependence of D^t for CH_4 and CF_4 diffusion in silicalite.[257] In this work, the authors assessed the validity of the Darken approximation, which assumes that D^c is independent of the gas concentration, such that D^t becomes a function of Γ and D^{self} in the limit of zero loading. For CH_4, the authors found that D^t increases monotonically with adsorbate concentration and can be accurately estimated using the Darken approximation at 298 K. In contrast, the Darken approximation generally overestimated D^t for CF_4, suggesting that there may be a number of related systems that show similar deviations and should be treated with caution when using this approximation.

One of the first MD simulations of gas diffusion in a MOF was reported by Skoulidas and Sholl, who calculated D^{self}, D^t and D^c values for the diffusion of Ar in **CuBTC**.[258] Like zeolites, diffusion in MOFs is found to be an activated process with activation energies ranging between 2.7 kJ mol^{-1} and 6.9 kJ mol^{-1} for Ar concentrations between infinite dilution and 80 atoms per unit cell. Analysis of D^{self}, D^t and D^c indicates that the diffusion of Ar in **CuBTC** is similar to diffusion in zeolites, both in magnitude and concentration dependence. As the concentration of Ar increases, the values of D^{self} and D^c are found to decrease, while D^t increases. At pressures between 10^3 bar and 10^4 bar, the difference between these values is as large as two orders of magnitude. Around the same time, Sarkisov *et al.* predicted D^{self} values for the diffusion of CH_4, *n*-pentane, *n*-hexane, *n*-heptane, cyclohexane and benzene in bipyridine molecular squares and **IRMOF-1**.[259] At low CH_4 concentration, the D^{self} value of 3.08×10^{-8} $m^2\,s^{-1}$ is found to be comparable to the self-diffusion of CH_4 in silicalite zeolites. For the *n*-alkanes, D^{self} values decrease as a function of chain length, ranging between 3.5×10^{-9} $m^2\,s^{-1}$ and 1.2×10^{-9} $m^2\,s^{-1}$. The large size of benzene and cyclohexane molecules results in a strong confinement effect where reasonable diffusion rates on the nanosecond timescale could not be recognised. In a subsequent study, Skoulidas and Sholl characterised the loading dependence of Ar, CH_4, CO_2, N_2 and H_2 diffusion in **IRMOF-1** to understand the relationships between molecular size, adsorption affinity and diffusivity.[260] In the limit of zero loading, D^{self} values are found to generally increase with decreasing molecular size. For all gases, D^{self} values decrease moderately as a function of gas loading due to increases in steric hinderance between the molecules. Analysis of D^c shows almost no dependence on the concentration for all gases indicating that Darken's approximation may be useful for estimating D^t. Therefore, the loading dependence of D^t is mainly influenced by Γ which increases monotonically with gas loading.

Comparably fewer studies have been performed to investigate the single-component diffusion properties in COFs. Garberoglio and Vallauri explored the adsorption and diffusion properties of H_2 and CH_4 in 2D COFs at 298 K.[261] For H_2, D^{self} values range between 7.5×10^{-8} $m^2\,s^{-1}$ and 1.60×10^{-6} $m^2\,s^{-1}$, whereas for CH_4, they range between 3.3×10^{-8} $m^2\,s^{-1}$ and 2.25×10^{-7} $m^2\,s^{-1}$. Many of these values are larger than the self-diffusion rates in MOFs but still orders of magnitude lower than in CNTs. Yang and Cao compared the diffusion and separation behaviour of H_2 and CH_4 in undoped and lithium-doped COFs.[262] In the undoped COFs, the H_2 and CH_4 D^{self} values decrease as a function of gas loading, ranging between 10^{-6} m^2 s^{-1} and 10^{-8} m^2 s^{-1}. Upon doping these COFs with Li^+ ions, D^{self} values for both gases decrease relative to the undoped COFs as the Li^+ ions provide a strong binding site which increases the activation energy of diffusion. After Li^+ doping, the activation energy for H_2 diffusion increases from approximately 1 kJ mol^{-1} to 9 kJ mol^{-1}. For CH_4, the activation energy increases from approximately 4 kJ mol^{-1} to a value between 25 and 40 kJ mol^{-1}, depending on the type of COF. As the gas loading is increased further, the Li sites become saturated and new molecules begin to adsorb at weaker sites which increases D^{self}. Finally, a further increase in pressure results in a lowering of the D^{self} values due to steric hinderance between diffusing molecules. While Li^+ doping is a favourable process for increasing the adsorption selectivity of CH_4 over H_2, it is unfavourable for kinetic separations due to a reduction in diffusion rates. For these types of process, undoped COFs are a potentially more promising material.

4.5.1.1 Transport Properties of Multi-component Mixtures

Many industrial applications involve gas mixtures. A number of mathematically equivalent formalisms can be used to treat gas mixtures such as the Fickian, Onsager and Maxwell–Stefan descriptions of mass transfer.[263] Once the diffusivities are known in one formalism, the other descriptions become fully defined making the choice of formalism a matter of convenience.[264] In the Fick formalism,[263] the flux of each component depends on the concentration gradient of the diffusing species, which can be generalised to:

$$(N) = -[\boldsymbol{D}](\nabla c)$$

where $[\boldsymbol{D}]$ is a square matrix of diffusivities which has the following elements for a binary mixture of components i and j:

$$[\boldsymbol{D}] = \begin{bmatrix} D_{ii} & D_{ij} \\ D_{ji} & D_{jj} \end{bmatrix}$$

The off-diagonal terms, D_{ij} and D_{ji}, describe the correlation effects between diffusing species and approach zero in the limit of zero loading. These terms can be positive or negative and in general are not symmetric, such that $D_{ij} \neq D_{ji}$. In the Onsager formalism,[263] the driving force for diffusion is recognised as the gradient of the chemical potential $\nabla\mu$, such that:

$$(N) = -\beta[\boldsymbol{L}]\,(\nabla\mu)$$

where $[\boldsymbol{L}]$ is a matrix of phenomenological coefficients that are functions of the concentration of each diffusing species and $\beta = 1/k_BT$. The advantage of this formalism is that the off-diagonal components of $[\boldsymbol{L}]$ are symmetric, providing a convenient check for the correctness of a simulation result.[265] Theodorou demonstrated that the phenomenological coefficients can be calculated using the VACF or from the MSD using linear response theory:[266]

$$L_{ij} = \frac{1}{2\mathrm{d}V} \lim_{t\to\infty} \frac{\mathrm{d}}{\mathrm{d}t} \left\langle \sum_{l=1}^{N_i} \left[\boldsymbol{r}_{li}(t) - \boldsymbol{r}_{li}(0) \right] \cdot \sum_{k=1}^{N_j} \left[\boldsymbol{r}_{kj}(t) - \boldsymbol{r}_{kj}(0) \right] \right\rangle$$

where V is the volume of the system, and $\boldsymbol{r}_{li}(t)$ is the position vector of the lth molecule of species i at time t. The relation between the Onsager coefficients and the Fickian diffusion matrix is expressed as:

$$[\boldsymbol{D}] = -\beta[\boldsymbol{L}] \left[\frac{\partial \mu}{\partial c} \right]$$

This expression can also be related to the MS formalism, where diffusion is driven by a chemical potential gradient that is assumed to be balanced by frictional drag forces exerted between diffusing species.[248] From the MS theory, the chemical potential gradients are written as linear functions of the fluxes through the following expression:

$$-\frac{c_i}{RT} \nabla \mu_i = \sum_{j=1;\, j\neq i}^{n} \frac{c_j N_i - c_i N_j}{c_{j,\mathrm{sat}} D_{ij}^{\mathrm{corr}}} + \frac{N_i}{D_i^{\mathrm{MS}}}$$

where R is the universal gas constant, D_i^{MS} is the MS diffusivity of component i, and D_{ij}^{corr} is the exchange diffusivity that describes correlation effects between components i and j. The MS diffusivities can be related to the Fickian diffusion matrix by:

$$[\boldsymbol{D}] = [\boldsymbol{B}]^{-1} [\Gamma]$$

where $[\Gamma]$ is a matrix of thermodynamic factors defined by the multi-component adsorption isotherm containing elements:

$$\Gamma_{ii} = \frac{\partial \ln(f_i)}{\partial \ln(c_i)}; \quad \Gamma_{ij} = \frac{c_i}{c_j} \frac{\partial \ln(f_i)}{\partial \ln(c_j)}$$

and $[\boldsymbol{B}]$ is a matrix of mixture MS diffusion coefficients with elements:

$$B_{ii} = \frac{1}{D_i^{\mathrm{MS}}} + \sum_{i\neq j} \frac{c_j}{D_{ij}^{\mathrm{corr}}}; \quad B_{ij} = -\frac{c_i}{D_{ij}^{\mathrm{corr}}}$$

In comparison with studies of single-component diffusion, there are fewer examples of simulations that aim to quantify the mixture diffusivities in $[\boldsymbol{D}]$ for a multi-component mixture. Sanborn and Snurr investigated the self and transport diffusivities of CF_4/n-alkane mixtures in faujasite using MD.[267] For all mixtures, the authors report that D^{t} is greater than D^{self} at the same loading and composition. Diffusivities obtained using the Kubo–Green or Einstein relation were found to be in good agreement with one another. Cross-term diffusivities were reported to be an order of magnitude lower than the main-term diffusivities. In a later study, they extended their investigation to higher gas loadings and found that the cross- and main-term diffusivities were of the same order of magnitude highlighting the importance of correlation effects.[265] The authors also predicted the flux of each component through a 1 μm membrane and found that neglecting the cross terms can lead to both quantitative and qualitative errors in the flux calculations.

Chen and Sholl investigated the diffusion of binary CH_4/H_2 mixtures in single-walled CNTs.[268] The self-diffusivities in the mixture are very similar to the single-component diffusivities and are found to be 2 to 3 orders of magnitude larger than in zeolite ZSM-12 across all studied gas loadings. Similar trends were also observed for the transport diffusivities with the cross- and main-term diffusivities being of comparable size, indicating that that correlation effects are highly important and that the net flux of one species is strongly affected by the concentration gradients of the other species. Skoulidas *et al.* employed detailed atomistic simulations to investigate CH_4/CF_4 mixture permeation through a silicalite membrane.[269] The ideal selectivity calculated from the ratio of the single-component fluxes predicts a CH_4 selective membrane. In contrast, calculations for the binary mixture show that the membrane is CF_4 selective, which is attributed to a high $S_{ads,CF_4/CH_4}$ indicating that CF_4 is selectively adsorbed over CH_4. The authors note that calculations on the binary mixture employ all elements of $[\boldsymbol{D}]$ and that setting off-diagonal elements to zero leads to quantitative inaccuracies in the predicted membrane selectivity.

One of the main advantages of the MS formalism is that it provides a framework for predicting multi-component diffusion effects using the information from single-component studies. This feature is highly desirable as many experimental studies investigate single-component behaviour due to the difficulties with studying mixtures. An example of such approach is presented by Skoulidas, Sholl and Krishna (SSK), who studied correlation effects in the diffusion of CH_4/CF_4 mixtures through an MFI zeolite.[270] To define the MS mixture diffusivities of a binary system, it is necessary to calculate the single-component D^c (referred to as D_i^{MS} in the MS formalism) and the binary exchange diffusivities for all components. In previous work, Paschek and Krishna derived the following expression to relate D^{self} to D_i^{MS} for a binary mixture containing tagged and untagged particles of the same species:[271]

$$\frac{1}{D_i^{self}} = \frac{1}{D_i^{MS}} + \frac{c_i}{D_{ii}^{corr}}$$

where D_{ii}^{corr} provides a measure of vacancy and geometry correlations as a result of confinement in a porous solid.[272] By estimating D_{ii}^{corr} from the single-component self- and corrected diffusion coefficients, the binary exchange diffusivities can be estimated using the following interpolation formula:[270]

$$c_{j,\mathrm{sat}}D_{ij}^{\mathrm{corr}} = \left[c_{j,\mathrm{sat}}D_{ii}^{\mathrm{corr}}\right]^{c_i/(c_i+c_j)}\left[c_{i,\mathrm{sat}}D_{jj}^{\mathrm{corr}}\right]^{c_i/(c_i+c_j)}$$

Now that the MS mixture diffusivities are determined, these can be recasted into Fickian or Onsager form to calculate the elements in [***D***] using [**Γ**]. In the SSK mixing theory, [**Γ**] is calculated from the multi-component adsorption isotherm which is predicted from single-component adsorption using IAST. The accuracy of the SSK mixing theory has been tested in a number of publications, showing good predictions for alkane diffusion in faujasite, mordenite and MFI zeolites,[273–276] *n*-alkane and H_2 diffusion in CNTs,[277] H_2 and CH_4 diffusion in **CuBTC**,[278] and for CO_2/CH_4 diffusion in **IRMOF-1**.[279] In contrast, the theory has been shown to be less accurate in materials with strong heterogeneous energy distributions.[280]

4.5.1.2 Modelling Rare Events with Transition State Theory

The maximum timestep of an MD simulation is defined by the fastest vibration that occurs in the system and typically lies on the femtosecond scale. This generally results in an accessible timeframe on the order of microseconds or lower. However, some diffusion processes in porous materials can occur outside this accessible timescale, such that alternative methods must be employed to accurately study their behaviour. For many of these slow or rare event processes, the adsorbate spends a large proportion of its time trapped in a local energy minimum due to favourable interactions with its surroundings. Therefore, diffusion can be characterised by a series of hops between local energy minima that are separated by large free energy barriers. In these cases, transition state theory (TST) can be employed to calculate the rate constant ($k_{\mathrm{a}\to\mathrm{b}}$) of a species hopping from state a to state b along a reaction coordinate, q. These states are partitioned by a dividing surface, q^*, which in many nanoporous materials represents a region of confinement, such as a window between two pores. Within the Bennett–Chandlers approach,[281,282] the rate constant is calculated from free energy sampling which estimates the probability of finding an adsorbate at the top of the free energy barrier, and a dynamical correction factor that accounts for adsorbates which re-cross the transition state when attempting a transition from state a to state b:

$$k_{\mathrm{a}\to\mathrm{b}} = \left(\kappa \times \sqrt{\frac{k_b T}{2\pi m}}\right) \times \frac{\mathrm{e}^{-\beta F(q^*)}}{\int_{q<q^*} \mathrm{e}^{-\beta F(q)}\mathrm{d}q}$$

where m is the mass of the diffusing adsorbate, $F(q)$ is the free energy profile along the reaction coordinate and κ is the dynamical correction factor or re-crossing coefficient ranging from 0 to 1 for a classical system. In practice, κ is usually determined from separate MD simulations with particles initially placed on top of the free energy barrier,[283] and $F(q)$ can be determined using histogram sampling (HS) methods from NVT Monte Carlo simulations.[284] In this method, a histogram of the particle positions across the reaction coordinate is converted into a free energy profile using:

$$\beta F(q) = -\ln\langle P(q)\rangle$$

where $P(q)$ denotes the probability of finding a molecule at a given position, q, according to the histogram. In systems with large free energy barriers, conventional HS may be impractical due to the large amount of time required to efficiently sample regions near the barriers and transition states. In these scenarios, importance sampling can be employed using the Widom insertion technique:[7]

$$\beta F(q) = -\ln\langle e^{-\beta\Delta U}\rangle_N$$

where $\langle e^{-\beta\Delta U}\rangle_N$ is the average Boltzmann factor over all positions in a perpendicular slice to the reaction coordinate. After evaluating $k_{a\to b}$, the D^{self} value for a diffusing adsorbate is calculated from:[285]

$$\begin{aligned} D^{self} &= k_{a\to b} \times \lambda^2 \\ &= \left(\kappa \times \sqrt{\frac{k_b T}{2\pi m}} \times \frac{e^{-\beta F(q^*)}}{\int_{q<q^*} e^{-\beta F(q)}\,dq}\right) \times \lambda^2 \end{aligned}$$

where λ is the lattice distance separating states a and b.

The application of TST to guest diffusion has mainly involved studies of zeolites, which have been previously reviewed by Dubbeldam and Snurr.[248] TST has played a large role in understanding the zeolite window effect; a phenomenon first observed in 1968 where ERI type zeolites are found to yield a bimodal product distribution when cracking long alkanes.[286] While C_{3-4} and C_{10-12} products are obtained, no products in the C_{5-8} range are observed. A few years later, Gorring showed that this product distribution has a very good correlation with the diffusion coefficient of each product.[287] For an increasing number of carbon atoms, the diffusion rates of these alkanes decreases and exhibits a minimum for C_8 chain lengths. For C_9 and larger, the diffusion coefficients begin to rise, suggesting that the lack of C_{5-8} products in ERI-type zeolites is because these chains diffuse too slowly to escape the zeolite without cracking into smaller products. Using TST simulations, Dubbeldam and Snurr predicted the D^{self} values of several linear alkanes in ERI-, CHA- and LTA-type zeolites.[288] The authors showed that the simulation results are able to reproduce the chain length dependence and location of the maxima observed by Gorring in ERI- and CHA-type zeolites. Analysis of the free energy difference as a function of chain length shows that the free energy minimum decreases for chain lengths that are commensurate with the size of the cage. By increasing the chain length further, part of the alkane is forced to pass through a window into a second

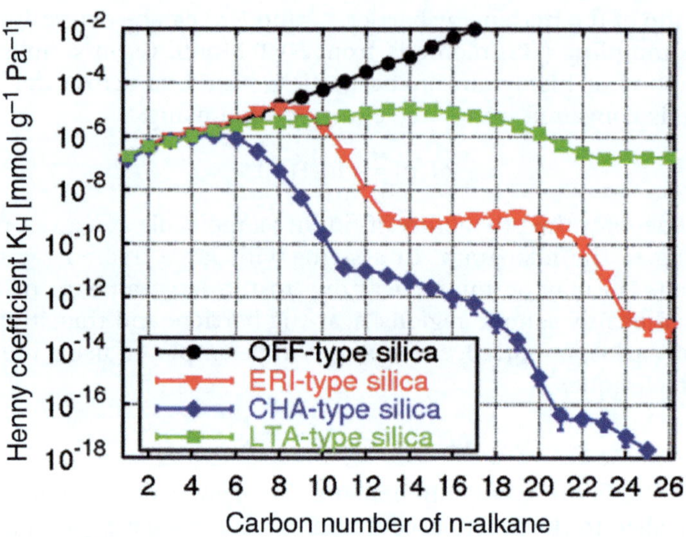

Figure 4.11 Henry coefficients in OFF-, ERI-, CHA- and LTA-type silica as a function of chain length at 600 K.
Reproduced from ref. 288 with permission from American Physical Society, Copyright 2003.

cage, resulting in an increased adsorption enthalpy and decreased Henry coefficient. It should be noted that this trend is characteristic for cage/window-type zeolites; results for the OFF-type zeolite (a channel) show a monotonic linear increase in the Henry coefficient as a function of chain length (Figure 4.11). In addition to a C_{10-12} product maximum, the cracking of C_{36} alkanes using ERI-type zeolites is found to exhibit additional maxima for products at C_{15-17} and C_{24-26} chain lengths.[286] Dubbeldam and co-workers suggested that very long chains adsorb partially into a cage near the outer surface and are split into smaller products. Afterwards, the non-adsorbed part can end up in the product or re-adsorb and undergo another cracking process. The maximum at C_{10-12} is because these products can rapidly diffuse away and escape complete consumption into smaller products.

TST has also been used to study other classes of nanoporous material. Liu *et al.* studied the impact of pore size and MOF interpenetration on the H_2 diffusivity.[289] A total of 10 MOFs were studied including six non-interpenetrated and four interpenetrated IRMOFs. MD simulations of the non-interpenetrated materials show that D^{self} values are 2 to 3 times larger than their interpenetrated counterparts and increase as a function of the pore size. This suggests that catenation imposes a large restriction on the motion of H_2. In addition, TST simulations were employed for **IRMOF-15** and **IRMOF-16**. In this work, the reaction coordinate describes the position of a H_2 molecule passing from one pore to another along the *z*-direction. Comparison of the free energy profile for **IRMOF-15** and **IRMOF-16** shows

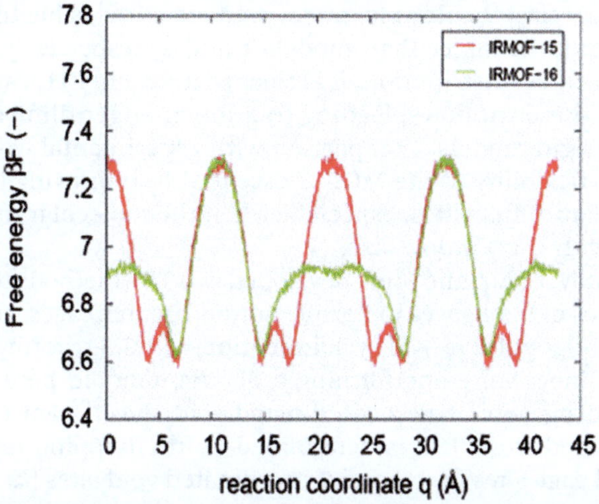

Figure 4.12 Free-energy profiles $F(q)$ of hydrogen in **IRMOF-15** and **IRMOF-16** at room temperature and infinite dilution. The reaction coordinate is chosen along the z-direction. For **IRMOF-15** the pore centres are located at 5.36, 16.09, 26.82 and 42.92 Å, while for **IRMOF-16** the pore centres are located at 0, 21.49 and 42.98 Å.
Reproduced from ref. 289 with permission from PCCP Owners Society.

that catenation has little impact on the overall barrier heights but introduces a large number of barriers per unit cell (Figure 4.12). From the MD simulations, it is shown that the H_2 diffusivity in **IRMOF-15** is about half of that in **IRMOF-16** in the limit of zero-loading. It is clear from the free energy profiles that the trend in diffusivities is a result of the larger number of barriers per unit cell in **IRMOF-15** due to catenation which decreases the lattice distance between the different pores along the reaction coordinate.

Haldoupis *et al.* employed molecular simulations to study the influence of framework flexibility on molecular diffusion in MOFs.[290] In particular, they focused on **ZIF-8** which possesses a flexible imidazolate linker that can rotate and alter the window diameter between two cages. Using *ab initio* MD (AIMD), the authors sampled 8000 window configurations over a short 1 ps length simulation. These were used to study the molecular diffusivities of H_2, He, CH_4, Ar and Xe in the limit of zero loading by calculating the TST hopping rates of each adsorbate through the pore windows. The average hopping rate across 8000 window configurations was used to calculate the molecular diffusivity. These were compared with molecular diffusivities calculated in three rigid models: (1) a structure based on neutron powder diffraction data, (2) a DFT-optimised structure and (3) a structure defined by the time-averaged atomic positions from the AIMD simulations. As H_2, He and Ar are all smaller than the size of the window, the differences in molecular diffusivity between the rigid and flexible models is fairly negligible. However, for CH_4 and Xe whose molecular sizes are commensurate with the average window size, the flexible model predicts a D^{self} value that is 5 and 13 orders of magnitude larger than models 1 and 2, respectively. By analysing over the full window distribution, it is clear that Xe and CH_4 can occasionally pass through these windows, leading to a much larger diffusivity than predicted by the rigid models. Comparison with experimental work shows that including the flexibility of the MOF is essential to improving the prediction of the molecular diffusivities, especially when the molecular dimensions are comparable with the window size.

In a later study, Camp and Sholl developed two TST methods to compute the self-diffusivities of light gases in flexible porous organic cages, such as **CC3**.[291] **CC3** is the product of a [4+6] cycloimination of 1,3,5-triformylbenzene and (*R*,*R*)-1,2 diaminocyclohexane forming a 3D diamondoid pore network with 4-coordinated cage sites and 2-coordinated void sites.[292] For **CC3**, the value of D^{self} calculated from TST is determined by the hopping rates out of the 4-coordinated cage sites ($k_{c\rightarrow v}$) and 2-coordinated void sites ($k_{v\rightarrow c}$):[291]

$$D^{\text{self}} = \frac{1}{6}\lambda^2[(4k_{c\rightarrow v}P_c) + (2k_{v\rightarrow c}P_v)]$$

where P_c and P_v are the equilibrium probabilities of occupying the cages or voids, and λ is the hopping distance between the cages and voids. As such, the 1D reaction coordinate lies along the vector connecting the centres of mass in two adjacent cages. Both P_c and P_v are directly proportional to the inverse of the hopping rate meaning that:[291]

$$D^{\text{self}} = \frac{1}{6}\lambda^2\left[\left(4\left(\frac{k_{v\rightarrow c}}{2k_{c\rightarrow v}+k_{v\rightarrow c}}\right)k_{c\rightarrow v}\right) + \left(2\left(\frac{k_{c\rightarrow v}}{2k_{c\rightarrow v}+k_{v\rightarrow c}}\right)k_{v\rightarrow c}\right)\right]$$

Initially, Camp and Sholl performed TST simulations to determine $k_{c\rightarrow v}$ and $k_{v\rightarrow c}$ in a rigid 2×2×2 homochiral **CC3-R** supercell. The free energy profile, $F(q)$, along the reaction coordinate, q, was determined from particle insertions on a number of regularly spaced square grids that were perpendicular to q. To study the influence of flexibility, two approaches were developed based on implicit ligand sampling (ILS) and umbrella sampling (US). In the ILS approach, the distribution of cage window sizes in empty **CC3** were determined from classical MD simulations employing the cage specific force-field (CSFF).[293] This is similar to the methodology of Haldoupis *et al.*, who studied TST hopping rates in flexible **ZIF-8**.[290] For each of the 1000 snapshots, independent cage to void ($k_{\text{cage 1}\rightarrow\text{void}}$ and $k_{\text{cage 2}\rightarrow\text{void}}$) and void to cage ($k_{\text{void}\rightarrow\text{cage 1}}$ and $k_{\text{void}\rightarrow\text{cage 2}}$) hopping rates were determined for each of the 16 cage–void–cage pairs. This is because the cages in each cage–void–cage pair may have different window sizes which changes $F(q)$ and hence, the hopping rate. In one snapshot, the free energy barrier for a Xe atom hopping between cage 1 and the void was found to be approximately 60 kJ mol^{-1} larger than the hop between cage 2 and the void. As such, $k_{\text{cage 1}\rightarrow\text{void}}$ was almost 10 orders of magnitude smaller than $k_{\text{cage 2}\rightarrow\text{void}}$, highlighting the impact which framework flexibility can have on diffusion.

Within the ILS approach, the flexible nature of **CC3** is captured by assessing the hopping rate across a large number of rigid structures. While these may cover a wide distribution of window sizes, they correspond to those found in an empty **CC3** framework. Therefore, the ILS approach assumes that the window dynamics are not strongly influenced by the presence of the adsorbates. To assess the validity of this assumption, it is necessary to calculate a free energy profile which accounts for the impact of host–guest interactions. In principle, this can be sampled from a long MD simulation of the flexible material, however, it is much more computationally efficient to employ the US approach, which places a series of harmonic springs (or umbrellas) along q. These restrain an adsorbate to a defined region of interest, allowing the nearby configurational space to be sampled more efficiently. For each umbrella, an adsorbate is placed at the centre of the spring and a short MD simulation is employed to sample the positions of the adsorbate (Figure 4.13a). The histograms from each umbrella can then be overlapped using the weighted histogram analysis method (WHAM)[294] to obtain $F(q)$. Figure 4.13b shows $F(q)$ for a Xe atom in **CC3** at 300 K. The cage

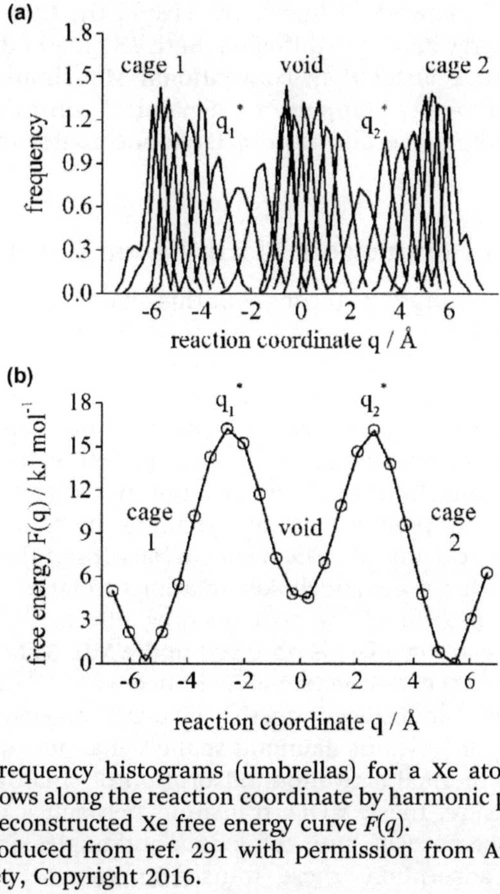

Figure 4.13 (a) Frequency histograms (umbrellas) for a Xe atom confined to 24 windows along the reaction coordinate by harmonic potentials at 300 K. (b) Reconstructed Xe free energy curve $F(q)$.
Reproduced from ref. 291 with permission from American Chemical Society, Copyright 2016.

to void free energy barrier of 16 kJ mol^{-1} is approximately 8 kJ mol^{-1} lower than predicted in the experimental structure. This result suggests that Xe hopping occurs predominantly in conformations where the **CC3** window is much wider than in the experimental structure, leading to a lower energy barrier.

In comparison with TST simulations employing the experimental structure of **CC3**, the ILS and US approaches predict higher D^{self} values for CH_4, CO_2, Xe, Rn, SF_6 and CS_2. This is most notable for CS_2 and SF_6, which are considered immobile in the experimental structure, but are able to freely diffuse in the flexible structure. For Kr, CH_4 and CO_2, the values of D^{self} are large enough to be studied directly from MD simulations. The diffusivities predicted by the ILS and US approaches are in good agreement with MD but generally overpredict D^{self} by factors between 1.2 and 2.4. This is attributed to the fact that the TST simulations are not corrected for re-crossing events, such that κ is assumed to be equal to 1. While the ILS and US methods generally predict similar values of D^{self} for Kr, CH_4, CO_2, Xe and Rn, noticeable differences are observed for CS_2 and SF_6. For these adsorbates, the US method predicts a significantly higher D^{self}, suggesting that their interactions with the framework influence the size of the **CC3** window and the corresponding energy barrier to diffusion. Both TST methods are found to be orders of magnitude faster than conventional MD simulations, and with additional simulations to compute the dynamical correction factors, show promise in describing the diffusion of large adsorbates in **CC3** and other porous materials.

4.5.2 Impact of Framework Flexibility on Guest Diffusion

One of the main assumptions throughout this chapter is that simulations of adsorption and diffusion in porous materials do not account for host dynamics or structural flexibility (*i.e.* the porous material is modelled as a rigid entity). This assumption drastically improves the computational cost of these calculations and is reasonable for many porous solids which are expected to have negligible changes to their structure in response to adsorption or diffusion. However, there are a number of porous materials that are characterised by their structural dynamics in response to different physical and chemical stimuli. The degree of structural flexibility can range from small local vibrations and linker rotations to large phase transitions such as swelling, 'breathing' and 'gate-opening' effects.[295–297] An archetypal example of the breathing effect is observed in the **MIL-53** MOF family. These contain a series of 1D chains where metals such as Al,[298] Ga,[299] In,[300] Fe,[301] V,[302] Cr[303] and Sc[304] are bridged together by μ-OH moieties and carboxylate ligands, affording a series of diamond-shaped channels in the shape of a wine rack. Upon thermal activation, adsorption of small molecules or application of pressure, these MOFs transition between a closed and open phase in which the porosity, unit cell length and angles drastically change. In the case of adsorption, these transitions can be observed on the adsorption isotherm by sharp inflections and hysteresis.[295] In recent years, molecular simulations have been employed to study these structural transitions using *ab initio* methods and classical MD.[305–309]

To study how structural flexibility can impact the guest adsorption and diffusion behaviour of porous solids, a number of groups have developed fully flexible models to describe the motion of these materials. Amirjalayer *et al.* extended the MM3 force-field to study self-diffusion of benzene in a flexible model of **IRMOF-1.**[310,311] At 300 K, the average D^{self} value of $(2.49 \pm 0.29) \times 10^{-9}$ m^2 s^{-1} is found to be in very good agreement with the experimental values of 1.8×10^{-9} m^2 s^{-1} and 2.0×10^{-9} m^2 s^{-1}. When employing a rigid model, an average D^{self} value of $(1.95 \pm 0.2) \times 10^{-8}$ m^2 s^{-1} is obtained, corresponding to an order of magnitude larger than the experimental values. These findings are consistent with the trend in diffusion activation energies which are determined to be 13.8 ± 1.3 kJ mol^{-1} and 4.01 ± 0.78 kJ mol^{-1} for the flexible and rigid models, respectively. An analysis of the probability densities in the flexible model indicates that benzene molecules typically occupy the corners of the MOF near the ZnO_4 clusters, whereas in the rigid model, a much higher density of benzene molecules is found in the centre of the pores. These differences are attributed to the correlated motion between the framework and adsorbates, which is not accounted for by the rigid model. A similar conclusion was later corroborated by Greathouse *et al.*, who used parameters from the CVFF and DREIDIING force-field to develop a flexible model for zinc carboxylate-based MOFs.[312]

Yang *et al.* studied the effects of framework flexibility on gas diffusion in **UiO-66.**[313] This MOF consists of $Zr_6O_4(OH)_4$ clusters that are coordinated by terephthalate ligands in an **fcu** topology. This contains a series of tetrahedral-shaped pores that surround a larger octahedral-shaped pore. A fully flexible model of **UiO-66** including bonded and non-bonded terms was developed using parameters from the CVFF and DREIDING force-fields. At a loading of 4 CH_4 molecules per unit cell, Yang and co-workers obtained D^{self} values of 2.8×10^{-10} m^2 s^{-1} and 1.0×10^{-9} m^2 s^{-1} for a rigid and a flexible framework, respectively. Calculations over several temperatures were used to compute activation energies, corresponding to 9.0 kJ mol^{-1} and 6.0 kJ mol^{-1} for the rigid and flexible frameworks, respectively. As such, the higher mobility in the flexible model is the result of a lower activation barrier for inter-cage hopping, resulting in lower cage residence times. In comparison with the experimental D^{self} value of 2.0×10^{-9} m^2 s^{-1}, the flexible model was in much better agreement than compared to the rigid model which under-predicted by nearly an order of magnitude.

Hertäg employed the AMBER[314] and DREIDING potentials to develop flexible models of **ZIF-8.**[315] This material is of particular interest due to its flexible window size controlled by the rotation of its 2-methylimidazolate ligands. In this work, the impact of the window flexibility on the diffusion rates of H_2 and CH_4 were studied using MD simulations. The authors find that the D^{self} values for H_2 and CH_4 are quite sensitive to the choice of force-fields, with the flexible DREIDING model generally predicting larger D^{self} values than the flexible AMBER model. This is attributed to two factors: (1) The DREIDING potential predicts a different angle around the Zn and N atoms than observed in XRD experiments, resulting in a larger window size. (2) The Zn-N force constants are very stiff in the DREIDING potential, resulting in slow rotational dynamics of the linkers. As for H_2, there is little difference between the diffusivities in the rigid and flexible models due to its small size relative to the pore window. As such, the energy barrier for inter-cage hopping is very small. As CH_4 is similar in size to the pore window diameter, a large difference in diffusivity is recorded for the flexible and rigid models. In the rigid model, the CH_4 molecules are trapped in the same cage for the entire MD simulation resulting in a D^{self} value smaller than 10^{-13} $m^2 s^{-1}$. When framework flexibility is modelled, the CH_4 molecules are able to diffuse through the windows and explore different cages, leading to a D^{self} value between 10^{-9} $m^2 s^{-1}$ and 10^{-11} $m^2 s^{-1}$. The authors conclude that lattice flexibility plays an important role in the diffusion of molecules that are commensurate in size to the pore windows.

In a later study, Zhang *et al.* computed the diffusivities and adsorption isotherms for CO_2 and CH_4 in **ZIF-8**.[316] Results from a rigid model are compared to a flexible model that describes the bonded and non-bonded terms using AMBER and UFF.[317] To model the flexibility of **ZIF-8** in the calculation of the adsorption isotherms, a hybrid approach using Gibbs ensemble Monte Carlo (GEMC) and MD was employed. For both gases, agreement with the experimental data is very good and the impact of flexibility is found to have a negligible effect on the adsorption isotherm at pressures up to 100 bar. In contrast, significant differences between the rigid and flexible models are found for the CO_2 and CH_4 diffusivities. In agreement with previous works,[315,318] CH_4 is found to be trapped inside a single cage and cannot pass through the pore windows of the rigid model. While the smaller CO_2 molecules are able to hop between different cages, it is found that a diffusive region of the Einstein relation is not observed within the timescale of the simulation. In the flexible model, the CH_4 and CO_2 D^{self} values are 6.2×10^{-10} $m^2 s^{-1}$ and 6.4×10^{-11} $m^2 s^{-1}$, respectively, at a loading of 3 molecules per unit cell. These match reasonably well with the experimental values of approximately 1.5×10^{-10} $m^2 s^{-1}$,[318] suggesting that the incorporation of flexibility is an important factor when the guest molecules are commensurate to the pore size.

Parkes *et al.* investigated Ar, Kr and Xe diffusion in both **ZIF-8** and **CuBTC** using flexible models that were parameterised from general force-fields and quantum mechanical calculations.[319,320] They conclude that D^{self} is highly dependent on a number of factors including the size of the adsorbate, the inter-cage window size and the inclusion of structural flexibility. These factors are very important for Kr diffusion, which is greatly enhanced by the use of a flexible model as the inter-cage windows can adapt to accommodate the larger size of the Kr atoms. Similar results were obtained by Kopelevich and Chang when studying all-silica sodalites which are isomorphic to **ZIF-8**.[321]

In addition, the authors discovered that the use of different rigid structures has a large impact on the resulting diffusion properties of these gases and advised that a range of initial structures should be employed to identify the importance of structural variation on guest diffusion properties.

Many studies from the Keskin group have explored the impact of structural flexibility on the membrane separation performance of HTS candidates.[322–324] Based on rigid structures, many of these screenings identify materials with narrow pores and low porosities to be excellent candidates for separations. When using a flexible model, the authors find that gas permeabilities are greatly affected while membrane selectivities remain fairly constant in comparison with the rigid structure results. This leads to the conclusion that the rigid framework approximation can provide accurate initial estimates of the selectivity in these materials. In comparison to membranes with large pores, structural flexibility is found to have negligible impact on both gas permeability and selectivity.

In summary, it is clear that the impact of flexibility on diffusion is an important issue even for structures that do not possess significant dynamical properties. This mainly affects materials with narrow pores or when the adsorbate is commensurate to the pore size. Many of these studies employ general force-field parameters to describe the bonded interactions in the framework. For more complex solids, it may be necessary to develop new parameters that are not currently described by the force-field which can be a time-consuming process. Therefore, the future development of general force-fields that are transferable to a range of flexible porous solids is highly desirable for the accurate description of guest transport properties.

4.5.3 Modelling Gas Permeability in Porous Membranes

Membrane separations offer a cheap and efficient alternative to traditional separation techniques such as cryogenic distillation, pressure swing adsorption and liquid absorption.[325] In these separations, a feed gas containing a mixture of adsorbates is driven through a membrane using a differential pressure drop. The feed gas is separated into permeate and retentate phases by exploiting differences in physical properties or membrane–guest interactions of its components. The permeate phase corresponds to the component of the feed gas that passes through the membrane, while the retentate is the component adsorbed and retained by the membrane. The overall separation performance of this process is governed by the permeability of each component and the selectivity of the membrane. The permeability describes how quickly a gas component passes through a membrane; it depends on the membrane's affinity for the gas component and the rate of which it diffuses through the pores. The selectivity is a measure of a membrane's relative affinity for one component over another. Membranes that simultaneously possess high selectivity and permeabilities are highly desirable. As such, the choice of material used to synthesise the membrane plays an important role. Polymer membranes are most commonly used at industry level as they are cheap, easily scalable and can be fabricated into hollow fibres and sheets.[326] Despite poor thermal and chemical stabilities, the main drawback of polymer membranes is their trade-off relationship between guest permeability and selectivity, *i.e.* polymers with high selectivity tend to possess low permeability and vice-versa. After plotting a large amount of experimental data for various small gas separations, Robeson observed an upper bound to this relationship, establishing a boundary to the overall performance of these materials.[327,328] An example of this can be observed in Figure 4.14 depicting the upper bound for O_2/N_2 separations.

To overcome the stability and trade-off limitations of polymer membranes, the development of inorganic membranes using carbonaceous, silica-based or zeolitic materials has proven valuable. These materials typically possess much larger permeabilities and selectivities and can be used in much harsher conditions. In addition to these, MOFs have received much attention due to their tuneable structural properties, making them versatile and a highly competitive choice for a range of gas separations. However, many of these materials are fragile and suffer from poor mechanical properties which increases their processing costs.[325] Approaches to combine the high permeability and selectivity of inorganic membranes with the processability of

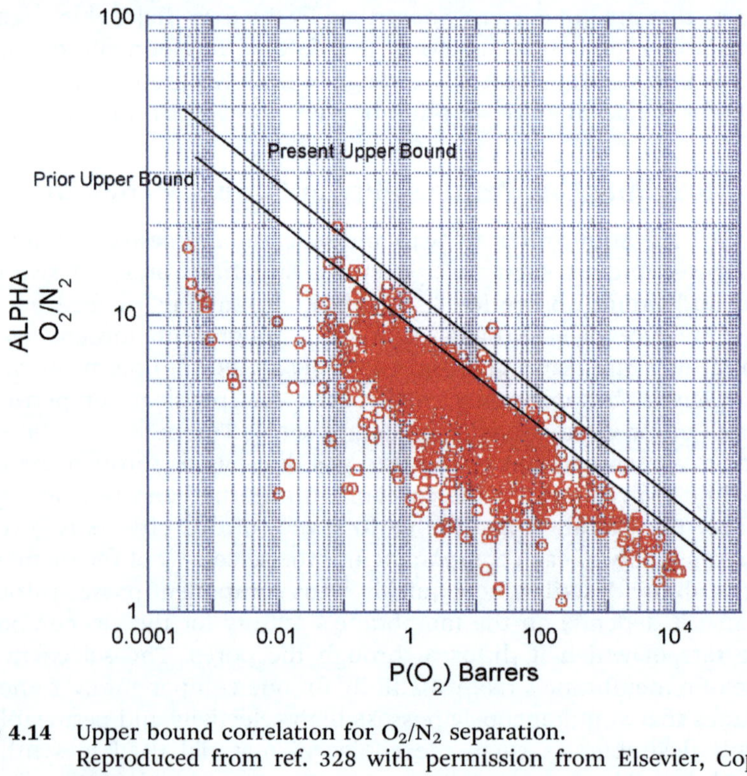

Figure 4.14 Upper bound correlation for O_2/N_2 separation.
Reproduced from ref. 328 with permission from Elsevier, Copyright 2008.

polymer membranes has recently been achieved by dispersing inorganic materials in polymer matrices, known as mixed matrix membranes (MMMs).[329] From an experimental view point, identifying promising materials for gas separation is a time-consuming and expensive task, requiring knowledge of the relationships between the structural and permeation properties of the membrane. Therefore, molecular simulations are often employed to improve our understanding of gaseous transport behaviours in membranes.

Many experimental studies of porous membranes report the single-component permeabilities of different gases where the ideal separation performance of a membrane is determined using ratios of the single-component data. The permeability of gas species, i, can be defined as:

$$P_i = \frac{N_i}{\Delta p L}$$

where N_i is the molar flux of species i, Δp is the pressure drop across the membrane and L is the membrane thickness. In many simulations, steady state fluxes are calculated using a shell model which assumes that the concentration gradient, ∇c_i, can be approximated from the gas loadings at the feed and permeate boundaries:[263,330,331]

$$\nabla c_i \cong \frac{c_i^{\text{permeate}} - c_i^{\text{feed}}}{L}$$

Assuming there are no surface resistances to the transport of gases entering or leaving the membrane, these gas loadings can be calculated from the equilibrium adsorption isotherm, defined from single-component GCMC simulations. Another approximation made by the shell model is that D^{t} is assumed to be constant at the adsorbate concentration defined by the arithmetic mean of the boundary loadings.

While single-component permeabilities can be accurately determined using this methodology,[330] separation applications typically involve mixtures of gases. It is well understood that the adsorption and diffusion properties in mixtures can be very different to their single-component behaviour.[332] As such, it is necessary to determine multi-component adsorption isotherms and use a theoretical framework that can accurately describe mixture diffusion and correlation effects between different species. GCMC simulations can be employed to determine the multi-component adsorption isotherms of a mixture or from single-component data using IAST. As mentioned previously, multi-component transport properties are defined by a matrix of diffusivities that can be calculated from MD simulations.[265,267–269,276,333] However, a number of studies have found it more convenient to employ the SSK mixing theory of Skoulidas, Sholl and Krishna[270] to predict the transport diffusivities from knowledge of the single-component D^{self} and D^{c} ($D^{\text{MS}}{}_i$) diffusivities.

4.5.3.1 Screening Membranes for Separation

In order to evaluate the separation performance of different porous solids, it is important to employ efficient methods that can rapidly estimate the permeability and selectivity of a membrane. In previous work, Krishna and van Baten employed the MS formalism to estimate the particle fluxes and membrane permeation selectivity (S_{perm}) of CO_2 and CH_4 across 12 zeolite membranes.[334] By assuming that the permeate side of a membrane is under vacuum conditions such that the downstream molecular loadings are negligibly small, Krishna and van Baten showed that for an equimolar mixture of components i and j, $S_{perm,i/j}$ can be approximated using the following equation:[334,335]

$$S_{perm,i/j} = S_{ads,i/j} \times S_{diff,i/j} = \frac{c_i}{c_j} \times \frac{D_i^{self}}{D_j^{self}}$$

where $S_{ads,i/j}$ and $S_{diff,i/j}$ are the adsorption and diffusion selectivities, respectively. $S_{ads,i/j}$ is calculated from the ratio of the adsorbate molar loadings of each species at the feed side of the membrane. $S_{diff,i/j}$ is calculated from the ratio of the mixture D^{self} values. In addition to S_{perm}, they derived an expression for the permeability of each species:

$$P_i = \phi D_i^{self} \frac{c_i}{f_i}$$

where ϕ is the materials void fraction and P_i, D_i^{self}, c_i and f_i are the permeability, mixture self-diffusion coefficient, equilibrium gas loading and bulk phase partial fugacity of component i. The main advantage of these approximations is that they avoid the calculation of the D^t which is more computationally expensive to evaluate than D^{self}. In a later study, Keskin and Sholl modified these expressions to account for mixtures of arbitrary molar ratios:[336]

$$S_{perm,i/j} = \frac{c_i/c_j}{y_i/y_j} \times \frac{D_i^{self}}{D_j^{self}}$$

where y_i and y_j are the mole fractions of species i and j. Keskin and Sholl validated this approximative method by comparing their predictions of $S_{perm,i/j}$ for CO_2/CH_4, CH_4/H_2 and CO_2/H_2 mixtures in **IRMOF-1** and **CuBTC** membranes with previous work which employs more detailed calculations that evaluate mixture-adsorption isotherms and D^t across various gas concentrations. While the approximative method accurately predicts membrane permeabilities for most mixtures, poor agreement is found for CO_2/H_2 mixtures in **CuBTC**. In general, the approximative method is anticipated to perform better for systems where IAST can accurately predict the isotherm of the binary mixture. If this is not the case, a more rigorous set of detailed calculations are required to study the permeation process, including the evaluation of mixture transport diffusivities across several gas loadings.

Over the past decade, the approximative methods of Krishna, van Baten, Keskin and Sholl have been employed to study various gas separations in MOF,[322,337–343] ZIF,[344–346] PCN[347] and COF[161,348] membranes. Atci and Keskin studied CH_4/H_2, CO_2/CH_4 and CO_2/H_2 separations over a comprehensive set of 15 ZIFs.[344] A direct comparison of the predicted gas mixture permeances for CO_2/H_2, N_2/H_2, CH_4/H_2, CO_2/CH_4 and CO_2/N_2 mixtures was made with experimental measurements on **ZIF-90** and **ZIF-69** membranes, showing reasonable agreement. For each of the CH_4/H_2, CO_2/CH_4 and CO_2/H_2 separations, a Robeson plot of the S_{perm} *versus* permeability was used to compare the ZIFs with other zeolite and MOF membranes (Figure 4.15). Two ZIFs and two zeolite membranes, **ZIF-65**, **ZIF-90**, LTA and CHA are found to possess $S_{perm,CH_4/H_2}$ less than 0.1 due to their molecular sieving properties (Figure 4.15a). While **ZIF-2** and **ZIF-79** show some of the highest S_{perm} and P_{CH_4} values among ZIF materials, they are still outclassed by CNT materials due to their specific PES which facilitates high P_{CH_4} and S_{perm}.[268] For CO_2/CH_4 separations, Robeson's 2008 upper bound was used to compare the separation performance of each membrane, relative to the best polymeric membranes (Figure 4.15b). Of the 15 ZIFs studied, **ZIF-90** was the only material which exceeds the upper bound due to its large S_{perm} value. A number of zeolitic and MOF membranes exhibit superior $S_{perm,CO_2/CH_4}$ and P_{CO_2} than the ZIF membranes, with **Mg-MOF-74** and **Zn-MOF-74** lying just above the upper bound. For the CO_2/H_2 separations, the molecular sieving nature of **ZIF-65** and **ZIF-90** is found to lead to high $S_{perm,CO_2/H_2}$. Out of the remaining ZIFs membranes, **ZIF-2** exhibits the largest $S_{perm,CO_2/H_2}$ and P_{CO_2}, surpassing several zeolite membranes, **MOF-177** and **CuBTC**. However, this material is outperformed by the **MOF-74** membranes which possess P_{CO2} greater than 100 Barrer and $S_{perm,CO_2/H_2}$ greater than 10. In summary, the authors noted that **ZIF-2** may be a very promising candidate for membrane separations of CH_4/H_2 and CO_2/H_2 mixtures.

Erucar and Keskin examined CH_4/H_2 separations in 173 MOFs from the CoRE MOF database.[323] An initial set of 250 structures was reduced to 159 structures by focusing on MOFs with a PLD greater than 4 Å. An additional 13 MOFs that possess narrow PLDs between 2.99 Å and 3.96 Å were also included as they were previously identified as top-performing adsorbents for separating CH_4/H_2 mixtures.[337] In all 172 MOF membranes, the $S_{perm,CH_4/H_2}$ values are found to be lower than the $S_{ads,CH_4/H_2}$ values. This is because each MOF interacts with CH_4 more strongly than H_2, leading to high $S_{ads,CH_4/H_2}$. However, as H_2 is smaller, lighter and interacts more weakly with the MOFs than CH_4, it can rapidly diffuse through the pores, leading to $S_{diff,CH_4/H_2}$ between 0.007 and 0.81 favouring H_2. The top 10 CH_4 selective candidates were identified as those possessing the highest $S_{perm,CH_4/H_2}$, ranging between 22 and 130. Some of these materials possess 1D pores which strongly confine the CH_4 molecules, leading to very large $S_{ads,CH_4/H_2}$ values up to 160. Interestingly, 16 MOFs were found to be H_2 selective with $S_{perm,H_2/CH_4}$ values as large as 7.92. These systems all possess 3D pore networks with high porosities that favour fast diffusion of the H_2

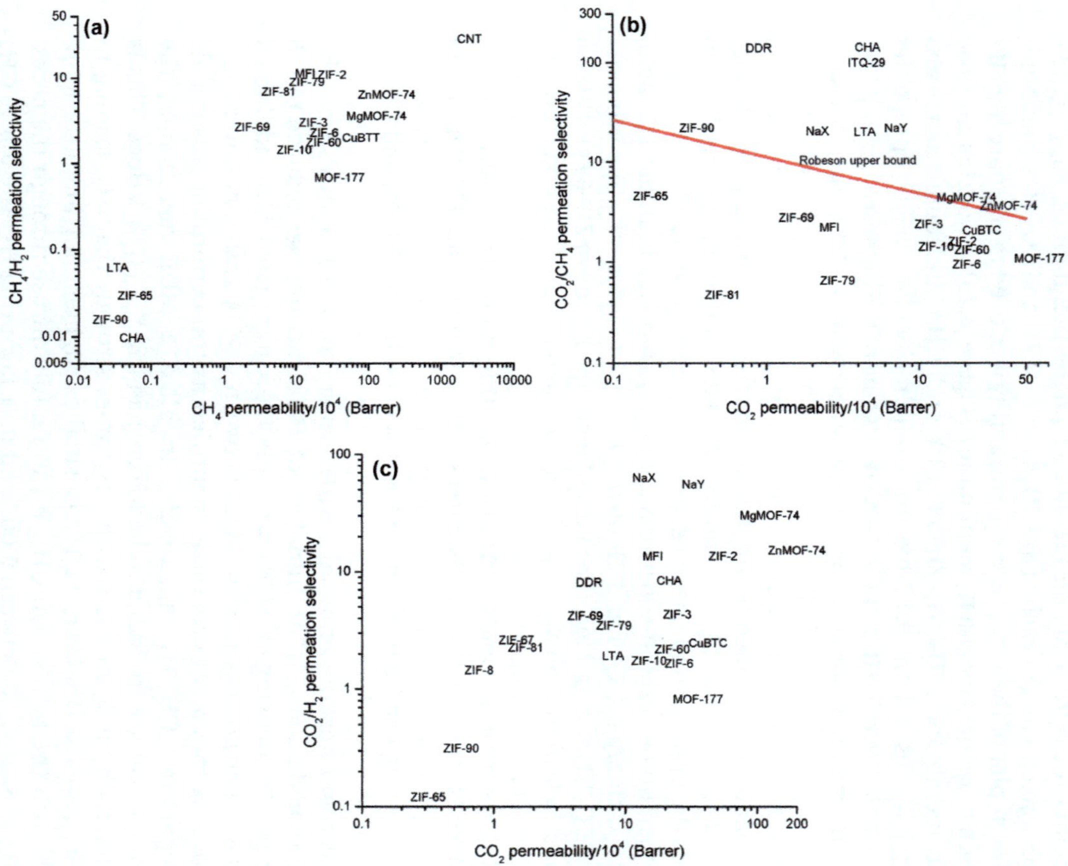

Figure 4.15 Permeation-based separation performance of ZIFs for (a) CH_4/H_2, (b) CO_2/CH_4 and (c) CO_2/H_2 mixtures. The bulk gas mixtures are CH_4/H_2 : 10/90, CO_2/CH_4 : 10/90 and CO_2/H_2 : 1/99 for ZIFs at 298 K and (a) 50/50, (b) 50/50 and (c) 15/85 for zeolites at 300 K. Each ZIF name represents a data point, which indicates its separation performance. Reproduced from ref. 344 with permission from American Chemical Society, Copyright 2012.

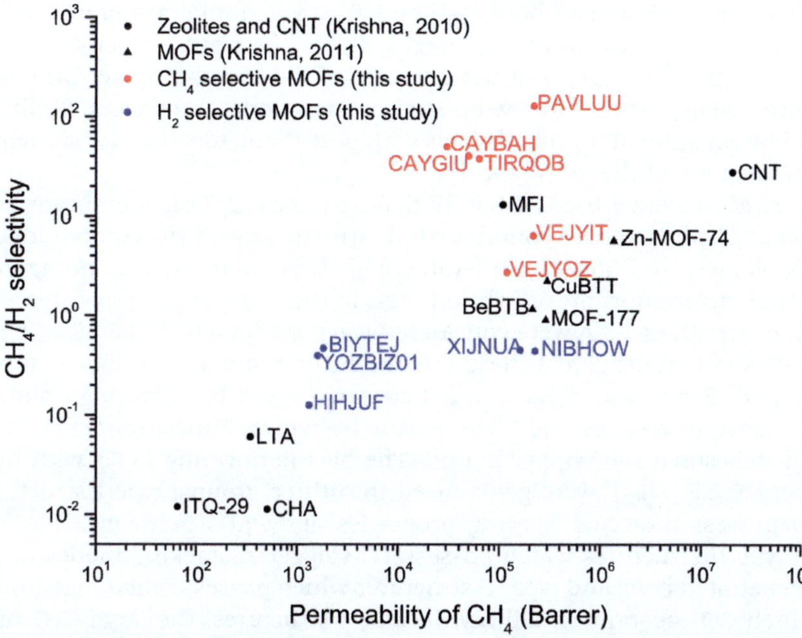

Figure 4.16 Comparison of selectivity and CH_4 permeability of top-performing MOF membranes with zeolite and CNT membranes. Bulk composition of CH_4/H_2 is 50/50, pressure is 10 bar, temperature is 298 K for this study and 300 K for others.
Reproduced from ref. 323 with permission from Elsevier, Copyright 2016.

molecules, *i.e.* D^{self} greater than $10^{-3}\,cm^2\,s^{-1}$. The top five CH_4 selective and H_2 selective MOFs were compared with calculations by Krishna *et al.* on other zeolites, MOF and CNT membranes (Figure 4.16).[335,341] MOF membranes **PAVLUU**, **CAYBAH**, **CAYGIU** and **TIRQOB** possess some of the highest $S_{perm,CH_4/H_2}$ values which outclass other studied zeolites, MOF and CNT membranes. However, the large permeabilities of these materials were still rivalled by **Zn-MOF-74** and CNT membranes, with values ranging between 10^6 and 10^7 Barrer. In the special case of CNT membranes, the smoothness of the PES leads to much faster diffusion of the guests than in MOFs and zeolites, with D^{self} values ranging between 0.1 $cm^2\,s^{-1}$ and 1.0 $cm^2\,s^{-1}$.[268] In the final part of their study, the authors studied the impact of framework flexibility on the separation performance of two MOFs: one containing small pore diameters and one with large pore diameters. For the large pore system (PLD and LCD equal to 13.64 Å and 23.98 Å, respectively), framework flexibility has a negligible effect on both gas P_{CH_4} and $S_{perm,CH_4/H_2}$. However, for the small pore system (PLD and LCD equal to 4.09 Å and 4.45 Å, respectively), the framework flexibility causes a sharp decrease in the PLD to 2.42 Å, resulting in a much smaller diffusion rate. This effect decreases both P_{CH_4} and P_{H_2} by a similar amount, causing the $S_{perm,CH_4/H_2}$ values to change negligibly. While rigid framework simulations can provide a good initial estimate of the membrane performance, the authors suggest that follow-up studies that model framework flexibility should be considered for membranes with pore diameters that are similar to the dimensions of the guest species.

Yan *et al.* screened the CoRE COF database for CO_2/CH_4 membrane separations.[161] This database contains 41 3D structures with the **ctn**, **bor**, **dia** or **pts** topologies, and 268 2D structures with triangular, square, hexagonal, octagonal, heteromorphic or hybrid pores. In this study, only structures with a PLD greater than 3.3 Å were considered leaving a total of 298 COFs. GCMC and MD simulations reveal these membranes possess P_{CO_2} values between 10^3 and 10^6 Barrer, and $S_{perm,CO_2/CH_4}$ between 0.1 and 10^4, showing a similar performance to zeolites and MOFs in the literature. Furthermore, 12 COFs exceed Robeson's 2008 upper bound. The best performing COFs with high P_{CO_2} and $S_{perm,CO_2/CH_4}$ were generalised into three groups: type I structures which possess molecular sieving properties and PLDs between 3.3 Å and 3.8 Å; type II structures which possess 2D staggered stacking modes or 3D interpenetrated nets; and type III structures which possess functional groups with high CO_2 adsorption affinity. In type I structures, the large CH_4 molecules cannot freely diffuse through the membrane due to their large kinetic diameter ($d = 3.8$ Å) whereas the smaller CO_2 molecules are able to rapidly diffuse through the pores. In type II structures, the 2D staggered conformations or 3D interpenetrated nets cause a large reduction in the pore size, which lead to large $S_{ads,CO_2/CH_4}$ and $S_{diff,CO_2/CH_4}$. In some cases, these materials possess $S_{perm,CO_2/CH_4}$ values that are between 3 and 18 times larger than their eclipsed/non-interpenetrated counterparts. Finally, the large P_{CO_2} and $S_{perm,CO_2/CH_4}$ values in type III materials are driven by their large adsorption affinities for CO_2, resulting in very large $S_{ads,CO_2/CH_4}$ values. While this leads to a lower CO_2 diffusion rate and $S_{diff,CO_2/CH_4}$, the relative size of the $S_{ads,CO_2/CH_4}$ contribution is much larger than the $S_{diff,CO_2/CH_4}$ contribution, leading to a large $S_{perm,CO_2/CH_4}$.

For the 298 COFs studied in this work, analysis of the relationship between $S_{perm,CO_2/CH_4}$, the PLD and P_{CO_2} indicate that COFs with PLDs smaller than 10 Å typically show good performance. With this as a criterion, Yan and co-workers designed 290 new COF materials by functionalising a selection of existing CoRE COFs. These included functional groups such as $-Br$, $-CF_3$, $-CH_3$, $-Cl$, $-COOH$, $-F$, $-NH_2$, $-NO_2$, $-OH$ or $-SO_3H$ (Figure 4.17). The inclusion of such groups was expected to improve $S_{ads,CO_2/CH_4}$ by creating smaller pore diameters and strong adsorption sites. While this reduces the diffusion rate of CO_2, the CH_4 diffusion rate is also expected to fall as a result of the smaller pore diameters and larger numbers of adsorbed CO_2 molecules. The separation performance of the newly designed COFs shows remarkable results, with 137 of the 290 COFs surpassing Robeson's upper bound. $S_{perm,CO_2/CH_4}$ values range between 3.0 and 194, and P_{CO_2} values range between 5.1×10^3 Barrer and 5.19×10^5 Barrer. Of the 26 materials with

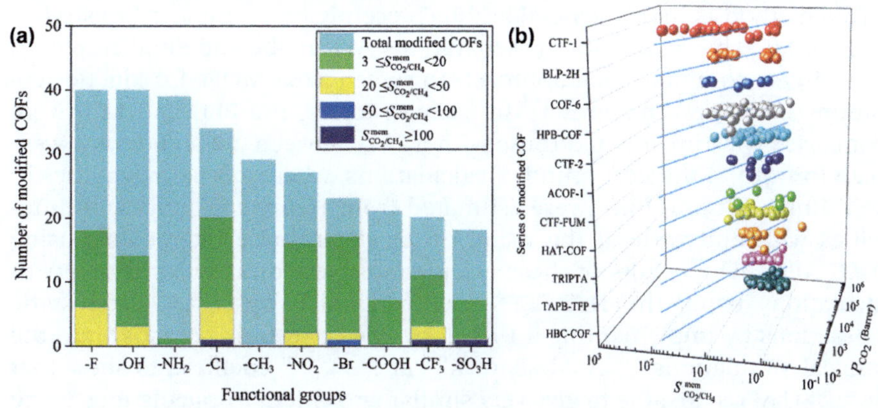

Figure 4.17 (a) Distribution of the modified COFs classified by the functional groups. The number of the functional groups is not the same for each column. It is the sum of the modified COFs with different numbers and sites of the corresponding functional group. (b) CO_2/CH_4 permeation selectivity and CO_2 permeability of the 290 modified COFs. The blue lines correspond to the 2008 Robeson's upper bound. Reproduced from ref. 161 with permission from American Chemical Society, Copyright 2019.

$S_{perm,CO_2/CH_4}$ greater than 20, 13 were found to possess a –F or –Cl functional group, highlighting the potential impact of these groups in the future design of COF-based membranes.

4.5.3.2 *Infinite Dilution Calculations*

As discussed in Section 4.4, a number of structural databases with tens of thousands of porous materials have been developed in recent years. These allow for much larger computational screenings to be performed which can evaluate structure–performance relationships over a diverse set of materials. With such a large number of potential candidates, it can be beneficial to determine an initial estimate of the membrane performance from calculations in the infinite dilution limit. This is an efficient way of rapidly screening materials and identifying top-performing candidates before employing a more detailed set of calculations. Haldoupis *et al.* demonstrated that in the infinite dilution limit, the permeability of a gas species, P_i^{inf}, can be estimated from its Henry coefficient, K_i^{inf}, and self-diffusivity, $D_i^{self,inf}$:[99]

$$P_i^{inf} = K_i^{inf} D_i^{self,inf}$$

Hence, the permeation selectivity in the limit of infinite dilution, $S_{perm,i/j}^{inf}$ is reduced to:

$$S_{perm,i/j}^{inf} = \frac{K_i^{inf} D_i^{self,inf}}{K_j^{inf} D_j^{self,inf}}$$

This method is highly general and has recently been employed to study a range of MOF-based membrane separations. Watanabe and Sholl employed a multiscale computational approach to screen 1163 MOFs for kinetic separations of CO_2/N_2 mixtures.[349] In each structure, the PLD and LCD were geometrically estimated and those with a PLD between 2.2 Å and 3.6 Å were taken forward to the next round of calculations, where gas permeabilities in the infinite dilution limit were estimated from Henry constants and diffusivities with and without the inclusion of electrostatic interactions using DDEC APCs. The results of these calculations were directly compared with calculations using the DFT-EPES method, which describes electrostatic interactions by implementing a DFT-derived electrostatic potential into the classical simulations.[350] Analysis of K^{inf} and $D^{\text{self,inf}}$ coefficients show that the DDEC APCs are able to give very similar agreement to calculations using the DFT-EPES. The dataset of structures was further refined to a total of 201 MOFs by including those with $D_{\text{CO}_2}^{\text{self,inf}}$ values greater than $10^{-8}\,\text{cm}^2\,\text{s}^{-1}$. In 90% of these materials, very large $S_{\text{ads,CO}_2/\text{N}_2}^{\text{inf}}$ values in excess of 400 are recorded, which is attributed to the larger quadrupole moment that exists in CO_2. Additionally, most MOFs possess $S_{\text{diff,CO}_2/\text{N}_2}^{\text{inf}}$ values between 0.1 and 10, and contain very narrow pores with LCDs smaller than 4 Å. Two exceptional candidates, **UFUNAK** and **OFIWIK**, are found to possess very large diffusion selectivities with $S_{\text{diff,CO}_2/\text{N}_2}^{\text{inf}}$ values of 441 and 104, respectively. However, the authors reiterated that these values may change when considering mixture effects or flexible frameworks that are not considered in these calculations. The predicted $S_{\text{perm,CO}_2/\text{N}_2}^{\text{inf}}$ values were used to form a Robeson plot and were compared against polymeric membranes using Robeson's 2008 upper bound. The top 10 identified materials possess $S_{\text{perm,CO}_2/\text{N}_2}^{\text{inf}}$ values between 1200 and 26 200 and $P_{\text{CO}_2}^{\text{inf}}$ values between 346 and 2.1×10^7 Barrer. To assess the impact of a real flue gas mixture, binary mixture adsorption isotherms were predicted for a 15/85 CO_2/N_2 mixture using single-component GCMC simulations and IAST. The results indicate that the high $S_{\text{ads,CO}_2/\text{N}_2}^{\text{inf}}$ values observed in the infinite dilution limit are also observed at higher pressures over a wide pressure range.

Altintas *et al.* employed similar calculations to estimate the H_2/CH_4 separation performance of 4240 MOF membranes.[351] These membranes were selected from the 54 808 structures in the MOF subset of the CSD as candidates with non-zero ASAs and PLDs greater than 3.8 Å. Of the 4240 structures, 3777 are identified to be selective towards CH_4, 68 possess little preference for either H_2 or CH_4 (*i.e.* $0.9 \leq S_{\text{perm,H}_2/\text{CH}_4}^{\text{inf}} \leq 1$), and 395 are selective towards H_2, with $S_{\text{perm,H}_2/\text{CH}_4}^{\text{inf}}$ values ranging between 0.0014 and 97, and $P_{\text{H}_2}^{\text{inf}}$ values ranging between 0.372 Barrer and 1.67×10^6 Barrer. A total of 1545 MOFs are found to exceed Robeson's 2008 upper bound highlighting their performance alongside polymeric membranes. Based on this rapid estimation of the separation performance, the top 10 materials with $P_{\text{H}_2}^{\text{inf}}$ greater than 10^4 Barrer and $S_{\text{perm,H}_2/\text{CH}_4}^{\text{inf}}$ greater than 10 were selected for a more detailed assessment of their separation performance using equimolar mixture GCMC and MD simulations at 1 bar and 298 K. The results from the calculations at infinite dilution conditions were in similar agreement to those obtained in the detailed calculations, which validates their use as a rapid tool for assessing the gas separation performance of porous solids. It should be noted that for separations at high pressures, the deviation between the detailed calculations and the infinite dilution predictions can become non-negligible as guest–guest interactions start to play a vital role.[337] Finally, the authors examined the structure–performance relationships of the materials in their study (Figure 4.18). These diagrams show that H_2/CH_4 separation performance is most promising in MOFs with PLDs between 3.8 Å and 6 Å, LCDs between 6 Å and 12 Å, ASAs smaller than 1000 $m^2 g^{-1}$, VFs between 0.5 and 0.75 and framework densities between 1.0 $g\,cm^{-3}$ and 1.5 $g\,cm^{-3}$. Furthermore, many of these MOFs are found to possess Cd, Cu and Zn.

In a later study, Daglar and Keskin employed a similar approach to screen 3806 MOF membranes for flue gas separations.[352] The top candidates were initially identified by rapidly estimating the separation performance of each membrane in the infinite dilution limit. Almost 77% of the total membranes were found to surpass Robeson's 2008 upper bound, with $S^{inf}_{perm,CO_2/N_2}$ values ranging between 1 and 2300, and $P^{inf}_{CO_2}$ values ranging between 10^3 Barrer and 10^9 Barrer. The top 15 membranes with $S^{inf}_{perm,CO_2/N_2}$ greater than 350 and $P^{inf}_{CO_2}$ greater than 10^6 Barrer were then studied using binary mixture GCMC and MD simulations for the industrial separation of a 15/85 CO_2/N_2 flue gas mixture at 1 bar and 298 K. In general, the selectivities and permeabilities at working conditions were lower than those predicted at infinite dilution conditions. However, these membranes are still able to compete with traditional zeolites and polymers under working conditions, with $S_{perm,CO_2/N_2}$ values ranging between 16.4 and 820, and P_{CO_2} values ranging between 1.19×10^5 Barrer and 1.95×10^6 Barrer. In the final round of calculations, H_2O was introduced into the mixture to create a 10/87/3 ternary $CO_2/N_2/H_2O$ separation. Although H_2O is a minor component, it is a highly competitive adsorbate leading to reductions in P_{CO_2} and P_{N_2} values when compared with the binary mixture and infinite dilution data. The inclusion of H_2O in these separations is crucial to obtain a realistic assessment of the separation performance under real working conditions. It is important to mention that these calculations do not provide any information about the relative stabilities of the membranes when exposed to H_2O, which should be further examined to evaluate the realistic performance of these materials. Analysis of the structure–performance relationships indicates that many of the top MOF membranes are found to possess narrow pore openings with low surface areas. Of the 3806 membranes that were studied, 190 are found to possess lanthanide metal centres. Although this is a relatively small percentage of the total dataset, it is surprising that 11 of the top 15 membranes are found to possess lanthanide metal centres in their structures. In summary, the results of the HTS indicate that lanthanide-containing MOFs with PLDs between 3.75 Å and 4.5 Å, and ASAs smaller than 1000 $m^2 g^{-1}$, are ideal candidates for membrane-based separations of flue gas.

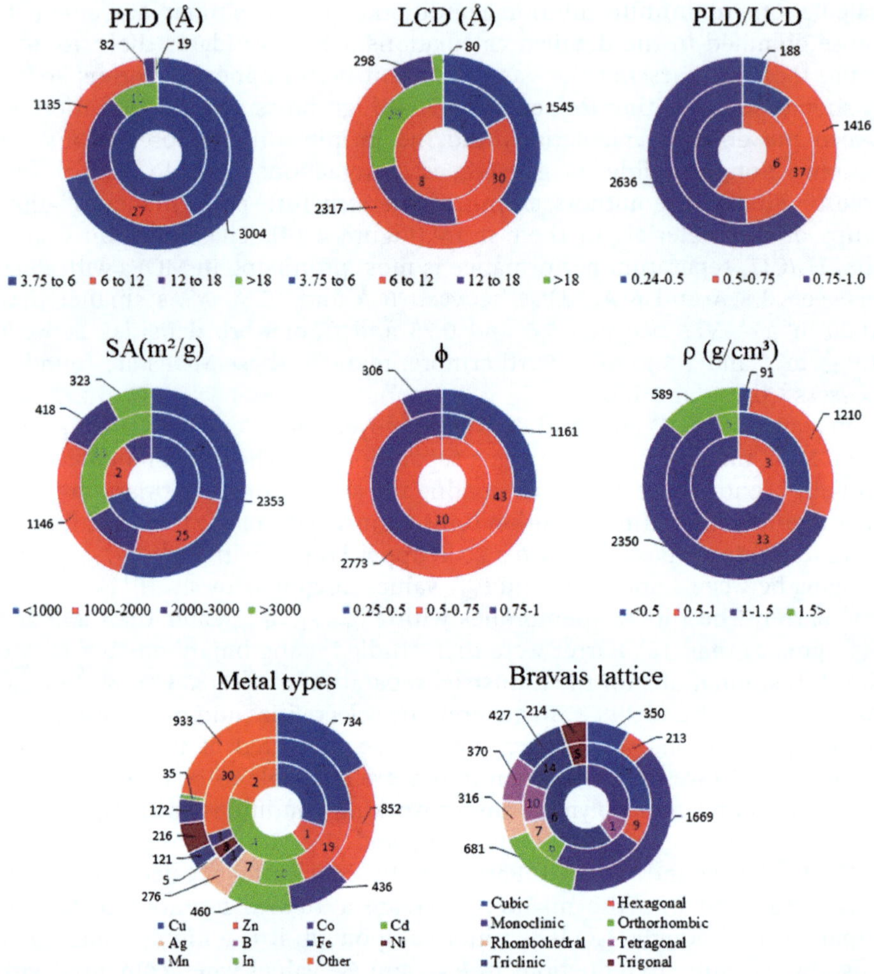

Figure 4.18 Effects of structural properties on the performance of MOF membranes. Numbers on the circles represent the number of MOFs. The outer circle represents all MOFs considered in this study (4240 MOFs), the middle circle represents the top 100 H_2 selective MOFs having $P_{H_2}^{\text{inf}} > 10^4$ Barrer and the inner circle represents the top 10 most promising MOFs, the ones having $S_{\text{perm},H_2/CH_4} > 10$ and $P_{H_2} > 10^4$ Barrer.
Reproduced from ref. 351 with permission from the Royal Society of Chemistry.

Qiao *et al.* screened 137 953 hypothetical MOFs for separations of CO_2 and N_2 from ternary mixtures containing CO_2, N_2 and CH_4.[353] Specific focus was given to MOFs with PLDs ranging between 3 and 4 Å, which refines the total number of structures to 17 257. Using calculations in the limit of infinite dilution, the permeabilities of CO_2, N_2 and CH_4 were estimated in each membrane and the top candidates were identified by forming Robeson plots for CO_2/CH_4 and N_2/CH_4 mixtures. A total of 5254 MOFs are found to lie above the CO_2/CH_4 upper bound, while 682 are found to lie above the CO_2/N_2 upper bound. By cross-referencing these structures, 24 materials with $S^{inf}_{perm,CO_2/CH_4}$ greater than 100, $P^{inf}_{CO_2}$ greater than 10^4 Barrer, $S^{inf}_{perm,N_2/CH_4}$ greater than 10 and $P^{inf}_{N_2}$ greater than 50 Barrer are found to exist above both upper bounds. These materials were subject to more detailed calculations using GCMC and MD simulations to probe their separation performance of a 10/70/20 $CO_2/N_2/CH_4$ mixture at 10 bar and 298 K. Similar to other screenings, MOFs with narrow pore diameters were frequently identified in the top candidates. These materials typically possess strong affinities for CO_2 leading to large selectivities between 260 and 10 557.

4.5.3.3 Modelling the Separation Performance of Mixed Matrix Membranes

MMMs combine the ideal processability and mechanical properties of polymers with the high selectivity and good thermal and chemical stabilities of many porous materials. A MMM is composed of a polymer matrix in which a series of porous filler particles such as CNTs, zeolites or MOFs are dispersed.[329] While a few studies have employed fully atomistic simulations to model the gas permeability properties of these materials,[354,355] the most common approach is to employ a theoretical permeation model such as the Maxwell, modified Maxwell, Bruggeman, Lewis–Nielson, Pal, Felske and modified Felske models.[356] These can be separated into two groups that consider ideal morphologies where the filler particles are perfectly dispersed within the polymer matrix (Maxwell, Bruggeman, Lewis–Nielsen, Pal), and non-ideal morphologies which possess interfacial voids or polymer rigidification around the filler particles (modified Maxwell, Felske and modified Felske).[357] These models require knowledge of the gas permeability in both the polymer matrix and porous filler particles, and the volume fraction of the dispersed filler particles. The mathematical form of each model is summarised in Table 4.2.

The choice of permeation model is very important in order to obtain accurate predictions of the separation performance in a MMM. In one of the earliest studies of MOF-based MMMs, Keskin and Sholl employed the Maxwell and Bruggeman models to study CO_2/CH_4 separations. To validate the accuracy of these models, the single-component and mixed gas permeabilities of CH_4, N_2, CO_2 and H_2 were predicted for **IRMOF-1**/Matrimid MMM.[358] Gas permeabilities for Matrimid were taken from experimental work, while molecular simulations were employed to predict the single-component and mixture permeabilities in **IRMOF-1** membranes. The authors' results show that both models are able to accurately reproduce the single-component permeabilities and selectivities of each gas. As for the mixture selectivities, both models were able to reproduce experimental trends where the CO_2/CH_4 mixture selectivity of the membrane decreases with increasing **IRMOF-1** content. With this validation, the authors proceeded to predict similar calculations for Matrimid membranes

Table 4.2 Summary of permeation models. Reproduced from ref. 356 with permission from John Wiley and Sons, Copyright 2013.

Models	Expressions	Parameters
Maxwell	$P_T = \frac{P}{P_m} = \left[\frac{2(1-\phi)+(1+2\phi)\lambda_{dm}}{(2+\phi)+(1-\phi)\lambda_{dm}}\right]$	$\boldsymbol{P_d}$: the permeability of dispersed phase $\boldsymbol{P_m}$: the permeability of continuous phase $\boldsymbol{P_r}$: the relative permeability $\boldsymbol{P}$: the permeability in MOF/polymer MMM $\boldsymbol{P_{eff}}$: the effective permeability of the pseudo-insert phase $\boldsymbol{P_I}$: the permeability of the interphase $\boldsymbol{\phi}$: the volume fraction of MOF particles $\boldsymbol{\phi_m}$: the maximum packing volume fraction of fillers $\boldsymbol{\phi_s}$: the volume fraction of the dispersed phase within the pseudo-insert phase $\boldsymbol{\phi_d}$: the volume fraction of dispersed phase $\boldsymbol{\phi_I}$: the volume fraction of the interphase $\boldsymbol{r_d}$: the insert radius $\boldsymbol{l_i}$: the interphase thickness $\boldsymbol{\delta}$: the ratio of outer radius of interfacial shell-to-core radius $\boldsymbol{\lambda_{Im}}$: P_I/P_m, $\boldsymbol{\lambda_{dI}}$: P_d/P_I, $\boldsymbol{\lambda_{dm}}$: P_d/P_m
Bruggeman	$P_t^{(1/3)}\left[\frac{\lambda_{dm}-1}{\lambda_{dm}-P_r}\right] = (1-\phi)^{-1}$	
Lewis–Nielsen	$P_T = \frac{P}{P_m} = \left[\frac{1+2((\lambda_{dm}-1)/(\lambda_{dm}+2)\phi)}{1-((\lambda_{dm}-1)/(\lambda_{dm}+2))\phi\varphi}\right]$	
Pal	$P_t^{(1/3)}\left[\frac{\lambda_{dm}-1}{\lambda_{dm}-P_r}\right] = \left(1-\frac{\phi}{\phi_m}\right)^{-\phi_m}$	
Modified Maxwell	$P_{eff} = P_I\left[\frac{2(1-\phi_s)+(1+2\phi_s)(P_d/P_I)}{(2+\phi_m)+(1-\phi_m)(P_d/P_I)}\right]$	
	$\phi_s = \frac{\phi_d}{\phi_d+\phi_I} = \frac{r_d^3}{(r_d+l_i)^3}$	
	$P_T = \frac{P}{P_m} = \left[\frac{2(1-\phi)+(1+2\phi)(P_{eff}/P_m)}{(2+\phi)+(1-\phi)(P_{eff}/P_m)}\right]$	
Felske	$P_T = \frac{P}{P_m} = \left[\frac{2(1-\phi)+(1+2\phi)(\beta/\gamma)}{(2+\phi)+(1-\phi)(\beta/\gamma)}\right]$	
	$\beta = (2+\delta^3)\lambda_{dm} - 2(1-\delta^3)\lambda_{Im}\gamma = (1+2\delta^3) - (1-\delta^3)\lambda_{dI}$	
Modified Felske	$P_T = \frac{P}{P_m} = \left[\frac{1+2((\beta-\gamma)/(\beta+2\gamma))\phi}{1-((\beta-\gamma)/(\beta+2\gamma))\phi\varphi}\right]$	

dispersed with **IRMOF-8, IRMOF-10, IRMOF-14, CuBTC** and Cu(hfipbb)(H_2hfipbb)$_{0.5}$.[359] Most MOFs are able to improve P_{CO_2} relative to the pure polymer but have negligible effect on $S_{perm,CO_2/CH_4}$ values. In contrast, simultaneous improvement of P_{CO_2} and $S_{perm,CO_2/CH_4}$ is found for increasing loadings of Cu(hfipbb)(H_2hfipbb)$_{0.5}$ due to the pure MOF membranes' large $S_{perm,CO_2/CH_4}$. These results demonstrate that selecting MOF membranes with very high membrane selectivities can lead to MMMs with significantly improved separation performance over pure polymeric membrane.

In a later study, the accuracy of several permeation models was evaluated by Erucar and Keskin for predicting CO_2 and CH_4 permeabilities in MOF-based MMMs.[357] To predict the performance of each model, the average absolute relative error (AARE) in the predicted CO_2 and CH_4 permeabilities was determined for **IRMOF-1**/Matrimid and **CuBTC**/Matrimid MMMs. The AARE values for these models decrease in size with the following order: Pal model > Bruggeman model > Lewis–Nielsen model > Maxwell model > modified Maxwell model > Felske > modified Felske. The modified Felske possesses the lowest AARE values of 8.37% and 13.6% for P_{CO_2} and P_{CH_4}, respectively. Models based on ideal morphologies tend to possess much larger AARE values with the Maxwell model possessing the lowest AARE values of 34.2 and 48.1% for CO_2 and CH_4, respectively. With these new insights, Erucar and Keskin employed the modified Felske and Maxwell models to study the CO_2/CH_4 separation performance in 80 new MMMs. The authors found that selecting MOF membranes that possess very high S_{perm} in their pure form can lead to MMMs with significantly improved separation performance over the pure polymeric membrane. In one example, the dispersion of a 30 wt% microporous metal–imidazolate framework (MMIF)[360] increased the $S_{perm,CO_2/CH_4}$ and P_{CO_2} values of Matrimid from 41.7 and 9 Barrer to 151 and 20 Barrer, respectively. For polymeric membranes that possess low $S_{perm,CO_2/CH_4}$ but high P_{CO_2}, the choice of MOF filler particle is critical to improve the separation performance. If the MOF membrane is relatively unselective, it can yield a MMM with a higher P_{CO_2} but lower $S_{perm,CO_2/CH_4}$. Alternatively, if the MOF membrane is highly selective, it can yield a MMM with a higher $S_{perm,CO_2/CH_4}$ but lower P_{CO_2}. These relationships are critical to understanding the link between the polymer, filler particles and separation performance of these MMMs. In recent years, a number of similar studies have been employed to screen MMMs for gas separations of small molecules, such as CO_2, N_2, CH_4 and H_2.[361–367]

4.6 Summary and Outlook

In this chapter, we have reviewed a number of modern computational approaches to evaluating the guest adsorption and transport properties in porous solids. We have shown that simulation techniques such as grand canonical Monte Carlo and molecular dynamics that employ classical force-fields can provide accurate predictions of the gas storage and separation properties in many porous adsorbents and membranes. While this area grows rapidly towards the high-throughput screening of very large structural databases, there are still a number of challenges that remain within the field. Although high-throughput studies are able to identify potential candidates for specific applications, it is important to recall that these simulations assume rigid, defect-free structure of membranes with little or no assessment of the stability of these materials under working conditions. As such, the real performance of these materials is likely to be worse than predicted by the simulations. Future studies assessing the stability, the effects of gas impurities and the flexibility of the host structure will be highly beneficial for improving the accuracy of these predictions. Many of the screening procedures employ general force-fields such as UFF, DREIDING and OPLS-aa. The accuracy of these force-fields is typically appreciable; however, we have shown that they are not always equipped to study complex interactions such as guest binding to co-ordinately unsaturated sites in MOFs. The future development of transferable force-fields that can describe the nature of these interactions across a wide array of materials is challenging but provides high reward and opportunities to progress the field.

References

1. H. v. Bekkum, E. M. Flanigen, P. A. Jacobs and J. C. Jansen, *Introduction to Zeolite Science and Practice*, Elsevier, 2nd edn, 2001.
2. R. Pophale, P. A. Cheeseman and M. W. Deem, *Phys. Chem. Chem. Phys.*, 2011, **13**, 12407–12412.
3. C. Baerlocher and L. B. McCusker, *Database of Zeolite Structures*, http://www.iza-structure.org/databases/, (accessed 1st May, 2020).
4. D.-Y. Peng and D. B. Robinson, *Ind. Eng. Chem. Fundam.*, 1976, **15**, 59–64.
5. S. Brandani, E. Mangano and L. Sarkisov, *Adsorption*, 2016, **22**, 261–276.
6. A. G. Bezus, A. V. Kiselev, A. A. Lopatkin and P. Q. Du, *J. Chem. Soc., Faraday Trans. 2*, 1978, **74**, 367–379.
7. B. S. D. Frenkel, *Understanding Molecular Simulation*, Academic Press, San Diego, 2nd edn, 2002.
8. D. Dubbeldam, S. Calero, T. J. H. Vlugt, R. Krishna, T. L. M. Maesen, E. Beerdsen and B. Smit, *Phys. Rev. Lett.*, 2004, **93**, 088302.
9. P. Pascual, P. Ungerer, B. Tavitian, P. Pernot and A. Boutin, *Phys. Chem. Chem. Phys.*, 2003, **5**, 3684–3693.
10. A. Martin-Calvo, J. J. Gutiérrez-Sevillano, J. B. Parra, C. O. Ania and S. Calero, *Phys. Chem. Chem. Phys.*, 2015, **17**, 24048–24055.
11. D. Dubbeldam, S. Calero, T. J. H. Vlugt, R. Krishna, T. L. M. Maesen and B. Smit, *J. Phys. Chem. B*, 2004, **108**, 12301–12313.
12. B. Liu, B. Smit, F. Rey, S. Valencia and S. Calero, *J. Phys. Chem. C*, 2008, **112**, 2492–2498.
13. A. García-Sánchez, C. O. Ania, J. B. Parra, D. Dubbeldam, T. J. H. Vlugt, R. Krishna and S. Calero, *J. Phys. Chem. C*, 2009, **113**, 8814–8820.
14. P. Bai, M. Tsapatsis and J. I. Siepmann, *J. Phys. Chem. C*, 2013, **117**, 24375–24387.

15. A. K. Rappe, C. J. Casewit, K. S. Colwell, W. A. Goddard and W. M. Skiff, *J. Am. Chem. Soc.*, 1992, **114**, 10024–10035.
16. S. L. Mayo, B. D. Olafson and W. A. Goddard, *J. Phys. Chem.*, 1990, **94**, 8897–8909.
17. W. L. Jorgensen, D. S. Maxwell and J. Tirado-Rives, *J. Am. Chem. Soc.*, 1996, **118**, 11225–11236.
18. H. A. Lorentz, *Ann. Phys.*, 1881, **248**, 127–136.
19. D. Berthelot, *Comptes Rendus Hebdomadaires des Séances de l'Académie des Sciences*, 1898, **126**, 1703–1855.
20. P. P. Ewald, *Ann. Phys.*, 1921, **369**, 253–287.
21. J. J. Potoff and J. I. Siepmann, *AIChE J.*, 2001, **47**, 1676–1682.
22. M. G. Martin and J. I. Siepmann, *J. Phys. Chem. B*, 1998, **102**, 2569–2577.
23. H. J. C. Berendsen, J. R. Grigera and T. P. Straatsma, *J. Phys. Chem.*, 1987, **91**, 6269–6271.
24. W. L. Jorgensen, J. Chandrasekhar, J. D. Madura, R. W. Impey and M. L. Klein, *J. Chem. Phys.*, 1983, **79**, 926–935.
25. M. W. Mahoney and W. L. Jorgensen, *J. Chem. Phys.*, 2000, **112**, 8910–8922.
26. J. G. Harris and K. H. Yung, *J. Phys. Chem.*, 1995, **99**, 12021–12024.
27. V. Buch, *J. Chem. Phys.*, 1994, **100**, 7610–7629.
28. F. Darkrim and D. Levesque, *J. Chem. Phys.*, 1998, **109**, 4981–4984.
29. R. S. Mulliken, *J. Chem. Phys.*, 1955, **23**, 1833–1840.
30. P. O. Löwdin, *J. Chem. Phys.*, 1950, **18**, 365–375.
31. A. E. Reed, R. B. Weinstock and F. Weinhold, *J. Chem. Phys.*, 1985, **83**, 735–746.
32. R. F. W. Bader, *Acc. Chem. Res.*, 1985, **18**, 9–15.
33. F. L. Hirshfeld, *Theor. Chim. Acta*, 1977, **44**, 129–138.
34. P. Bultinck, C. Van Alsenoy, P. W. Ayers and R. Carbó-Dorca, *J. Chem. Phys.*, 2007, **126**, 144111.
35. C. M. Breneman and K. B. Wiberg, *J. Comput. Chem.*, 1990, **11**, 361–373.
36. C. I. Bayly, P. Cieplak, W. Cornell and P. A. Kollman, *J. Phys. Chem.*, 1993, **97**, 10269–10280.
37. U. C. Singh and P. A. Kollman, *J. Comput. Chem.*, 1984, **5**, 129–145.
38. C. Campañá, B. Mussard and T. K. Woo, *J. Chem. Theory Comput.*, 2009, **5**, 2866–2878.
39. T. A. Manz and D. S. Sholl, *J. Chem. Theory Comput.*, 2010, **6**, 2455–2468.
40. A. K. Rappe and W. A. Goddard, *J. Phys. Chem.*, 1991, **95**, 3358–3363.
41. C. E. Wilmer, K. C. Kim and R. Q. Snurr, *J. Phys. Chem. Lett.*, 2012, **3**, 2506–2511.
42. A. Oda and O. Takahashi, *Chem. Phys. Lett.*, 2010, **495**, 155–159.
43. B. A. Wells, C. De Bruin-Dickason and A. L. Chaffee, *J. Phys. Chem. C*, 2015, **119**, 456–466.
44. S. Ramachandran, T. G. Lenz, W. M. Skiff and A. K. Rappé, *J. Phys. Chem.*, 1996, **100**, 5898–5907.
45. D. Ongari, P. G. Boyd, O. Kadioglu, A. K. Mace, S. Keskin and B. Smit, *J. Chem. Theory Comput.*, 2019, **15**, 382–401.

46. C. Altintas and S. Keskin, *Mol. Syst. Des. Eng.*, 2020, **5**, 532–543.
47. Q. Xu and C. Zhong, *J. Phys. Chem. C*, 2010, **114**, 5035–5042.
48. E. Argueta, J. Shaji, A. Gopalan, P. Liao, R. Q. Snurr and D. A. Gómez-Gualdrón, *J. Chem. Theory Comput.*, 2018, **14**, 365–376.
49. M. Mezei, *Mol. Phys.*, 1980, **40**, 901–906.
50. R. Q. Snurr, A. T. Bell and D. N. Theodorou, *J. Phys. Chem.*, 1993, **97**, 13742–13752.
51. J. I. Siepmann and D. Frenkel, *Mol. Phys.*, 1992, **75**, 59–70.
52. T. J. H. Vlugt, R. Krishna and B. Smit, *J. Phys. Chem. B*, 1999, **103**, 1102–1118.
53. S. P. Bates, W. J. M. van Well, R. A. van Santen and B. Smit, *J. Am. Chem. Soc.*, 1996, **118**, 6753–6759.
54. B. Smit and J. I. Siepmann, *Science*, 1994, **264**, 1118–1120.
55. E. J. Maginn, A. T. Bell and D. N. Theodorou, *J. Phys. Chem.*, 1995, **99**, 2057–2079.
56. W. Shi and E. J. Maginn, *J. Chem. Theory Comput.*, 2007, **3**, 1451–1463.
57. F. Wang and D. P. Landau, *Phys. Rev. Lett.*, 2001, **86**, 2050–2053.
58. H. Zhang and R. Q. Snurr, *J. Phys. Chem. C*, 2017, **121**, 24000–24010.
59. S. Brunauer, P. H. Emmett and E. Teller, *J. Am. Chem. Soc.*, 1938, **60**, 309–319.
60. I. Langmuir, *J. Am. Chem. Soc.*, 1918, **40**, 1361–1403.
61. M. Thommes, K. Kaneko, A. V. Neimark, J. P. Olivier, F. Rodriguez-Reinoso, J. Rouquerol and K. S. W. Sing, *Pure Appl. Chem.*, 2015, **87**, 1051–1069.
62. J. Rouquerol, P. Llewellyn and F. Rouquerol, in *Studies in Surface Science and Catalysis*, ed. F.R. Reinoso, J. Rouqerol, P. Llewellyn and N. Seaton, Elsevier, 2007, vol. 160, pp. 49–56.
63. P. H. Emmett and S. Brunauer, *J. Am. Chem. Soc.*, 1937, **59**, 1553–1564.
64. F. Rouquerol, P. Llewellyn, G. Maurin and K. S. Sing, *Adsorption by Powders and Porous Solids: Principles, Methodology and Applications*, Academic Press, Kidlington, Oxford, 2nd edn, 2014.
65. K. Sing, *Colloids Surf., A*, 2001, **187–188**, 3–9.
66. J. L. C. Rowsell and O. M. Yaghi, *J. Am. Chem. Soc.*, 2006, **128**, 1304–1315.
67. K. S. Walton and R. Q. Snurr, *J. Am. Chem. Soc.*, 2007, **129**, 8552–8556.
68. Y.-S. Bae, A. Ö. Yazaydın and R. Q. Snurr, *Langmuir*, 2010, **26**, 5475–5483.
69. M. Eddaoudi, J. Kim, N. Rosi, D. Vodak, J. Wachter, M. Keeffe and O. M. Yaghi, *Science*, 2002, **295**, 469.
70. T. C. Wang, W. Bury, D. A. Gómez-Gualdrón, N. A. Vermeulen, J. E. Mondloch, P. Deria, K. Zhang, P. Z. Moghadam, A. A. Sarjeant, R. Q. Snurr, J. F. Stoddart, J. T. Hupp and O. K. Farha, *J. Am. Chem. Soc.*, 2015, **137**, 3585–3591.
71. D. A. Gomez-Gualdron, O. V. Gutov, V. Krungleviciute, B. Borah, J. E. Mondloch, J. T. Hupp, T. Yildirim, O. K. Farha and R. Q. Snurr, *Chem. Mater.*, 2014, **26**, 5632–5639.
72. D. A. Gómez-Gualdrón, P. Z. Moghadam, J. T. Hupp, O. K. Farha and R. Q. Snurr, *J. Am. Chem. Soc.*, 2016, **138**, 215–224.

73. J. Glover and E. Besley, *Phys. Chem. Chem. Phys.*, 2018, **20**, 23616–23624.
74. O. K. Farha, C. E. Wilmer, I. Eryazici, B. G. Hauser, P. A. Parilla, K. O'Neill, A. A. Sarjeant, S. T. Nguyen, R. Q. Snurr and J. T. Hupp, *J. Am. Chem. Soc.*, 2012, **134**, 9860–9863.
75. O. K. Farha, I. Eryazici, N. C. Jeong, B. G. Hauser, C. E. Wilmer, A. A. Sarjeant, R. Q. Snurr, S. T. Nguyen, A. Ö. Yazaydın and J. T. Hupp, *J. Am. Chem. Soc.*, 2012, **134**, 15016–15021.
76. T. Düren, F. Millange, G. Férey, K. S. Walton and R. Q. Snurr, *J. Phys. Chem. C*, 2007, **111**, 15350–15356.
77. B. Lee and F. M. Richards, *J. Mol. Biol.*, 1971, **55**, 379-IN374.
78. A. Shrake and J. A. Rupley, *J. Mol. Biol.*, 1973, **79**, 351–371.
79. L. D. Gelb and K. E. Gubbins, *Langmuir*, 1998, **14**, 2097–2111.
80. F. M. Richards, *Annu. Rev. Biophys. Bioeng.*, 1977, **6**, 151–176.
81. M. Connolly, *J. Appl. Crystallogr.*, 1983, **16**, 548–558.
82. A. L. Myers and P. A. Monson, *Langmuir*, 2002, **18**, 10261–10273.
83. O. Talu and A. L. Myers, *AIChE J.*, 2001, **47**, 1160–1168.
84. D. Ongari, P. G. Boyd, S. Barthel, M. Witman, M. Haranczyk and B. Smit, *Langmuir*, 2017, **33**, 14529–14538.
85. H. Giesche, *Part. Part. Syst. Charact.*, 2006, **23**, 9–19.
86. J. Mitchell, J. B. W. Webber and J. H. Strange, *Phys. Rep.*, 2008, **461**, 1–36.
87. A. P. Radlinski, M. Mastalerz, A. L. Hinde, M. Hainbuchner, H. Rauch, M. Baron, J. S. Lin, L. Fan and P. Thiyagarajan, *Int. J. Coal Geol.*, 2004, **59**, 245–271.
88. C.-S. Tsao, M.-S. Yu, T.-Y. Chung, H.-C. Wu, C.-Y. Wang, K.-S. Chang and H.-L. Chen, *J. Am. Chem. Soc.*, 2007, **129**, 15997–16004.
89. E. P. Barrett, L. G. Joyner and P. P. Halenda, *J. Am. Chem. Soc.*, 1951, **73**, 373–380.
90. J. C. P. Broekhoff and J. H. de Boer, *J. Catal.*, 1967, **9**, 8–14.
91. G. Horváth and K. Kawazoe, *J. Chem. Eng. Jpn.*, 1983, **16**, 470–475.
92. A. Saito and H. C. Foley, *AIChE J.*, 1991, **37**, 429–436.
93. L. S. Cheng and Y. Ralph T, *Chem. Eng. Sci.*, 1994, **49**, 2599–2609.
94. C. Lastoskie, K. E. Gubbins and N. Quirke, *J. Phys. Chem.*, 1993, **97**, 4786–4796.
95. N. A. Seaton, J. P. R. B. Walton and N. quirke, *Carbon*, 1989, **27**, 853–861.
96. A. V. Neimark, Y. Lin, P. I. Ravikovitch and M. Thommes, *Carbon*, 2009, **47**, 1617–1628.
97. J. Landers, G. Y. Gor and A. V. Neimark, *Colloids Surf., A*, 2013, **437**, 3–32.
98. L. D. Gelb and K. E. Gubbins, *Langmuir*, 1999, **15**, 305–308.
99. E. Haldoupis, S. Nair and D. S. Sholl, *J. Am. Chem. Soc.*, 2010, **132**, 7528–7539.
100. L. Sarkisov and A. Harrison, *Mol. Simul.*, 2011, **37**, 1248–1257.
101. J. Hoshen and R. Kopelman, *Phys. Rev. B*, 1976, **14**, 3438–3445.
102. T. F. Willems, C. H. Rycroft, M. Kazi, J. C. Meza and M. Haranczyk, *Microporous Mesoporous Mater.*, 2012, **149**, 134–141.
103. M. P. Suh, H. J. Park, T. K. Prasad and D.-W. Lim, *Chem. Rev.*, 2012, **112**, 782–835.

104. United States Driving Research and Innovation for Vehicle efficiency and Energy sustainability, Target Explanation Document: Onboard Hydrogen Storage for Light-Duty Fuel Cell Vehicles, https://www.energy.gov/sites/prod/files/2017/05/f34/fcto_targets_onboard_hydro_storage_explanation.pdf, (accessed 1st May, 2020).
105. H. W. Langmi, A. Walton, M. M. Al-Mamouri, S. R. Johnson, D. Book, J. D. Speight, P. P. Edwards, I. Gameson, P. A. Anderson and I. R. Harris, *J. Alloys Compd.*, 2003, **356–357**, 710–715.
106. J. G. Vitillo, G. Ricchiardi, G. Spoto and A. Zecchina, *Phys. Chem. Chem. Phys.*, 2005, **7**, 3948–3954.
107. K. A. Williams and P. C. Eklund, *Chem. Phys. Lett.*, 2000, **320**, 352–358.
108. M. Rzepka, P. Lamp and M. A. de la Casa-Lillo, *J. Phys. Chem. B*, 1998, **102**, 10894–10898.
109. Q. Wang and J. K. Johnson, *J. Phys. Chem. B*, 1999, **103**, 4809–4813.
110. S. K. Bhatia and A. L. Myers, *Langmuir*, 2006, **22**, 1688–1700.
111. Y. Basdogan and S. Keskin, *CrystEngComm*, 2015, **17**, 261–275.
112. A. Dailly and E. Poirier, *Energy Environ. Sci.*, 2011, **4**, 3527–3534.
113. H. Frost, T. Düren and R. Q. Snurr, *J. Phys. Chem. B*, 2006, **110**, 9565–9570.
114. H. Frost and R. Q. Snurr, *J. Phys. Chem. C*, 2007, **111**, 18794–18803.
115. J. L. C. Rowsell and O. M. Yaghi, *Angew. Chem., Int. Ed.*, 2005, **44**, 4670–4679.
116. S. S. Han and W. A. Goddard, *J. Am. Chem. Soc.*, 2007, **129**, 8422–8423.
117. D. Cao, J. Lan, W. Wang and B. Smit, *Angew. Chem., Int. Ed.*, 2009, **48**, 4730–4733.
118. D. H. Jung, D. Kim, T. B. Lee, S. B. Choi, J. H. Yoon, J. Kim, K. Choi and S.-H. Choi, *J. Phys. Chem. B*, 2006, **110**, 22987–22990.
119. P. Ryan, L. J. Broadbelt and R. Q. Snurr, *Chem. Commun.*, 2008, 4132–4134.
120. O. K. Farha, A. Özgür Yazaydın, I. Eryazici, C. D. Malliakas, B. G. Hauser, M. G. Kanatzidis, S. T. Nguyen, R. Q. Snurr and J. T. Hupp, *Nat. Chem.*, 2010, **2**, 944–948.
121. T. A. Makal, J.-R. Li, W. Lu and H.-C. Zhou, *Chem. Soc. Rev.*, 2012, **41**, 7761–7779.
122. Y. He, W. Zhou, G. Qian and B. Chen, *Chem. Soc. Rev.*, 2014, **43**, 5657–5678.
123. V. C. Menon and S. Komarneni, *J. Porous Mater.*, 1998, **5**, 43–58.
124. T. Düren, L. Sarkisov, O. M. Yaghi and R. Q. Snurr, *Langmuir*, 2004, **20**, 2683–2689.
125. S. Ma, D. Sun, J. M. Simmons, C. D. Collier, D. Yuan and H.-C. Zhou, *J. Am. Chem. Soc.*, 2008, **130**, 1012–1016.
126. S. Wang, *Energy Fuels*, 2007, **21**, 953–956.
127. S.-J. Lee and Y.-S. Bae, *J. Phys. Chem. C*, 2014, **118**, 19833–19841.
128. G. Garberoglio, *Langmuir*, 2007, **23**, 12154–12158.
129. H. Li, K. Wang, Y. Sun, C. T. Lollar, J. Li and H.-C. Zhou, *Mater. Today*, 2018, **21**, 108–121.
130. J.-R. Li, J. Sculley and H.-C. Zhou, *Chem. Rev.*, 2012, **112**, 869–932.

131. J.-R. Li, R. J. Kuppler and H.-C. Zhou, *Chem. Soc. Rev.*, 2009, **38**, 1477–1504.
132. A. L. Myers and J. M. Prausnitz, *AIChE J.*, 1965, **11**, 121–127.
133. E. García-Pérez, J. B. Parra, C. O. Ania, A. García-Sánchez, J. M. van Baten, R. Krishna, D. Dubbeldam and S. Calero, *Adsorption*, 2007, **13**, 469–476.
134. L. Huang, L. Zhang, Q. Shao, L. Lu, X. Lu, S. Jiang and W. Shen, *J. Phys. Chem. C*, 2007, **111**, 11912–11920.
135. S. Cavenati, C. A. Grande and A. E. Rodrigues, *J. Chem. Eng. Data*, 2004, **49**, 1095–1101.
136. A. Vishnyakov, P. I. Ravikovitch and A. V. Neimark, *Langmuir*, 1999, **15**, 8736–8742.
137. Q. Yang and C. Zhong, *J. Phys. Chem. B*, 2006, **110**, 17776–17783.
138. B. Liu and B. Smit, *Langmuir*, 2009, **25**, 5918–5926.
139. Y. Liu, D. Liu, Q. Yang, C. Zhong and J. Mi, *Ind. Eng. Chem. Res.*, 2010, **49**, 2902–2906.
140. J. H. Cavka, S. Jakobsen, U. Olsbye, N. Guillou, C. Lamberti, S. Bordiga and K. P. Lillerud, *J. Am. Chem. Soc.*, 2008, **130**, 13850–13851.
141. Q. Yang, A. D. Wiersum, P. L. Llewellyn, V. Guillerm, C. Serre and G. Maurin, *Chem. Commun.*, 2011, **47**, 9603–9605.
142. R. Babarao, J. Jiang and S. I. Sandler, *Langmuir*, 2009, **25**, 5239–5247.
143. D. Wu, C. Wang, B. Liu, D. Liu, Q. Yang and C. Zhong, *AIChE J.*, 2011, **58**, 2078–2084.
144. D. Wu, Q. Yang, C. Zhong, D. Liu, H. Huang, W. Zhang and G. Maurin, *Langmuir*, 2012, **28**, 12094–12099.
145. D. Dubbeldam, R. Krishna, S. Calero and A. Ö. Yazaydın, *Angew. Chem., Int. Ed.*, 2012, **51**, 11867–11871.
146. M. W. Deem and J. M. Newsam, *Nature*, 1989, **342**, 260–262.
147. M. W. Deem, R. Pophale, P. A. Cheeseman and D. J. Earl, *J. Phys. Chem. C*, 2009, **113**, 21353–21360.
148. D. J. Earl and M. W. Deem, *Ind. Eng. Chem. Res.*, 2006, **45**, 5449–5454.
149. M. W. Deem and J. M. Newsam, *J. Am. Chem. Soc.*, 1992, **114**, 7189–7198.
150. Y. G. Chung, J. Camp, M. Haranczyk, B. J. Sikora, W. Bury, V. Krungleviciute, T. Yildirim, O. K. Farha, D. S. Sholl and R. Q. Snurr, *Chem. Mater.*, 2014, **26**, 6185–6192.
151. C. E. Wilmer, M. Leaf, C. Y. Lee, O. K. Farha, B. G. Hauser, J. T. Hupp and R. Q. Snurr, *Nat. Chem.*, 2012, **4**, 83–89.
152. R. L. Martin, L.-C. Lin, K. Jariwala, B. Smit and M. Haranczyk, *J. Phys. Chem. C*, 2013, **117**, 12159–12167.
153. J. Goldsmith, A. G. Wong-Foy, M. J. Cafarella and D. J. Siegel, *Chem. Mater.*, 2013, **25**, 3373–3382.
154. Y. J. Colón, D. Fairen-Jimenez, C. E. Wilmer and R. Q. Snurr, *J. Phys. Chem. C*, 2014, **118**, 5383–5389.
155. D. A. Gómez-Gualdrón, Y. J. Colón, X. Zhang, T. C. Wang, Y.-S. Chen, J. T. Hupp, T. Yildirim, O. K. Farha, J. Zhang and R. Q. Snurr, *Energy Environ. Sci.*, 2016, **9**, 3279–3289.

156. Y. J. Colón, D. A. Gómez-Gualdrón and R. Q. Snurr, *Cryst. Growth Des.*, 2017, **17**, 5801–5810.
157. P. Z. Moghadam, A. Li, S. B. Wiggin, A. Tao, A. G. P. Maloney, P. A. Wood, S. C. Ward and D. Fairen-Jimenez, *Chem. Mater.*, 2017, **29**, 2618–2625.
158. Y. G. Chung, E. Haldoupis, B. J. Bucior, M. Haranczyk, S. Lee, H. Zhang, K. D. Vogiatzis, M. Milisavljevic, S. Ling, J. S. Camp, B. Slater, J. I. Siepmann, D. S. Sholl and R. Q. Snurr, *J. Chem. Eng. Data*, 2019, **64**, 5985–5998.
159. R. L. Martin, C. M. Simon, B. Medasani, D. K. Britt, B. Smit and M. Haranczyk, *J. Phys. Chem. C*, 2014, **118**, 23790–23802.
160. R. Mercado, R.-S. Fu, A. V. Yakutovich, L. Talirz, M. Haranczyk and B. Smit, *Chem. Mater.*, 2018, **30**, 5069–5086.
161. T. Yan, Y. Lan, M. Tong and C. Zhong, *ACS Sustainable Chem. Eng.*, 2019, 7, 1220–1227.
162. R. L. Martin, C. M. Simon, B. Smit and M. Haranczyk, *J. Am. Chem. Soc.*, 2014, **136**, 5006–5022.
163. P. Z. Moghadam, T. Islamoglu, S. Goswami, J. Exley, M. Fantham, C. F. Kaminski, R. Q. Snurr, O. K. Farha and D. Fairen-Jimenez, *Nat. Commun.*, 2018, **9**, 1378.
164. J. K. Schnobrich, O. Lebel, K. A. Cychosz, A. Dailly, A. G. Wong-Foy and A. J. Matzger, *J. Am. Chem. Soc.*, 2010, **132**, 13941–13948.
165. C. E. Wilmer, O. K. Farha, T. Yildirim, I. Eryazici, V. Krungleviciute, A. A. Sarjeant, R. Q. Snurr and J. T. Hupp, *Energy Environ. Sci.*, 2013, **6**, 1158–1163.
166. J. Kim, L.-C. Lin, R. L. Martin, J. A. Swisher, M. Haranczyk and B. Smit, *Langmuir*, 2012, **28**, 11914–11919.
167. J. Kim, A. Maiti, L.-C. Lin, J. K. Stolaroff, B. Smit and R. D. Aines, *Nat. Commun.*, 2013, **4**, 1694.
168. X. Lin, I. Telepeni, A. J. Blake, A. Dailly, C. M. Brown, J. M. Simmons, M. Zoppi, G. S. Walker, K. M. Thomas, T. J. Mays, P. Hubberstey, N. R. Champness and M. Schröder, *J. Am. Chem. Soc.*, 2009, **131**, 2159–2171.
169. C. E. Wilmer, O. K. Farha, Y.-S. Bae, J. T. Hupp and R. Q. Snurr, *Energy Environ. Sci.*, 2012, **5**, 9849–9856.
170. B. J. Sikora, C. E. Wilmer, M. L. Greenfield and R. Q. Snurr, *Chem. Sci.*, 2012, **3**, 2217–2223.
171. N. S. Bobbitt, J. Chen and R. Q. Snurr, *J. Phys. Chem. C*, 2016, **120**, 27328–27341.
172. A. Ahmed, S. Seth, J. Purewal, A. G. Wong-Foy, M. Veenstra, A. J. Matzger and D. J. Siegel, *Nat. Commun.*, 2019, **10**, 1568.
173. A. Ahmed, Y. Liu, J. Purewal, L. D. Tran, A. G. Wong-Foy, M. Veenstra, A. J. Matzger and D. J. Siegel, *Energy Environ. Sci.*, 2017, **10**, 2459–2471.
174. T. K. Prasad and M. P. Suh, *Chem. – Eur. J.*, 2012, **18**, 8673–8680.
175. K. Koh, J. D. Van Oosterhout, S. Roy, A. G. Wong-Foy and A. J. Matzger, *Chem. Sci.*, 2012, **3**, 2429–2432.

176. C. M. Simon, R. Mercado, S. K. Schnell, B. Smit and M. Haranczyk, *Chem. Mater.*, 2015, **27**, 4459–4475.
177. L. Chen, P. S. Reiss, S. Y. Chong, D. Holden, K. E. Jelfs, T. Hasell, M. A. Little, A. Kewley, M. E. Briggs, A. Stephenson, K. M. Thomas, J. A. Armstrong, J. Bell, J. Busto, R. Noel, J. Liu, D. M. Strachan, P. K. Thallapally and A. I. Cooper, *Nat. Mater.*, 2014, **13**, 954–960.
178. J. Pérez-Pellitero, H. Amrouche, F. R. Siperstein, G. Pirngruber, C. Nieto-Draghi, G. Chaplais, A. Simon-Masseron, D. Bazer-Bachi, D. Peralta and N. Bats, *Chem. – Eur. J.*, 2010, **16**, 1560–1571.
179. D. Fairen-Jimenez, R. Galvelis, A. Torrisi, A. D. Gellan, M. T. Wharmby, P. A. Wright, C. Mellot-Draznieks and T. Düren, *Dalton Trans.*, 2012, **41**, 10752–10762.
180. S. Vandenbrande, T. Verstraelen, J. J. Gutiérrez-Sevillano, M. Waroquier and V. Van Speybroeck, *J. Phys. Chem. C*, 2017, **121**, 25309–25322.
181. M. J. Lennox, M. Bound, A. Henley and E. Besley, *Mol. Simul.*, 2017, **43**, 828–837.
182. S. S.-Y. Chui, S. M.-F. Lo, J. P. H. Charmant, A. G. Orpen and I. D. Williams, *Science*, 1999, **283**, 1148–1150.
183. N. C. Jeong, B. Samanta, C. Y. Lee, O. K. Farha and J. T. Hupp, *J. Am. Chem. Soc.*, 2012, **134**, 51–54.
184. V. K. Peterson, Y. Liu, C. M. Brown and C. J. Kepert, *J. Am. Chem. Soc.*, 2006, **128**, 15578–15579.
185. H. Wu, W. Zhou and T. Yildirim, *J. Am. Chem. Soc.*, 2009, **131**, 4995–5000.
186. S. R. Caskey, A. G. Wong-Foy and A. J. Matzger, *J. Am. Chem. Soc.*, 2008, **130**, 10870–10871.
187. E. D. Bloch, W. L. Queen, R. Krishna, J. M. Zadrozny, C. M. Brown and J. R. Long, *Science*, 2012, **335**, 1606–1610.
188. Q. Yang and C. Zhong, *ChemPhysChem*, 2006, **7**, 1417–1421.
189. H. Wu, J. M. Simmons, G. Srinivas, W. Zhou and T. Yildirim, *J. Phys. Chem. Lett.*, 2010, **1**, 1946–1951.
190. N. Lamia, M. Jorge, M. A. Granato, F. A. Almeida Paz, H. Chevreau and A. E. Rodrigues, *Chem. Eng. Sci.*, 2009, **64**, 3246–3259.
191. A. Ö. Yazaydın, R. Q. Snurr, T.-H. Park, K. Koh, J. Liu, M. D. LeVan, A. I. Benin, P. Jakubczak, M. Lanuza, D. B. Galloway, J. J. Low and R. R. Willis, *J. Am. Chem. Soc.*, 2009, **131**, 18198–18199.
192. L. Hamon, N. Heymans, P. L. Llewellyn, V. Guillerm, A. Ghoufi, S. Vaesen, G. Maurin, C. Serre, G. De Weireld and G. D. Pirngruber, *Dalton Trans.*, 2012, **41**, 4052–4059.
193. M. Fischer, F. Hoffmann and M. Fröba, *ChemPhysChem*, 2009, **10**, 2647–2657.
194. M. Fischer, J. R. B. Gomes, M. Fröba and M. Jorge, *Langmuir*, 2012, **28**, 8537–8549.
195. G. Maurin, S. Bourrelly, P. L. Llewellyn and R. G. Bell, *Microporous Mesoporous Mater.*, 2006, **89**, 96–102.
196. G. Maurin, P. L. Llewellyn and R. G. Bell, *J. Phys. Chem. B*, 2005, **109**, 16084–16091.

197. D. F. Plant, G. Maurin, I. Deroche and P. L. Llewellyn, *Microporous Mesoporous Mater.*, 2007, **99**, 70–78.
198. G. Maurin, Y. Belmabkhout, G. Pirngruber, L. Gaberova and P. Llewellyn, *Adsorption*, 2007, **13**, 453–460.
199. L. Liu, L. Zhao and H. Sun, *J. Phys. Chem. C*, 2009, **113**, 16051–16057.
200. S. S. Han, W.-Q. Deng and W. A. Goddard III, *Angew. Chem., Int. Ed.*, 2007, **46**, 6289–6292.
201. J. L. Mendoza-Cortés, S. S. Han, H. Furukawa, O. M. Yaghi and W. A. Goddard, *J. Phys. Chem. A*, 2010, **114**, 10824–10833.
202. S. S. Han, H. Furukawa, O. M. Yaghi and W. A. Goddard, *J. Am. Chem. Soc.*, 2008, **130**, 11580–11581.
203. S. S. Han, S.-H. Choi and W. A. Goddard, *J. Phys. Chem. C*, 2010, **114**, 12039–12047.
204. J. G. McDaniel, K. Yu and J. R. Schmidt, *J. Phys. Chem. C*, 2012, **116**, 1892–1903.
205. J. G. McDaniel and J. R. Schmidt, *J. Phys. Chem. C*, 2012, **116**, 14031–14039.
206. J. G. McDaniel and J. R. Schmidt, *J. Phys. Chem. A*, 2013, **117**, 2053–2066.
207. A. L. Dzubak, L.-C. Lin, J. Kim, J. A. Swisher, R. Poloni, S. N. Maximoff, B. Smit and L. Gagliardi, *Nat. Chem.*, 2012, **4**, 810–816.
208. O. Engkvist, P.-O. Åstrand and G. Karlström, *Chem. Rev.*, 2000, **100**, 4087–4108.
209. J. Borycz, L.-C. Lin, E. D. Bloch, J. Kim, A. L. Dzubak, R. Maurice, D. Semrouni, K. Lee, B. Smit and L. Gagliardi, *J. Phys. Chem. C*, 2014, **118**, 12230–12240.
210. W. Zhou, H. Wu and T. Yildirim, *J. Am. Chem. Soc.*, 2008, **130**, 15268–15269.
211. E. Haldoupis, J. Borycz, H. Shi, K. D. Vogiatzis, P. Bai, W. L. Queen, L. Gagliardi and J. I. Siepmann, *J. Phys. Chem. C*, 2015, **119**, 16058–16071.
212. M. Fischer, B. Kuchta, L. Firlej, F. Hoffmann and M. Fröba, *J. Phys. Chem. C*, 2010, **114**, 19116–19126.
213. M. Fischer, F. Hoffmann and M. Fröba, *ChemPhysChem*, 2010, **11**, 2220–2229.
214. S. Grimme, *Wiley Interdiscip. Rev.: Comput. Mol. Sci.*, 2011, **1**, 211–228.
215. M. Dion, H. Rydberg, E. Schröder, D. C. Langreth and B. I. Lundqvist, *Phys. Rev. Lett.*, 2004, **92**, 246401.
216. J. D. Weeks, D. Chandler and H. C. Andersen, *J. Chem. Phys.*, 1971, **54**, 5237–5247.
217. M. Jorge, M. Fischer, J. R. B. Gomes, C. Siquet, J. C. Santos and A. E. Rodrigues, *Ind. Eng. Chem. Res.*, 2014, **53**, 15475–15487.
218. C. Campbell, C. A. Ferreiro-Rangel, M. Fischer, J. R. B. Gomes and M. Jorge, *J. Phys. Chem. C*, 2017, **121**, 441–458.
219. C. Campbell, J. R. B. Gomes, M. Fischer and M. Jorge, *J. Phys. Chem. Lett.*, 2018, **9**, 3544–3553.

220. L. Chen, C. A. Morrison and T. Düren, *J. Phys. Chem. C*, 2012, **116**, 18899–18909.
221. S. Carrà and D. D. Konowalow, *Il Nuovo Cimento (1955–1965)*, 1964, **34**, 205–214.
222. L. Chen, L. Grajciar, P. Nachtigall and T. Düren, *J. Phys. Chem. C*, 2011, **115**, 23074–23080.
223. H. Fang, P. Kamakoti, J. Zang, S. Cundy, C. Paur, P. I. Ravikovitch and D. S. Sholl, *J. Phys. Chem. C*, 2012, **116**, 10692–10701.
224. R. T. Cygan, J.-J. Liang and A. G. Kalinichev, *J. Phys. Chem. B*, 2004, **108**, 1255–1266.
225. O. Bludský, M. Rubeš, P. Soldán and P. Nachtigall, *J. Chem. Phys.*, 2008, **128**, 114102.
226. H. S. Antila and E. Salonen, in *Biomolecular Simulations: Methods and Protocols*, ed. L. Monticelli and E. Salonen, Humana Press, Totowa, NJ, 2013, pp. 215–241, https://doi.org10.1007/978-1-62703-017-5_9.
227. S. W. Rick and S. J. Stuart, in *Reviews in Computational Chemistry*, 2002, pp. 89–146, https://doi.org/10.1002/0471433519.ch3.
228. V. Lachet, A. Boutin, B. Tavitian and A. H. Fuchs, *J. Phys. Chem. B*, 1998, **102**, 9224–9233.
229. J. L. Belof, A. C. Stern, M. Eddaoudi and B. Space, *J. Am. Chem. Soc.*, 2007, **129**, 15202–15210.
230. T. M. Becker, D. Dubbeldam, L.-C. Lin and T. J. H. Vlugt, *J. Comput. Sci.*, 2016, **15**, 86–94.
231. T. M. Becker, J. Heinen, D. Dubbeldam, L.-C. Lin and T. J. H. Vlugt, *J. Phys. Chem. C*, 2017, **121**, 4659–4673.
232. D. Franz, K. A. Forrest, T. Pham and B. Space, *Cryst. Growth Des.*, 2016, **16**, 6024–6032.
233. K. A. Forrest, T. Pham, K. McLaughlin, J. L. Belof, A. C. Stern, M. J. Zaworotko and B. Space, *J. Phys. Chem. C*, 2012, **116**, 15538–15549.
234. T. Pham, K. A. Forrest, R. Banerjee, G. Orcajo, J. Eckert and B. Space, *J. Phys. Chem. C*, 2015, **119**, 1078–1090.
235. L. Riekert, in *Advances in Catalysis*, ed. D. D. Eley, H. Pines and P. B. Weisz, Academic Press, 1970, vol. 21, pp. 281–322.
236. J. Kärger, D. M. Ruthven and D. N. Theodorou, in *Diffusion in Nanoporous Materials*, 2012, pp. 111–142, https://doi.org/10.1002/9783527651276.ch5.
237. P. A. Fedders, *Phys. Rev. B*, 1978, **17**, 40–46.
238. J. Kärger, M. Petzold, H. Pfeifer, S. Ernst and J. Weitkamp, *J. Catal.*, 1992, **136**, 283–299.
239. V. Gupta, S. S. Nivarthi, A. V. McCormick and H. Ted Davis, *Chem. Phys. Lett.*, 1995, **247**, 596–600.
240. K. Hahn, J. Kärger and V. Kukla, *Phys. Rev. Lett.*, 1996, **76**, 2762–2765.
241. V. Kukla, J. Kornatowski, D. Demuth, I. Girnus, H. Pfeifer, L. V. C. Rees, S. Schunk, K. K. Unger and J. Kärger, *Science*, 1996, **272**, 702–704.
242. K. Hahn and J. Kärger, *J. Phys. Chem.*, 1996, **100**, 316–326.
243. D. S. Sholl and K. A. Fichthorn, *Phys. Rev. Lett.*, 1997, **79**, 3569–3572.

244. D. S. Sholl, *Chem. Phys. Lett.*, 1999, **305**, 269–275.
245. D. S. Sholl and C. K. Lee, *J. Chem. Phys.*, 2000, **112**, 817–824.
246. P. H. Nelson and S. M. Auerbach, *J. Chem. Phys.*, 1999, **110**, 9235–9243.
247. S. Keskin, J. Liu, R. B. Rankin, J. K. Johnson and D. S. Sholl, *Ind. Eng. Chem. Res.*, 2009, **48**, 2355–2371.
248. D. Dubbeldam and R. Q. Snurr, *Mol. Simul.*, 2007, **33**, 305–325.
249. A. I. Skoulidas and D. S. Sholl, *J. Phys. Chem. A*, 2003, **107**, 10132–10141.
250. R. Krishna and L. J. P. van den Broeke, *Chem. Eng. J. Biochem. Eng. J.*, 1995, **57**, 155–162.
251. D. Paschek and R. Krishna, *Langmuir*, 2001, **17**, 247–254.
252. P. Demontis, E. S. Fois, G. B. Suffritti and S. Quartieri, *J. Phys. Chem.*, 1990, **94**, 4329–4334.
253. L. Leherte, G. C. Lie, K. N. Swamy, E. Clementi, E. G. Derouane and J. M. André, *Chem. Phys. Lett.*, 1988, **145**, 237–241.
254. R. L. June, A. T. Bell and D. N. Theodorou, *J. Phys. Chem.*, 1990, **94**, 8232–8240.
255. G. Sastre, C. R. A. Catlow and A. Corma, *J. Phys. Chem. B*, 1999, **103**, 5187–5196.
256. A. I. Skoulidas, D. M. Ackerman, J. K. Johnson and D. S. Sholl, *Phys. Rev. Lett.*, 2002, **89**, 185901.
257. A. I. Skoulidas and D. S. Sholl, *J. Phys. Chem. B*, 2001, **105**, 3151–3154.
258. A. I. Skoulidas, *J. Am. Chem. Soc.*, 2004, **126**, 1356–1357.
259. L. Sarkisov, T. Düren and R. Q. Snurr, *Mol. Phys.*, 2004, **102**, 211–221.
260. A. I. Skoulidas and D. S. Sholl, *J. Phys. Chem. B*, 2005, **109**, 15760–15768.
261. G. Garberoglio and R. Vallauri, *Microporous Mesoporous Mater.*, 2008, **116**, 540–547.
262. Z. Yang and D. Cao, *J. Phys. Chem. C*, 2012, **116**, 12591–12598.
263. J. A. Wesselingh and R. Krishna, *Mass Transfer in Multicomponent Mixtures*, Delft University Press, Delft, The Netherlands, 2000.
264. D. S. Sholl, *Acc. Chem. Res.*, 2006, **39**, 403–411.
265. M. J. Sanborn and R. Q. Snurr, *AIChE J.*, 2001, **47**, 2032–2041.
266. D. N. Theodorou, R. Q. Snurr and A. T. Bell, *Comprehensive Supramolecular Chemistry*, Oxford, 1996.
267. M. J. Sanborn and R. Q. Snurr, *Sep. Purif. Technol.*, 2000, **20**, 1–13.
268. H. Chen and D. S. Sholl, *J. Am. Chem. Soc.*, 2004, **126**, 7778–7779.
269. A. I. Skoulidas, T. C. Bowen, C. M. Doelling, J. L. Falconer, R. D. Noble and D. S. Sholl, *J. Membr. Sci.*, 2003, **227**, 123–136.
270. A. I. Skoulidas, D. S. Sholl and R. Krishna, *Langmuir*, 2003, **19**, 7977–7988.
271. D. Paschek and R. Krishna, *Chem. Phys. Lett.*, 2001, **333**, 278–284.
272. S. M. Auerbach, K. A. Carrado and P. K. Dutta, in *Handbook of Zeolite Science and Technology*, Marcel Dekker, New York, 2003, ch. 10.
273. S. Chempath, R. Krishna and R. Q. Snurr, *J. Phys. Chem. B*, 2004, **108**, 13481–13491.
274. R. Krishna and J. M. van Baten, *J. Phys. Chem. B*, 2005, **109**, 6386–6396.
275. J. M. van Baten and R. Krishna, *Microporous Mesoporous Mater.*, 2005, **84**, 179–191.

276. R. Krishna and J. M. van Baten, *Chem. Phys. Lett.*, 2005, **407**, 159–165.
277. R. Krishna and J. M. van Baten, *Ind. Eng. Chem. Res.*, 2006, **45**, 2084–2093.
278. S. Keskin, J. Liu, J. K. Johnson and D. S. Sholl, *Langmuir*, 2008, **24**, 8254–8261.
279. R. Babarao and J. Jiang, *Langmuir*, 2008, **24**, 5474–5484.
280. D. S. Sholl, *Langmuir*, 2006, **22**, 3707–3714.
281. D. Chandler, *J. Chem. Phys.*, 1978, **68**, 2959–2970.
282. C. H. Bennett, in *Diffusion in Solids: Recent Developments*, ed. A. Nowick and J. Burton, Academic, New York, 1975, pp. 73–113.
283. Q. Yang, D. Liu, C. Zhong and J.-R. Li, *Chem. Rev.*, 2013, **113**, 8261–8323.
284. D. Dubbeldam, E. Beerdsen, T. J. H. Vlugt and B. Smit, *J. Chem. Phys.*, 2005, **122**, 224712.
285. E. Beerdsen, B. Smit and D. Dubbeldam, *Phys. Rev. Lett.*, 2004, **93**, 248301.
286. N. Y. Chen, S. J. Lucki and E. B. Mower, *J. Catal.*, 1969, **13**, 329–332.
287. R. L. Gorring, *J. Catal.*, 1973, **31**, 13–26.
288. D. Dubbeldam, S. Calero, T. L. M. Maesen and B. Smit, *Phys. Rev. Lett.*, 2003, **90**, 245901.
289. B. Liu, Q. Yang, C. Xue, C. Zhong and B. Smit, *Phys. Chem. Chem. Phys.*, 2008, **10**, 3244–3249.
290. E. Haldoupis, T. Watanabe, S. Nair and D. S. Sholl, *ChemPhysChem*, 2012, **13**, 3449–3452.
291. J. S. Camp and D. S. Sholl, *J. Phys. Chem. C*, 2016, **120**, 1110–1120.
292. T. Tozawa, J. T. A. Jones, S. I. Swamy, S. Jiang, D. J. Adams, S. Shakespeare, R. Clowes, D. Bradshaw, T. Hasell, S. Y. Chong, C. Tang, S. Thompson, J. Parker, A. Trewin, J. Bacsa, A. M. Z. Slawin, A. Steiner and A. I. Cooper, *Nat. Mater.*, 2009, **8**, 973–978.
293. D. Holden, K. E. Jelfs, A. I. Cooper, A. Trewin and D. J. Willock, *J. Phys. Chem. C*, 2012, **116**, 16639–16651.
294. S. Kumar, J. M. Rosenberg, D. Bouzida, R. H. Swendsen and P. A. Kollman, *J. Comput. Chem.*, 1992, **13**, 1011–1021.
295. A. Schneemann, V. Bon, I. Schwedler, I. Senkovska, S. Kaskel and R. A. Fischer, *Chem. Soc. Rev.*, 2014, **43**, 6062–6096.
296. S. Horike, S. Shimomura and S. Kitagawa, *Nat. Chem.*, 2009, **1**, 695–704.
297. K. Uemura, R. Matsuda and S. Kitagawa, *J. Solid State Chem.*, 2005, **178**, 2420–2429.
298. T. Loiseau, C. Serre, C. Huguenard, G. Fink, F. Taulelle, M. Henry, T. Bataille and G. Férey, *Chem. - Eur. J.*, 2004, **10**, 1373–1382.
299. M. Vougo-Zanda, J. Huang, E. Anokhina, X. Wang and A. J. Jacobson, *Inorg. Chem.*, 2008, **47**, 11535–11542.
300. E. V. Anokhina, M. Vougo-Zanda, X. Wang and A. J. Jacobson, *J. Am. Chem. Soc.*, 2005, **127**, 15000–15001.
301. T. R. Whitfield, X. Wang, L. Liu and A. J. Jacobson, *Solid State Sci.*, 2005, **7**, 1096–1103.

302. K. Barthelet, J. Marrot, D. Riou and G. Férey, *Angew. Chem., Int. Ed.*, 2002, **41**, 281–284.
303. C. Serre, F. Millange, C. Thouvenot, M. Noguès, G. Marsolier, D. Louër and G. Férey, *J. Am. Chem. Soc.*, 2002, **124**, 13519–13526.
304. J. P. S. Mowat, S. R. Miller, A. M. Z. Slawin, V. R. Seymour, S. E. Ashbrook and P. A. Wright, *Microporous Mesoporous Mater.*, 2011, **142**, 322–333.
305. L. Chen, J. P. S. Mowat, D. Fairen-Jimenez, C. A. Morrison, S. P. Thompson, P. A. Wright and T. Düren, *J. Am. Chem. Soc.*, 2013, **135**, 15763–15773.
306. F. Salles, S. Bourrelly, H. Jobic, T. Devic, V. Guillerm, P. Llewellyn, C. Serre, G. Ferey and G. Maurin, *J. Phys. Chem. C*, 2011, **115**, 10764–10776.
307. F. Salles, A. Ghoufi, G. Maurin, R. G. Bell, C. Mellot-Draznieks and G. Férey, *Angew. Chem.*, 2008, **120**, 8615–8619.
308. D. S. Coombes, R. G. Bell, C. Mellot-Draznieks, N. A. Ramsahye and G. Maurin, in *Studies in Surface Science and Catalysis* ed. R. Xu, Z. Gao, J. Chen and W. Yan, Elsevier, 2007, vol. 170, pp. 918–925.
309. L. Vanduyfhuys, T. Verstraelen, M. Vandichel, M. Waroquier and V. Van Speybroeck, *J. Chem. Theory Comput.*, 2012, **8**, 3217–3231.
310. S. Amirjalayer, M. Tafipolsky and R. Schmid, *Angew. Chem., Int. Ed.*, 2007, **46**, 463–466.
311. M. Tafipolsky, S. Amirjalayer and R. Schmid, *J. Comput. Chem.*, 2007, **28**, 1169–1176.
312. J. A. Greathouse and M. D. Allendorf, *J. Phys. Chem. C*, 2008, **112**, 5795–5802.
313. Q. Yang, H. Jobic, F. Salles, D. Kolokolov, V. Guillerm, C. Serre and G. Maurin, *Chem. – Eur. J.*, 2011, **17**, 8882–8889.
314. Y. Duan, C. Wu, S. Chowdhury, M. C. Lee, G. Xiong, W. Zhang, R. Yang, P. Cieplak, R. Luo, T. Lee, J. Caldwell, J. Wang and P. Kollman, *J. Comput. Chem.*, 2003, **24**, 1999–2012.
315. L. Hertäg, H. Bux, J. Caro, C. Chmelik, T. Remsungnen, M. Knauth and S. Fritzsche, *J. Membr. Sci.*, 2011, **377**, 36–41.
316. L. Zhang, G. Wu and J. Jiang, *J. Phys. Chem. C*, 2014, **118**, 8788–8794.
317. L. Zhang, Z. Hu and J. Jiang, *J. Am. Chem. Soc.*, 2013, **135**, 3722–3728.
318. E. Pantatosaki, G. Megariotis, A.-K. Pusch, C. Chmelik, F. Stallmach and G. K. Papadopoulos, *J. Phys. Chem. C*, 2012, **116**, 201–207.
319. L. Zhao, Q. Yang, Q. Ma, C. Zhong, J. Mi and D. Liu, *J. Mol. Model.*, 2011, **17**, 227–234.
320. B. Zheng, M. Sant, P. Demontis and G. B. Suffritti, *J. Phys. Chem. C*, 2012, **116**, 933–938.
321. D. I. Kopelevich and H.-C. Chang, *J. Chem. Phys.*, 2001, **115**, 9519–9527.
322. I. Erucar and S. Keskin, *Chem. Eng. Sci.*, 2015, **130**, 120–128.
323. I. Erucar and S. Keskin, *J. Membr. Sci.*, 2016, **514**, 313–321.
324. C. Altintas and S. Keskin, *ACS Sustainable Chem. Eng.*, 2019, 7, 2739–2750.

325. X. Y. Chen, H. Vinh-Thang, A. A. Ramirez, D. Rodrigue and S. Kaliaguine, *RSC Adv.*, 2015, **5**, 24399–24448.
326. S. Basu, A. L. Khan, A. Cano-Odena, C. Liu and I. F. J. Vankelecom, *Chem. Soc. Rev.*, 2010, **39**, 750–768.
327. L. M. Robeson, *J. Membr. Sci.*, 1991, **62**, 165–185.
328. L. M. Robeson, *J. Membr. Sci.*, 2008, **320**, 390–400.
329. T.-S. Chung, L. Y. Jiang, Y. Li and S. Kulprathipanja, *Prog. Polym. Sci.*, 2007, **32**, 483–507.
330. E. Adatoz and S. Keskin, *J. Nanomater.*, 2015, **2015**, 136867.
331. H. Chen and D. S. Sholl, *J. Membr. Sci.*, 2006, **269**, 152–160.
332. R. Krishna and D. Paschek, *Sep. Purif. Technol.*, 2000, **21**, 111–136.
333. R. Krishna, *Chem. Soc. Rev.*, 2012, **41**, 3099–3118.
334. R. Krishna and J. M. van Baten, *Chem. Eng. J.*, 2007, **133**, 121–131.
335. R. Krishna and J. M. van Baten, *J. Membr. Sci.*, 2010, **360**, 323–333.
336. S. Keskin and D. S. Sholl, *Langmuir*, 2009, **25**, 11786–11795.
337. Y. Basdogan, K. B. Sezginel and S. Keskin, *Ind. Eng. Chem. Res.*, 2015, **54**, 8479–8491.
338. Z. Sumer and S. Keskin, *Chem. Eng. Sci.*, 2017, **164**, 108–121.
339. Z. Sumer and S. Keskin, *Ind. Eng. Chem. Res.*, 2017, **56**, 8713–8722.
340. Y. Gurdal and S. Keskin, *J. Phys. Chem. C*, 2013, **117**, 5229–5241.
341. R. Krishna and J. M. van Baten, *Phys. Chem. Chem. Phys.*, 2011, **13**, 10593–10616.
342. C. Altintas and S. Keskin, *Chem. Eng. Sci.*, 2016, **139**, 49–60.
343. S. Keskin, *J. Phys. Chem. C*, 2010, **114**, 13047–13054.
344. E. Atci and S. Keskin, *J. Phys. Chem. C*, 2012, **116**, 15525–15537.
345. J. Liu, S. Keskin, D. S. Sholl and J. K. Johnson, *J. Phys. Chem. C*, 2011, **115**, 12560–12566.
346. S. Keskin, *J. Phys. Chem. C*, 2011, **115**, 800–807.
347. T. N. Ozturk and S. Keskin, *J. Phys. Chem. C*, 2014, **118**, 13988–13997.
348. M. Tong, Y. Lan, Q. Yang and C. Zhong, *Chem. Eng. Sci.*, 2017, **168**, 456–464.
349. T. Watanabe and D. S. Sholl, *Langmuir*, 2012, **28**, 14114–14128.
350. T. Watanabe, T. A. Manz and D. S. Sholl, *J. Phys. Chem. C*, 2011, **115**, 4824–4836.
351. C. Altintas, G. Avci, H. Daglar, E. Gulcay, I. Erucar and S. Keskin, *J. Mater. Chem. A*, 2018, **6**, 5836–5847.
352. H. Daglar and S. Keskin, *J. Phys. Chem. C*, 2018, **122**, 17347–17357.
353. Z. Qiao, C. Peng, J. Zhou and J. Jiang, *J. Mater. Chem. A*, 2016, **4**, 15904–15912.
354. L. Zhang, Z. Hu and J. Jiang, *J. Phys. Chem. C*, 2012, **116**, 19268–19277.
355. X. Kong and J. Liu, *J. Phys. Chem. C*, 2019, **123**, 15113–15121.
356. I. Erucar, G. Yilmaz and S. Keskin, *Chem. – Asian J.*, 2013, **8**, 1692–1704.
357. I. Erucar and S. Keskin, *Ind. Eng. Chem. Res.*, 2011, **50**, 12606–12616.
358. S. Keskin and D. S. Sholl, *Energy Environ. Sci.*, 2010, **3**, 343–351.
359. T. Watanabe, S. Keskin, S. Nair and D. S. Sholl, *Phys. Chem. Chem. Phys.*, 2009, **11**, 11389–11394.

360. S.-S. Chen, M. Chen, S. Takamizawa, M.-S. Chen, Z. Su and W.-Y. Sun, *Chem. Commun.*, 2011, **47**, 752–754.
361. J. D. Evans, D. M. Huang, M. R. Hill, C. J. Sumby, A. W. Thornton and C. J. Doonan, *J. Phys. Chem. C*, 2014, **118**, 1523–1529.
362. G. Yilmaz and S. Keskin, *Ind. Eng. Chem. Res.*, 2012, **51**, 14218–14228.
363. C. Altintas and S. Keskin, *Mol. Simul.*, 2015, **41**, 1396–1408.
364. Z. Sumer and S. Keskin, *J. Nanomater.*, 2016, **2016**, 12.
365. H. Daglar and S. Keskin, *Adv. Theory Simul.*, 2019, **2**, 1900109.
366. S. Budhathoki, O. Ajayi, J. A. Steckel and C. E. Wilmer, *Energy Environ. Sci.*, 2019, **12**, 1255–1264.
367. I. Erucar and S. Keskin, *J. Membr. Sci.*, 2012, **407–408**, 221–230.

CHAPTER 5

Spectroscopic and Catalytic Properties

JORDI MORALES-VIDAL AND MANUEL A. ORTUÑO*

Institute of Chemical Research of Catalonia (ICIQ), Barcelona Institute of Science and Technology (BIST), Av. Països Catalans 16, 43007 Tarragona, Spain
*Email: mortuno@iciq.es

5.1 Introduction

Porous materials, in general terms, encompass any compounds with voids or pores within their structure. Zeolites have been the flagship of these materials for adsorption and catalytic applications, but recent developments on organic/inorganic porous solids, including porous molecules, have made a breakthrough in the field. Focusing on reticular chemistry,[1] the present chapter mainly covers metal–organic frameworks (MOFs), a family of porous materials that are formed by inorganic centres (nodes or secondary building units) connected through organic ligands (linkers),[2] and covalent organic frameworks (COFs), assembled by connecting organic modules *via* covalent bonds.[3] These building blocks could be hypothetically assembled in almost any possible way, thus creating the opportunity to explore and finely tune pore sizes and framework topologies at will.[4] Moreover, these porous materials can be further tailored by post-synthetic procedures *via* engineering defects or anchoring species to linkers and nodes. Such versatility has proven to be outstanding for a myriad of applications, ranging from gas adsorption, gas separation, sensing, drug delivery, and catalysis, among others.[2,5–7] In this scenario, theoretical techniques, assisted by the

Inorganic Materials Series No. 8
Computer Simulation of Porous Materials: Current Approaches and Future Opportunities
Edited by Kim Jelfs

Published by the Royal Society of Chemistry, www.rsc.org

development of up-to-date databases and computation-ready experimental structures,[8–13] offer huge opportunities for material characterisation, examination of active sites, elucidation of reaction mechanisms, high-throughput screening, and *in silico* catalyst design (see Figure 5.1).

The two elements that one needs to address in any kind of computational study are the method and the model. Regarding the theoretical method, density functional theory (DFT) within the Kohn–Sham formalism[14,15] is the most employed methodology to compute spectroscopic and catalytic properties of solid materials and molecules. Very briefly, the different approaches of the exchange–correlation density functional (DF) are the local density approximation (LDA), the generalised gradient approximation (GGA), the meta-gradient approximation (meta-GGA), and the hybrid approximation, which incorporates a percentage of nonlocal Hartree–Fock exchange. This type of hierarchy is usually referred to as the Jacob's ladder of DFs.[16] As a caveat, the majority of them cannot properly account for long-range electron correlation. However, there are several approaches to take into consideration these dispersion forces and the combination of these procedures with DFT has become common practice nowadays. The reader is referred to specialised literature for a more comprehensive view of the different dispersion correction approaches.[17,18] Despite the huge success of DFT in materials simulations, complex electronic structure situations still pose a great challenge. The most common example related to spectroscopic and catalytic properties regards the accurate calculation of molecules and materials containing 3d transition metals.[19] This feature does not only apply to metallic frameworks, but to any porous material that can anchor transition metals: zeolites, COFs, and even other MOFs. Wave function post-Hartree–Fock and multireference (MR) methods, although very costly, are sometimes mandatory in these situations. For further information on these topics, the reader is referred to specialised literature in methodologies and applications.[19–22]

Depending on the nature of the chemical system and the properties under investigation, the choice of DF can be decisive to obtain reliable results. In that sense, DF benchmarks have been reported for structural parameters of MOFs. An initial study showed that most GGA DFs can reproduce the structural features within the accuracy error;[23] a later contribution, focused on PBE and PBEsol with different dispersion corrections, highlighted PBE-D2 as a suitable method to describe both rigid and flexible porous materials;[24] a recent report recommended revM06-L and SCAN to reproduce both structural parameters and band gaps.[25] However, these guidelines may not be fully transferable when computing nontrivial electronic structures, redox-active materials, or bond-breaking and bond-forming processes present in catalysis. When appropriate, recommendations on DFs will be discussed in the pertinent sections throughout the chapter.

As for the computational model, the ideal system must contain as much information as possible while maintaining a balance between computational cost and chemical accuracy. On that note, one should be careful with over-simplifications (or excessive reductionism) since they could convey

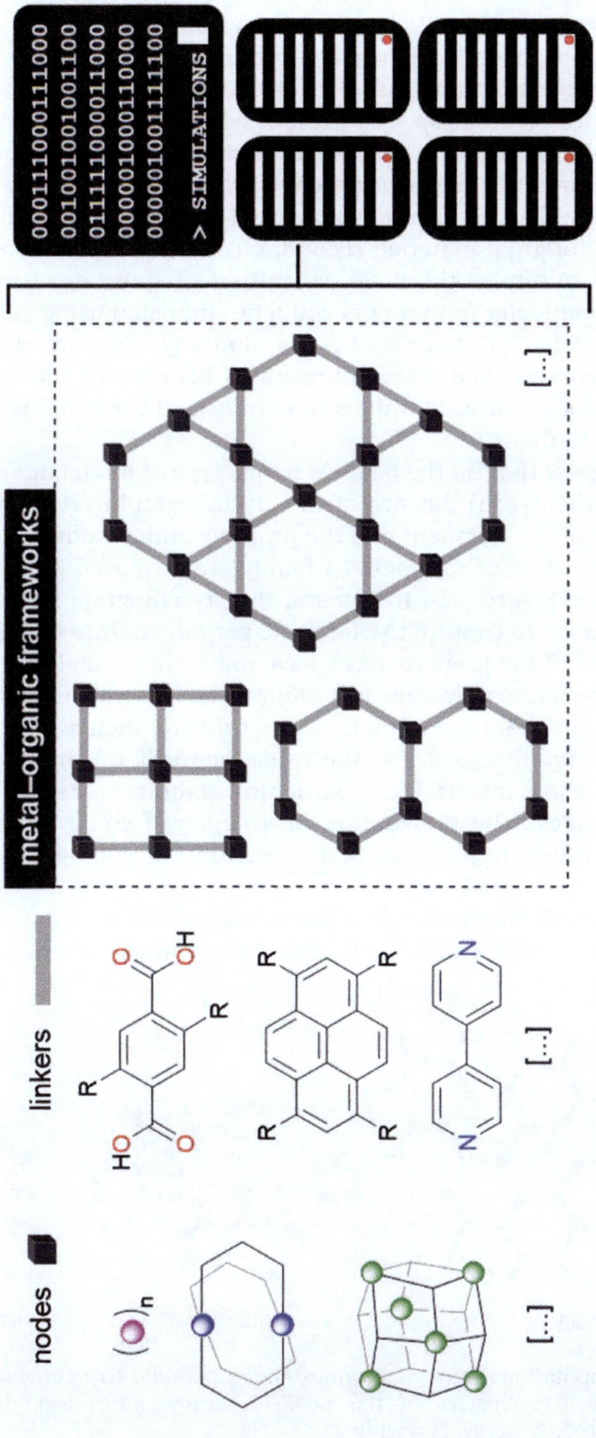

Figure 5.1 Simulation of metal–organic frameworks.

misleading information.[26] To represent porous materials, there are two general approaches that should be carefully evaluated and tested to make an educated decision: extended periodic structures and finite-size molecular clusters. Periodic simulations usually take refined X-ray structures where solvent molecules or other species present within the pores have been removed.[10,11] Molecular clusters are usually extracted from periodic structures by capping the organic linkers in different ways. Such finite-size approach has a few caveats: first, some atoms may need to be fixed to mimic the bulk structure of the original material; second, charge neutrality should be usually ensured to minimise the potential impact of spurious charge effects. Zeolites and 3D reticular frameworks could be simulated using either cluster or periodic models. 2D materials behave similarly, although an extended approach is recommended when interactions between layers are relevant. Porous molecules are already finite-size entities and the capping procedure does not apply to them.

The main factors that tip the balance to one type of model or another are: (i) the level of theory, (ii) the size of the crystallographic cell, and (iii) the local behaviour of the fragment and the property under study. Both periodic and cluster structures of the MOF-74 family are displayed below as prototypical cases (see Figure 5.2). In general, the crystallographic unit cell is a good starting point to capture the intrinsic periodic nature of the material. On the one hand, if the property has a local nature (*e.g.*, catalysis occurs in a well-defined site at the node), one can safely isolate fragments of the MOF by capping the linkers. The use of high levels of theory, such as MR, may need more drastic simplifications. On the other hand, if the property has an impact on the whole material (*e.g.*, structural stability by removing linkers that connect nodes), the preparation of a supercell could be required to avoid spurious results (*e.g.*, high defect concentration leading to the collapse of the pores).

Once the methods and models have been selected and their advantages and drawbacks have been identified, one can move towards the next step: the

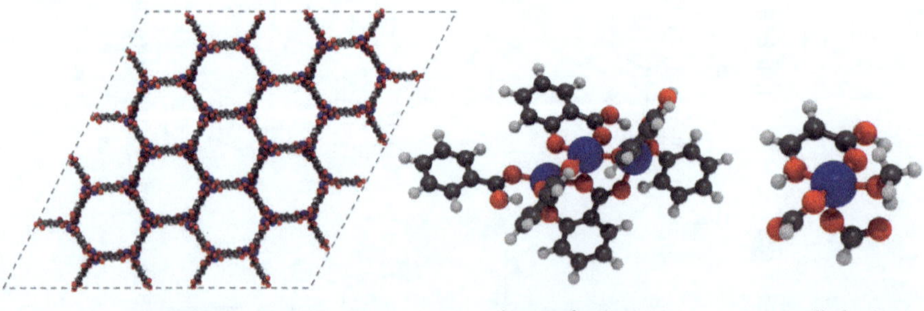

Figure 5.2 Computational models: from extended periodic structures to finite-size molecular clusters of the MOF-74 family. Atom legend: M = blue, O = red, C = grey, H = white.

simulation of spectroscopic and catalytic properties. As such, the chapter is divided into two sections: spectroscopy, band gap, and magnetism (see Section 5.2) and catalysis (see Section 5.3). The former will explain the theoretical protocols to compute a variety of spectroscopic properties, band gaps, and magnetic features by illustrating a few representative examples; the latter will discuss the elucidation of active sites and reaction mechanisms of selected catalytic processes, organised by reaction type. When necessary, the impact of the method and the model on the computational results will be discussed accordingly. The chapter will close with a summary of the contents surveyed and future perspectives in the field.

Comprehensive reviews on the computational simulation of porous materials at different levels of theory and atomic scales can be consulted in the literature: MOFs,[27–32] COFs,[31,33] zeolites,[34–36] and porous molecules.[37]

5.2 Spectroscopy, Band Gap, and Magnetism

The characterisation of materials is the first and foremost step prior to catalytic testing (as a matter of fact, to any type of application).[38] This section presents computational procedures to simulate vibrational spectroscopy with periodic and molecular approaches alike, while UV–visible spectroscopy is mainly tackled through cluster models; band gaps are computed for periodic structures and are particularly sensitive to the DF of choice; finally, magnetism is exemplified with a case study.

5.2.1 Vibrational Spectroscopy

Vibrational spectroscopy is widely used to characterise MOF structures, as well as to track probe molecules attached to them. It entails IR and Raman spectroscopies, where the former provides information about the functional groups and structural features of the framework and the latter is particularly useful to understand metal–linker interactions.[39] General trends in DFT-computed vibrational frequencies within the harmonic approximation are often accurate, but GGA and hybrid DFs tend to underestimate and overestimate the values, respectively.[28] This issue is commonly addressed by applying scale factors for vibrational frequencies according to the level of theory employed.[40]

Structural studies using DFT were performed on periodic and cluster models of pristine MOF-5 (or IRMOF-1), a material with Zn_4O nodes and benzene-1,4-dicarboxylate (BDC) linkers (see Figure 5.3).[41] The IR and Raman spectra were computed at periodic B3LYP level with atom-centred basis sets and the resulting vibrational modes, although slightly overestimated, were in agreement with experiments. Of important note was the region concerning the vibrations of the Zn_4O node, where the only experimental Raman signal at 578 cm^{-1} was initially assigned to a totally symmetric mode.[42] However, according to computed vibrations with both

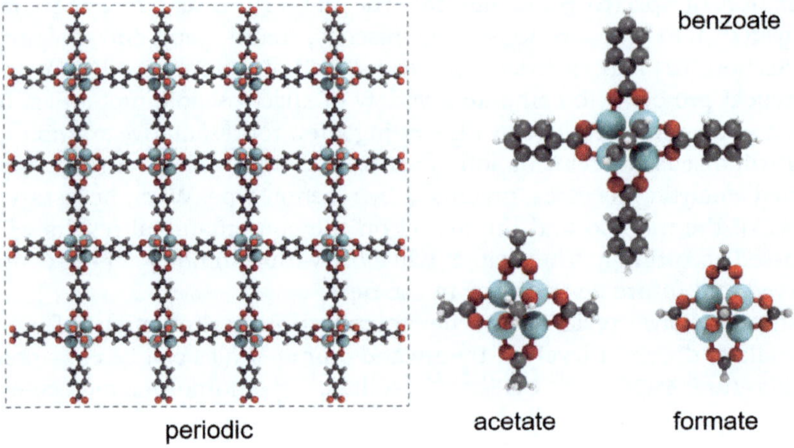

Figure 5.3 Computational models for MOF-5: periodic structure (left) and cluster models capped with different carboxylate groups (right). Atom legend: Zn = cyan, O = red, C = grey, H = white.

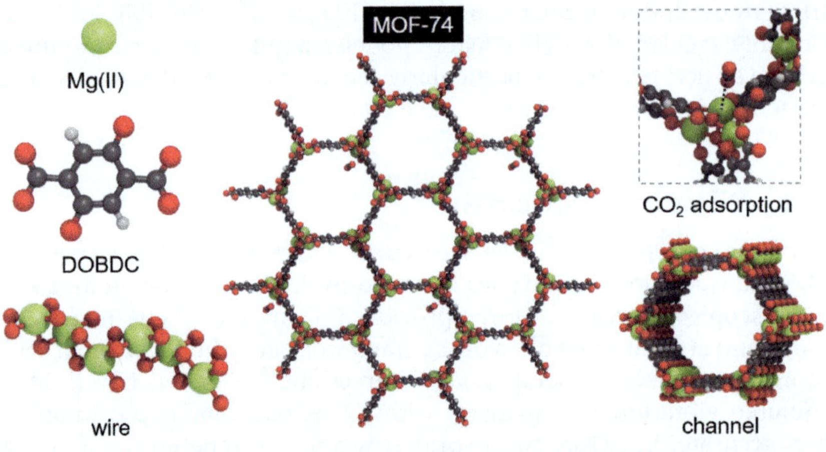

Figure 5.4 Periodic structure of MOF-74 with adsorbed CO_2 and its different components around. Atom legend: Mg = pale green, O = red, C = grey, H = white.

periodic and cluster models, that signal was better assigned to a Zn–O stretching mode.

The presence of open metal sites within the pores of MOFs allows the adsorption of gases, thus the comparison between the vibrational frequencies of isolated and adsorbed guest molecules became common practice. A good example would be the adsorption of CO_2 on MOF-74, a material formed by Mg(II) ions and 2,5-dihydroxybenzene-1,4-dicarboxylate (DOBDC) linkers (see Figure 5.4). Periodic DFT calculations with B3LYP and dispersion corrections predicted a blue shift of 10–15 cm^{-1} in the asymmetric stretching mode of an adsorbed CO_2 compared to the isolated molecule,[43] which was in agreement with experiments. A slight overestimation of this shift was later reported using the GGA vDW-DF2 approach.[44] In contrast to previous experimental and theoretical results, LDA predicted a red shift of 13 cm^{-1}, manifesting the drawbacks of this method.[45] Similarly, the open metal sites of HKUST-1 (or Cu-BTC), formed by Cu(II) ions and 1,3,5-benzene-tricarboxylate (BTC), can also bind a variety of gas molecules. Cluster calculations with B3LYP and dispersion corrections overestimated the vibrational frequencies and, for CO_2, only a minor red shift of 7 cm^{-1} was observed upon adsorption.[46]

The UiO family of porous materials are formed by $Zr_6O_4(OH)_4$ nodes connected through twelve ditopic carboxylates; UiO-66 contains BDC linkers and UiO-67 presents 4,4-biphenyl-dicarboxylate (BPDC) ligands. Representations of the periodic structures of these materials can be consulted in Figure 5.5.

Under thermal treatment, $Zr_6O_4(OH)_4$ nodes can undergo dehydroxylation to form Zr_6O_6 derivatives.[47,48] Vibrational modes were computed by means of periodic DFT at B3LYP level with atomic-centred basis sets to elucidate the structural differences between the two states. Simulated IR band positions and intensities were in line with experiments and vibrational modes were unambiguously assigned. The main changes observed in the IR spectra were originated by the cluster distortions upon dehydroxylation (see Figure 5.6). The (HO)–Zr–(OH) bending mode at 470 cm^{-1} was only present in the original cluster, and the μ_3-O stretching mode at 672 cm^{-1} was blue-shifted by

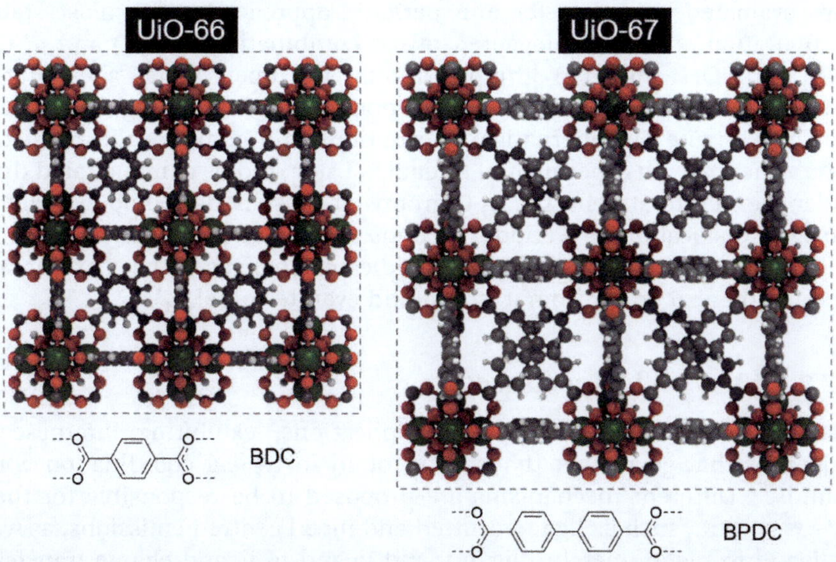

Figure 5.5 Periodic structures and organic linkers of UiO-66 (left) and UiO-67 (right). Atom legend: Zr = green, O = red, C = grey, H = white.

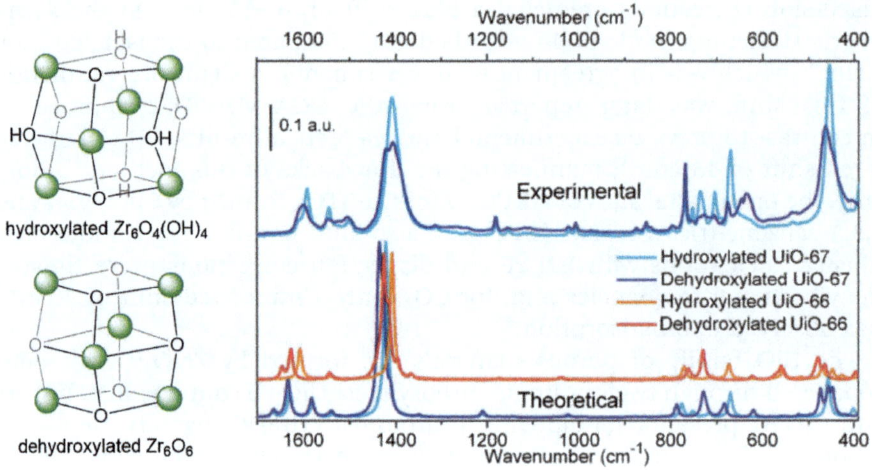

Figure 5.6 Representation of hydroxylated and dehydroxylated UiO-66 nodes (left) and experimental and computed IR spectra (right).
Adapted from ref. 48 with permission from the Royal Society of Chemistry.

58 cm^{-1} in the dehydroxylated node. Using the same protocol, similar results were found for the related UiO-67 material containing BPDC linkers.[48] Later, complete mechanistic studies were computed for dehydroxylation processes in pristine[49] and defective[50] UiO-66 nodes.

NU-1000 comprises $Zr_6O_{16}H_{16}$ clusters connected through eight 1,3,6,8-tetrakis(*p*-benzoate)pyrene linkers. Several proton distributions of the node were evaluated using cluster and periodic approaches. The most stable configuration entailed a staggered mixed combination of μ_3-O and μ_3-OH units with OH and H_2O groups capping the open metal sites of the 8-connected node (see Figure 5.7). Computed O–H stretching frequencies were key to rule out alternative distributions and compare the predicted node structure with experimental IR data.[51] Later studies demonstrated that, under thermal treatment, the μ_3-O groups of these nodes could experience transient dislocations.[52] Periodic *ab initio* molecular dynamics followed by cluster-based geometry optimisations indicated changes in the coordination mode of the oxo group from μ_3 to μ_2 (and even terminal μ_1).

5.2.2 UV–visible Spectroscopy

MOFs show potential for optical applications, exhibiting luminescent properties that go further than those found in typical coordination compounds.[53] Different mechanisms are proposed to be responsible for their UV-vis spectra, such as ligand-centred and metal-centred emissions, as well as ligand to metal, metal to ligand, and ligand to ligand charge transfers. Computational modelling of UV–vis spectra needs the proper description of both ground and excited states. These calculations are thus carried out on

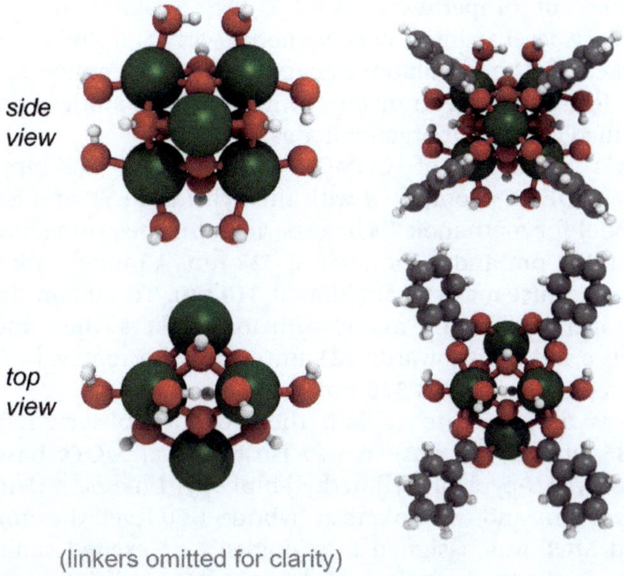

Figure 5.7 Side and top views of the mixed proton topology of the NU-1000 node.[51] Atom legend: Zr = green, O = red, C = grey, H = white.

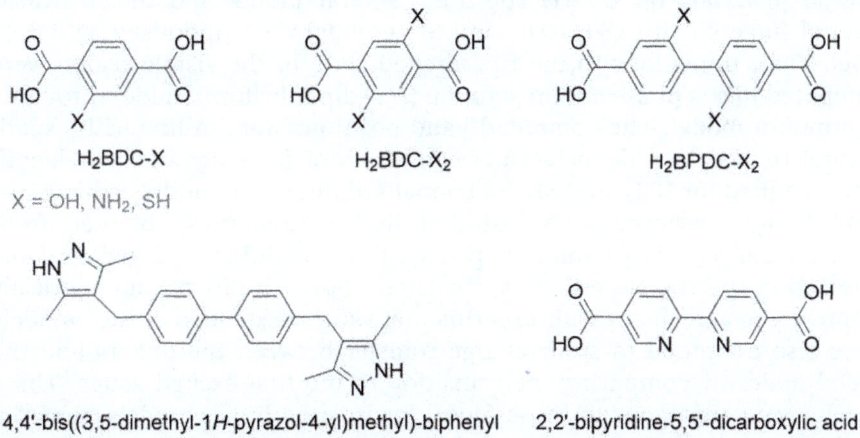

Scheme 5.1

cluster models using time-dependent density functional theory, or TD-DFT. Hybrid and long-range-corrected DFs are usually required to explain excitations to relevant Rydberg states and long-range charge-transfer character.[54,55] These simulations are extremely useful to understand the origin of energy transitions as showcased in the following contributions. Schematic representations of the linkers used in these examples can be consulted in Scheme 5.1.

The luminescent properties of MOF-5 were explored using relativistic TD-DFT at BLYP level using a cluster model with one ZnO_4 node with six benzoate linkers.[56] The calculations suggested that the emission came from a ligand to ligand charge transfer, while previous studies pointed to a mechanism involving the inorganic nodes.[42]

The UV–vis spectrum of COMOC-4, a gallium 2,2′-bipyridine-5,5′-dicarboxylate MOF was computed with the hybrid B3LYP and a continuum solvation model for methanol.[57] The experimental spectrum showed a main absorption at 307 nm and a shoulder at 328 nm. A model with two linkers was able to reproduce a π–π* transition at 310 nm. To explain the shoulder, defects were introduced in a model with four linkers: the removal of one linker yielded a red shift towards 321 nm, while the removal of a bridging OH led to a red shift towards 326 nm.

TD-DFT was employed to explain the red shift observed from Cd to Zn (from 335 nm to 441 nm) in two isostructural MOFs based on 4,4′-bis((3,5-dimethyl-1*H*-pyrazol-4-yl)methyl)-biphenyl linkers.[58] Using a model with two metal ions and two linkers at hybrid PBE0 level, the emission from the Cd-based MOF was assigned to a singlet π–π* excited state, while the luminescence of the Zn-based material might have a different origin, likely from a triplet π–π* excited state. Similar results were obtained when using CAM-B3LYP, a long-range corrected DF.

TD-DFT simulations explored the impact of linker functionalisation in UiO-66 materials on UV–vis spectra.[59] Several mono- and di-substituted neutral linkers (with OH, NH_2, and SH groups) were optimised at B3LYP level. Their transitions to the first excited state in the visible region were computed in gas phase and in solution (*N*,*N*-dimethylformamide) through a continuum model. The computed band positions were in line with experimental trends, but higher accuracy in terms of intensity and wavelength were required for NH_2- and SH-functionalised linkers. To address this issue, TD-DFT was performed on distorted linker geometries obtained from *ab initio* molecular dynamics snapshots, thus providing averaged-dynamic absorption spectra. Nevertheless, the latter approach did not automatically improve the agreement with experimental values. Extended cluster models were also examined to study charge transfer between the linkers and the metal nodes by computing the transition to the first excited states. These results were analogous to those found for isolated linker models, thus the limited extension of ligand to metal processes and its high dependence on linker functionalisation.

Additional spectroscopic studies were carried out on functionalised UiO-66 and UiO-67 to examine how vibrational modes can impact absorption properties.[60] Following a similar protocol, computational models of the organic linkers were obtained from gas phase optimisations and *ab initio* molecular dynamics simulations (periodic structures and molecular clusters) at PBE level supplemented with the D3 dispersion approach. The absorption spectra were then simulated with TD-DFT at B3LYP level and a solvation continuum model (*N*,*N*-dimethylformamide). Non-negligible differences were found between the averaged-dynamic absorption spectra and the ones derived from gas phase optimisations, in particular for the models extracted from periodic frameworks. Overall, it was found that excitation energies were influenced by low-frequency modes of the organic linkers, *e.g.*, 'butterfly' motions of BDC in UiO-66 and biphenyl rotations of BPDC in UiO-67.

5.2.3 Band Gap

The band gap is typically defined as the energy difference between the top valence bands and the bottom conduction bands of a solid. Depending on the gap width, the materials are classified as conductors, semiconductors, or insulators. In this regard, MOFs typically show large band gaps and most of them are classified as insulators, although the use of redox-active ligands can yield conductive materials.[61]

The accurate computation of band gaps for MOFs (or for any other solid) is not trivial. The values are obtained from density of states (DOS) and band structure calculations, but these values must be taken with extreme caution as they highly depend on the DF employed.[62,63] While structural parameters computed at GGA level are usually well described and in agreement with experiments, this approximation is no longer reliable for band gap predictions, as it tends to over-delocalise valence electrons due to the self-interaction error (SIE).[64] This situation derives in a systematic underestimation of computed band gaps *cf.* experiments. To address this issue, one must rely on (i) hybrid DFs which contain a certain percentage of Hartree–Fock exchange, or (ii) corrections to localise the electron density through parameterisation.

Hybrid DFs correct SIE to some extent and are therefore suitable for electronic structure calculations. However, they are computationally costly and geometry optimisations at this level are almost prohibited when using plane-wave periodic DFT. The typical procedure then involves geometry optimisations at an affordable GGA level of theory followed by single-point energy calculations with the hybrid DF of choice. In that regard, HSE03/06 are commonly employed on porous materials as they were initially designed for solids. Nevertheless, hybrid DFs usually lead to overestimated band gaps as discussed below.

The electronic properties of the Zn-based MOF-5 and its isostructural analogues (Be, Mg, Ca, Cd) were computed at GGA level with periodic models.[65] In good agreement with experiments,[42] a band gap of 3.50 eV was predicted for MOF-5 and similar values were obtained for its derivatives. These studies show the small influence of metals in the band gap and the ability of GGA to describe the electronic properties of these materials. Simulations using the hybrid B3LYP yielded an overestimated band gap of 5.00 eV.[41]

Band gaps for the hydroxylated and dehydroxylated states of UiO-66[47] and UiO-67[48] were computed at B3LYP level with atom-centred basis sets. In line with experimental data, the band gap of the hydroxylated node was larger than that for the dehydroxylated one, although the absolute values were overestimated. According to computed DOS and band structures for the different UiO materials, the origin of the band gap modulation upon dehydroxylation was attributed to electronic changes in the organic linkers, where the Zr atoms from the node were not involved.

DFT simulations were also employed to rationalise the band gap modulation in UiO-66 due to linker functionalisation[59,66] and lanthanide doping.[67] Although both PBE and HSE03/06 provided qualitatively good results, the band gaps computed with hybrid DFs were quantitatively closer to experimental values. On the one hand, the linker functionalisation led to a decrease on the band gap due to a valence band shift to higher energy states, which was highly influenced by the functional group. The computed band gap for pristine UiO-66 was 4.03 eV while those for UiO-66-X (X = NO_2, OH, NH_2, SH) presented values between 3.01 eV and 2.10 eV (see Figure 5.8). On the other hand, the Ce doping significantly decreased the band gap by bringing down the conduction band energy levels; that is, the Ce 4f empty orbitals generated a broad band within the pristine UiO-66. Furthermore, defective structures of UiO-66 were optimised with PBEsol followed by single-point calculations with HSE06,[68] but their corresponding band gaps did not significantly change by the presence of point defects.

A comprehensive screening of 20 MOFs pertaining to four distinct structural families pointed to Fe as the most effective metal to promote electrical conductivity.[69] The systems were optimised with PBEsol and their electronic properties were computed with the hybrid HSE06. The calculations showed that the highest energy of the valence electrons of high-spin

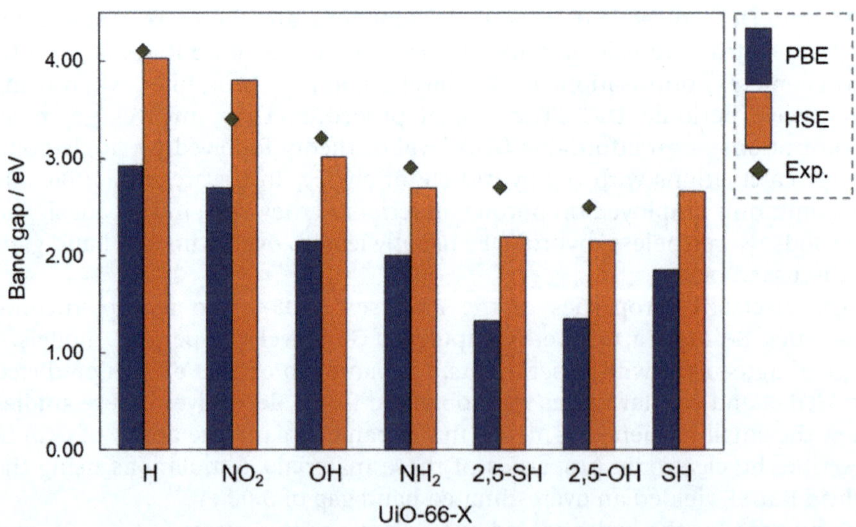

Figure 5.8 Computed and experimental band gaps for UiO-66-X with respect to the functional group of the linker.[59,66]

Fe(II) yielded smaller band gaps than other transition metal derivatives. However, while materials with low-spin Fe(II) centres behaved as insulators, the presence of Fe(III) created a mixed-valence system that promoted electrical conductivity.

Another way to tackle the over-delocalisation issue concerns the use of the Hubbard correction, namely DFT + U. In this approach, on-site Coulomb interaction term (U) and site exchange term (J) are added to electrons in highly correlated orbitals. J and U are combined in the U effective parameter, or $U^{eff} = |U - J|$, which can be understood as an energy toll to force the electron occupation of selected orbitals to avoid unphysical over-delocalisation.[70,71] This methodology provides a computational affordable way to model solids with extended periodic structures; unfortunately, there is no universal way to choose the U^{eff} values and these are not easily transferable between systems. U^{eff} parameters are usually fitted to experimental band gaps;[62] thus this approach lacks predictive usage for this property. More interestingly, they can be obtained at *ab initio* level *via* linear response to variations in density related to changes in occupation of d orbitals, *e.g.*, for Ti–Cu metal atoms within the MOF-74 family.[72]

As a representative example of the impact of DFT + U on band gaps, Mn- and Co-MOF-74 electronic properties were examined with two GGA DFs (PBE and PW91) including dispersion corrections and U^{eff} parameters ranging between 0.00 and 8.00 eV.[73] The general trends were similar for both DFs, thus only PBE data will be discussed herein. The computed band gaps of these open-shell MOFs exhibited a clear dependence with the U^{eff} value of choice. The band gap values were underestimated at DFT level, while an excellent agreement with experiments was found when including the Hubbard approach. The optimal U^{eff} values were 3.0 and 5.3 eV for Co-MOF-74 (see Figure 5.9) and >5.5 eV for Mn-MOF-74.

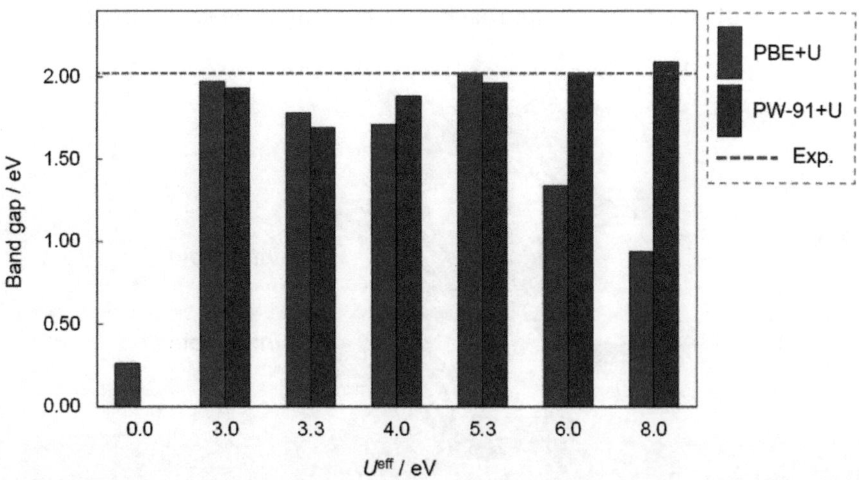

Figure 5.9 Computed and experimental band gaps for Co-MOF-74 with respect to the U^{eff} value.[73]

5.2.4 Magnetism

MOFs can exhibit magnetic properties due to the presence of unpaired electrons in their structure, either from open-shell metal centres or from interactions with guest molecules.[74] Quantum mechanical methods can describe the interaction between electrons from different centres, or isotropic magnetic coupling, and the interaction between electrons at the same centre, or anisotropic magnetic coupling. The isotropic magnetic coupling can be described at DFT level using either periodic or cluster models. The anisotropic magnetic coupling needs high level wave function methods and calculations are thus carried out on cluster models.

Some materials within the MOF-74 family have shown promising potential as 1D linear magnets due to their structural features.[75] Their magnetic properties have been widely explored from a computational standpoint, and we will therefore review these studies to exemplify how to model magnetism in MOFs in a general way. J coupling constants representing the interaction between metal centres in different chains are labelled as J_{inter}, while those interactions within the same chain are labelled as J_{intra} (see Figure 5.10).

Periodic DFT was employed to compute both isotropic couplings, J_{intra} and J_{inter}. The magnetic properties of Co-MOF-74, Ni-MOF-74, and Fe-MOF-74 were studied at PBE level.[76] Similar results were obtained for the three isostructural MOFs: ferromagnetic J_{intra} couplings were one order of magnitude larger than antiferromagnetic J_{inter} ones. Although this trend was in line with experiments, the ferromagnetic J_{intra} values were overestimated. The origin of this discrepancy is again the SIE, as in the band gap calculations (see Section 5.2.3). Thus, GGA + U and hybrid DFs were invoked to mitigate this problem.[77] The two methods led to very similar J_{intra} values for Fe-MOF-74, but in better agreement with experiments than the ones obtained at GGA level. The best results in terms of both accuracy and computational cost were achieved with the DFT + U approach. Furthermore, the screening of the U^{eff}

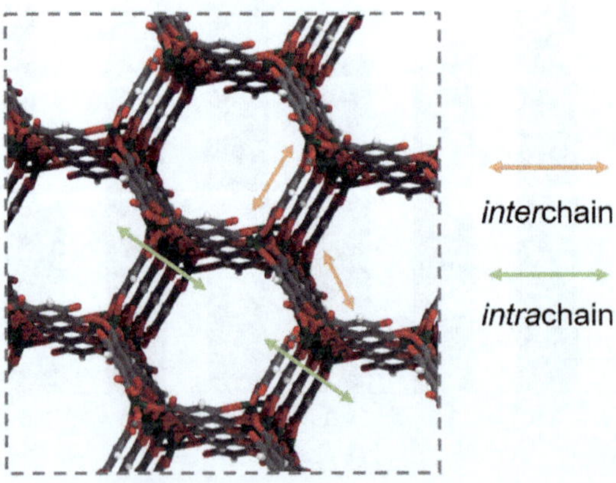

Figure 5.10 Intrachain and interchain interactions between the wires of Fe-based MOF-74. Atom legend: Fe = dark green, O = red, C = grey, H = white.

parameter showed that a value of 2.0 eV provided the best match with experiments.[78]

Regarding cluster models, an 88-atom cluster[79] was designed for Fe-MOF-74 containing three Fe atoms in the same chain (see Figure 5.11). This model cannot account for J_{inter}, but allows the computation of J_{intra} using a wide variety of DFs. Then, J_{intra} values were calculated with M06-L (meta-GGA) and M06 (hybrid) DFs, which provided a better agreement with experiments than those computed at GGA level. The same protocol was also employed to account for the small (but not negligible) variation of magnetic properties caused by the adsorption of either N_2[80] or hydrocarbons[78] on the open metal sites of Fe-MOF-74. The sensitivity of J_{intra} values with respect to the DF of choice was benchmarked by testing 25 DFs on the 88-atom cluster. Hybrid DFs performed best compared to experimental data, particularly those with 20–28% Hartree–Fock exchange. Moreover, the impact of the basis set and the cluster size (88-atom and 106-atom, see Figure 5.11) on J_{intra} values was evaluated, and the 88-atom model with a triple-ζ plus polarisation basis set was the best cost–accuracy set up.

Anisotropic couplings arising from ligand field and spin–orbit interactions were studied for bare Fe-MOF-74[79] and with adsorbed hydrocarbons[78] using cluster models. Single-ion anisotropy was computed on the 88-atom cluster by means of wave-function-based methods with state-averaged complete active space and multireference perturbation theory including spin–orbit coupling. Anisotropic couplings can be affected by the presence of paramagnetic ions in the second coordination sphere. However, since this influence is mostly of electrostatic character, the substitution of paramagnetic ions by diamagnetic ones of similar ionic radius and same nominal charge was reasonable. Therefore, two peripheral Fe(II) were substituted by Zn(II). The single-ion anisotropy computed values placed Fe-MOF-74 in the weak-exchange regime. This and the noncollinear character of the Fe(II) ions anisotropy suggested that the MOF would present fast magnetic relaxation pathways, which was consistent with experiments.

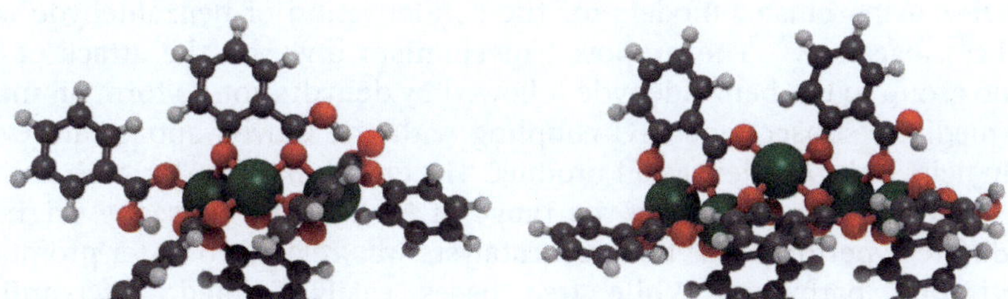

Figure 5.11 Molecular models for Fe-based MOF-74 containing three metal atoms (88-atom cluster on the left) and five metal atoms (106-atom cluster on the right).[78] Atom legend: Fe = dark green, O = red, C = grey, H = white.

Furthermore, Fe-MOF-74 magnetic anisotropy was predicted to be moderately affected when considering hydrocarbons' adsorption.

5.3 Catalysis

After a proper characterisation of electronic and structural properties of MOFs (see Section 5.2), it is now time to address their applications in catalysis. The key feature of MOF catalysts is the presence of homogeneously-distributed and well-defined metal–organic moieties within an extended porous solid. These sites have the potential to combine the high activity and selectivity of homogeneous catalysts with the facile separation and recycling properties of heterogeneous systems.[81–85] In this scenario, MOFs can play several roles: (i) encapsulating catalytic species, (ii) hosting nanoparticles, (iii) supporting catalytic sites, and (iv) promoting reactivity themselves.[86] The latter two types will be covered herein, with particular emphasis on computational studies reporting reaction mechanisms, rate-determining steps (RDS), and energy profiles. The catalytic applications of COFs are not as widely developed as those for MOFs and fewer computational mechanistic studies are available. The opposite is true for zeolites, where the literature is immense; we therefore refer to computational reviews for detailed information,[34–36] including specialised contributions on methanol to olefin[87] and methane oxidation[88] processes.

5.3.1 Condensation Reactions

Due to the Lewis acidic character of some MOFs bearing open metal sites, they have been used as heterogeneous catalysts for condensation reactions. The general reaction mechanism for the MOF-catalysed aldol and Knoevenagel condensations (see Scheme 5.2) entails coordination of the electrophile to the node, nucleophilic attack on the activated substrate, and dehydration followed by production release.

Early studies proved that IRMOF functionalised with NH_2 groups performed better for the Knoevenagel condensation than the parent material.[89] DFT simulations evaluated the activity of free amines and the amino-based IRMOF-3 using cluster models for the condensation of benzaldehyde with ethyl-cyanoacetate.[90] The proposed mechanism involved the attack of the amino group to the benzaldehyde followed by dehydration to form an imine intermediate; subsequent C–C coupling with the ethyl-cyanoacetate eventually yielded the condensation product. The energy barriers for both aniline and IRMOF-3 were similar in the range of 2.0–2.2 eV. Thus, the origin of the different performance between catalysts was related to water-promoted deactivation pathways. While free bases easily formed water–aniline complexes, the nodes of IRMOF-3 captured water molecules and protected aniline linkers from deactivation.

The open metal sites in Cu-BTC could also act as Lewis acid catalysts for the Knoevenagel reaction between benzaldehyde and propanedinitrile.[91] Periodic and cluster DFT simulations were employed to unravel the

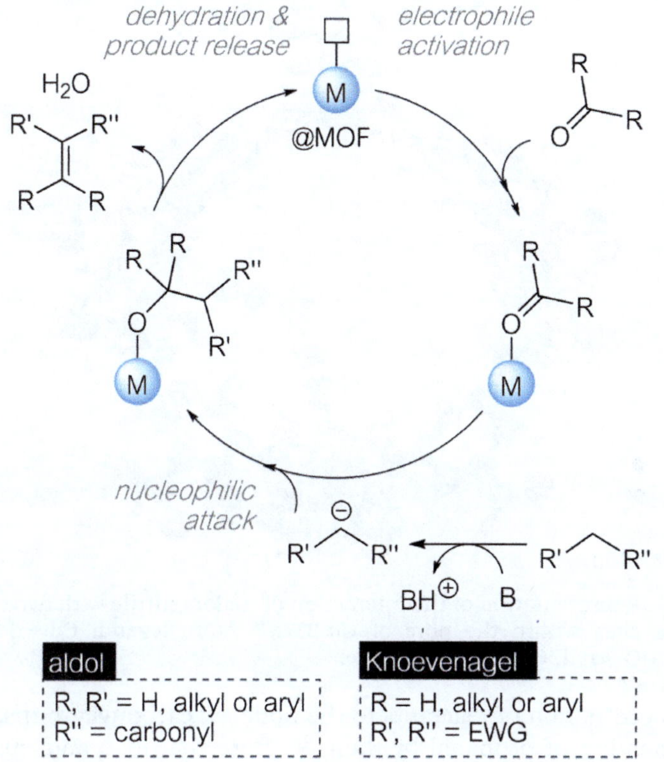

Scheme 5.2

condensation mechanism.[92] The initial deprotonation of propanedinitrile to form the reactive carbanion took place in two steps: first, the oxygen from benzaldehyde abstracted an acidic hydrogen while one nitrile attacked the carbonyl group forming a transient C–N bond; then, the proton was transferred to a carboxylate linker, creating a transient defect and recovering the nitrile moiety. The resulting carbanion attacked the carbonyl group of benzaldehyde and, at the same time, the linker transferred back the proton and healed the defect. Subsequent dehydration yielded the condensation product. The RDS concerned the last dehydration process and entailed 2.28 eV above the initial reactants. As a final remark, the use of periodic models was recommended due to the interaction of the substrate with several nodes within the MOF pore (see Figure 5.12). Similar roles of metal–linker moieties in deprotonation steps were also proposed for the Knoevenagel reaction catalysed by the MOF-74 family of materials.[93]

The UiO family is well-known for the dual role of the $Zr_6O_4(OH)_4$ nodes containing hydroxo and oxo groups as Brønsted bases and open metal sites as Lewis acids. As such, UiO-66 and UiO-66-NH_2 showed catalytic activity towards the cross-aldol condensation of benzaldehyde with propanal as well as the self-aldol condensation of propanal.[94] The reaction pathway started

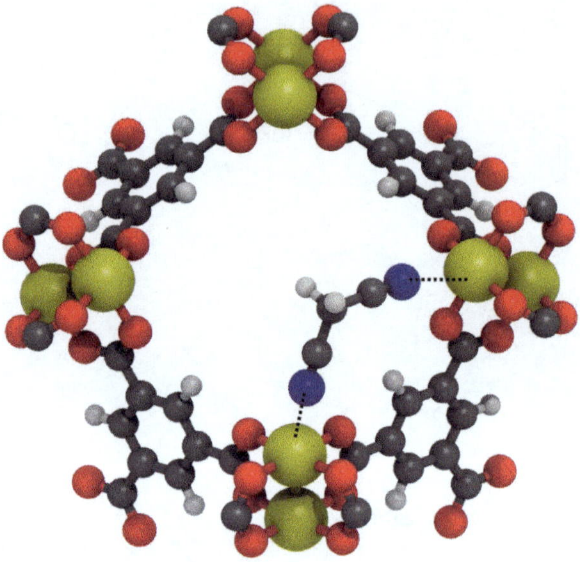

Figure 5.12 Representation of the interaction of malononitrile with two open metal sites within the pore of Cu-BTC.[92] Atom legend: Cu = dark yellow, O = red, C = grey, H = white.

with the coordination of reactants to the node *via* carbonyl groups, followed by deprotonation of propanal by the μ_3-O from the node and nucleophilic attack of the enolate to the aldehyde to form a new C–C bond. Subsequent proton transfer and second μ_3-O-mediated deprotonation provided the condensation product and a hydrolysed node. Final dehydration regenerated the catalytic active site. The extended cluster model (capped with benzoate and *o*-NH_2-benzoate linkers) predicted an activation free energy of 1.04 eV for the second deprotonation RDS, while the periodic model (with one missing linker) gave an activation free energy of 1.24 eV for the dehydration RDS. Although some intermediates were stabilised in the periodic structure due to pore confinement effects, both models provided similar qualitative results. Regarding catalyst functionalisation, the activation free energies for the aldol condensation steps in the UiO-66-NH_2 model were similar to those found in the UiO-66 model, thus no direct role during the mechanism could be assigned to the amino groups.

UiO-66 also catalysed the esterification of carboxylic acids with alcohols, particularly when defective nodes were present. Periodic DFT simulations (with two missing linkers) were performed to identify the reaction mechanism between propionic acid and methanol.[95] Both hydrated and dehydrated nodes were computationally evaluated, which corresponded to materials without and with heat treatment by experiments, respectively. The general mechanism entailed coordination of the acid to Zr to activate the C=O group, nucleophilic attack of the methanol/methoxy, and rate-determining water removal to generate the ester product. The relative activation free energies for the dehydration step involving hydrated and dehydrated nodes were 0.32 eV and 0.56 eV, respectively, which agreed with a decrease in catalytic activity observed after thermal treatment. Finally, and in line with previous reports,[94] the NH_2 groups in UiO-66-NH_2 models only played a spectator role.

An imine-based COF decorated with acid and base molecules through supramolecular non-covalent functionalisation exhibited activity in the Knoevenagel condensation.[96] Periodic DFT simulations at the PBE-D2 level explored the interactions between CH_3COOH, NEt_3, and COF. The most important bonding patterns entailed the formation of a $CH_3COOH\cdots NEt_3$ adduct and the H-bonding between the acid and the imine group. Later, the effect of the confinement in the Knoevenagel reaction was evaluated by computing transition states at molecular level (M06) and then performing constrained geometry optimisations of these structures within the pores (PBE-D2).

5.3.2 Hydrolysis of Nerve Agents

The tuneable porosity and Lewis acid properties of heterogeneous MOF materials make them suitable for adsorption, separation, and conversion of gases. These features have been successfully exploited for the degradation of organophosphorus nerve agents (see Scheme 5.3).[97–99] As discussed below, most efforts are centred around materials containing Zr-based nodes due to their high thermal, mechanical, and hydrolytic stabilities.[100,101] Only recently has the post-synthetic functionalisation of COFs found applications in this field.[102]

An efficient degradation protocol to mitigate toxicity comprises the hydrolysis of P–F bonds in organophosphorus nerve agents (P–OR bonds in nerve agent simulants). The general mechanism in MOFs entails coordination of the substrate to the node *via* the O atom of P=O, nucleophilic attack of Zr–OH (intramolecular) or H_2O (intermolecular) to the P atom, Berry pseudorotations followed by elimination of the leaving group (LG), and product desorption from the node (see Scheme 5.4).

Experimental studies reported a fast degradation of GD and simulant DMNP promoted by NU-1000, which features 8-connected nodes compared to the 12-connected ones in the UiO family.[103] Interestingly, a heat treatment

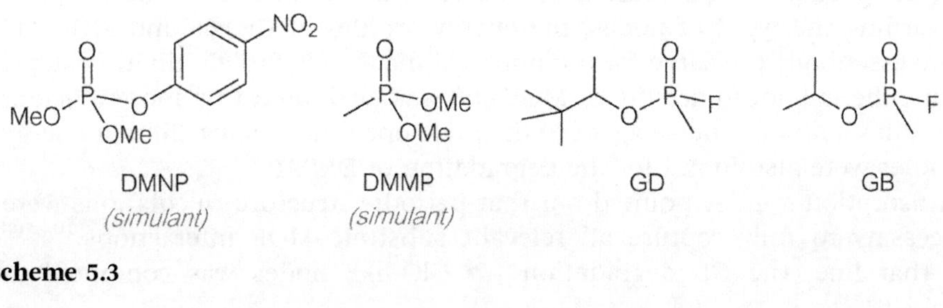

Scheme 5.3

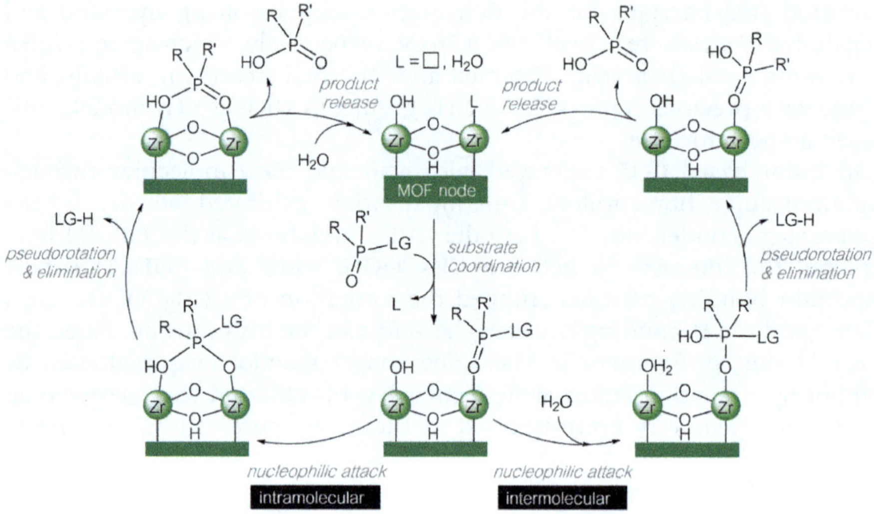

Scheme 5.4

of the NU-1000 MOF increased the reaction rates. DFT calculations on cluster models shed light into the coordination modes of different nerve agents as well as their reaction energetics. This computational work was later complemented by a mechanistic study on DMNP hydrolysis using pristine (before thermal treatment) and dehydrated distorted (after thermal treatment) NU-1000 nodes.[104] Apart from the initial dehydration step, which is not necessary in the distorted node, both systems presented similar free energy profiles with RDS of 0.45 eV and 0.41 eV, respectively.

The GB degradation *via* defective UiO-66 and MOF-808 was studied using cluster models capped with formate ligands.[105] The catalytic site of the UiO-66 model contained one Zr–OH and one Zr with a vacant site. The cycle started with an exothermic coordination of GB to the Zr open metal site *via* its oxygen atom. The subsequent nucleophilic attack of Zr–OH to the P atom required 0.57 eV and gave rise to five-coordinate P intermediates bridging two Zr sites. Berry pseudorotations were required to place the leaving groups (F or ^{i}PrO) in the apical position of the trigonal bipyramid. Overall, the catalyst regeneration *via* product desorption was the RDS. The inclusion of one explicit water molecule increased the energy barrier of the nucleophilic attack by 0.15 eV, but decreased the catalyst inhibition by the product. Regarding the types of nodes, the energy profiles of UiO-66 and MOF-808 were essentially the same for the nucleophilic attack and elimination steps. Thus, the enhanced activity of MOF-808 was attributed to a more efficient coordination of the nerve agent to the node open metal sites. Similar energy profiles were also found for the degradation of DMMP.[106]

Adsorption studies pointed out that periodic structure calculations were necessary to fully capture all relevant substrate–MOF interactions.[107,108] In that line, the GB degradation *via* UiO-like nodes was computed at periodic DFT level using different solvent molecules as nucleophiles: H_2O, MeOH, and ^iPrOH.[109] All three systems shared a similar energy profile where the rate-determining pseudorotations required 0.46 eV, 0.50 eV, and 0.60 eV, respectively.

Coming back to the coordination site availability, the different connectivity patterns of Zr-based nodes play an important role in substrate binding and reactivity. Free energy profiles for GB hydrolysis were reported using both cluster and periodic models for a variety of nodes with different amounts of Zr coordination sites: 10-connected defective UiO-66, 8-connected NU-1000 (two different locations), and 6-connected MOF-808 (see Figure 5.13).[110]

In contrast to previous studies,[105] these simulations employed nodes bearing Zr–OH and Zr–OH_2 groups, thus the initial coordination of substrates was now considered in the reaction path.[110] Using models with benzoate linkers, the substitution of the Zr-bound water by GB was the RDS for UiO-66 and NU-1000, while the nucleophilic attack was the limiting step for MOF-808 (see Figure 5.14). The computed activation free energies followed the trend MOF-808 $<$ NU-1000 $\approx$ UiO-66, in line with the highest activity observed for MOF-808. Additional DFT calculations predicted a higher activity for defective Zr_{12} nodes compared to Zr_6 nodes in MOF-808.[111]

Not only node connectivity, but also linker functionalisation can have an impact on degradation kinetics. DFT simulations (cluster models with explicit water molecules) evaluated the role of NH_2 groups in UiO-66-NH_2 clusters for the water nucleophilic attack on DMNP.[112] Water deprotonation by NH_2 demanded 0.74 eV, while the previously-discussed deprotonation by Zr–OH only required 0.41 eV (see Figure 5.15). Thus, the NH_2 groups did not behave as Brønsted bases, but rather participated in tuning the coordination shell around the catalytic site. Moreover, the free energy barriers for UiO-66-NH_2 (0.41 eV), UiO-66 (0.49 eV), and UiO-66-OH (0.58 eV) followed the experimental reactivity trend. The functionalisation of linkers with NH_2 and F groups was also evaluated for the hydrolysis of GB.[110] For UiO-66, the NH_2 groups decreased the rate-determining water displacement by 0.21 eV.

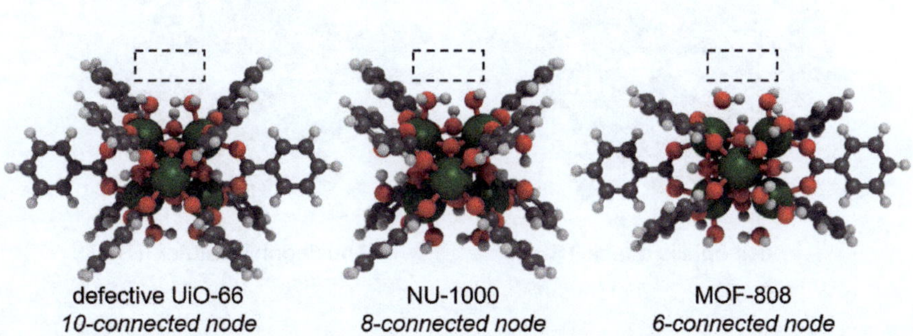

Figure 5.13 Cluster models for Zr_6-based nodes of defective UiO-66, NU-1000, and MOF-808.[110] The absence of linkers, which creates coordination sites, is indicated with dashed lines. Atom legend: Zr = green, O = red, C = grey, H = white.

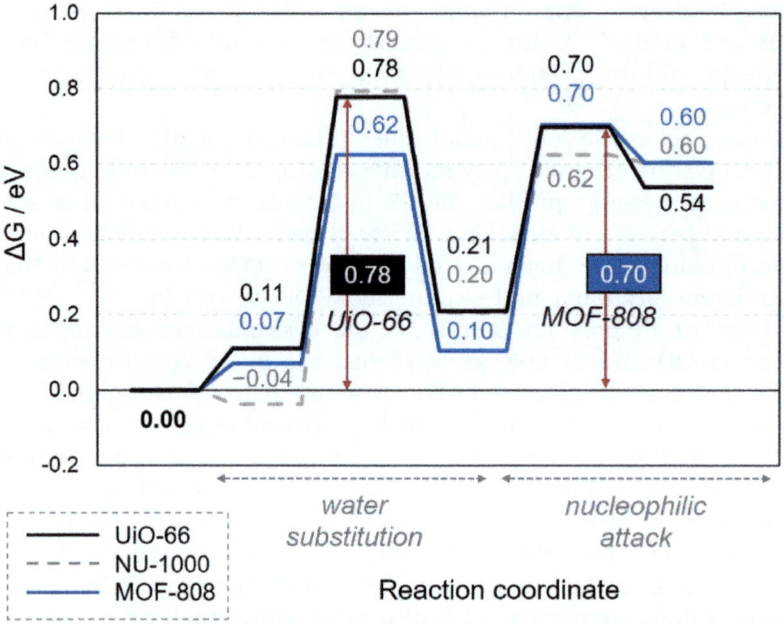

Figure 5.14 Free energy profiles for the hydrolysis of GB catalysed by UiO-66, NU-1000, and MOF-808 nodes.[110]

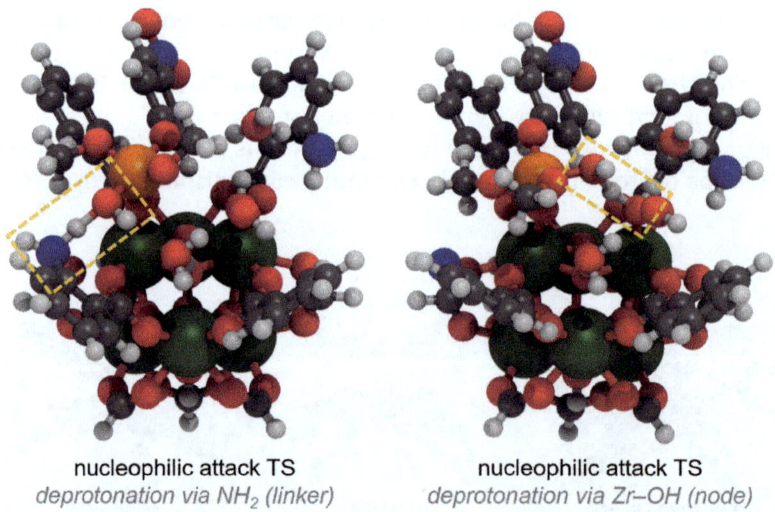

Figure 5.15 Transition state structures of water nucleophilic attacks in UiO-66-NH_2 cluster models.[112] Atom legend: Zr = green, P = orange, O = red, N = blue, C = grey, H = white.

For MOF-808, the F and NH_2 groups had a minor impact on the rate-determining nucleophilic attack, reducing the activation free energy by only 0.05 eV. In both cases, perfluorination of the linkers had an overall detrimental effect on the reaction kinetics.

UiO-66-X materials functionalised with halogen atoms in the linkers (X = F, Cl, Br, I) were prepared and tested for the decomposition of DMNP.[113] Among them, only UiO-66-I showed higher catalytic activity compared to the parent material. As in previous works, the water nucleophilic attack was computed on UiO-66 and UiO-66-I, where the halogen atom was placed in both *ortho* and *meta* positions when using benzoate as a model for the real BDC linker. The computed activation free energies for UiO-66 and UiO-66-*o*-I were 0.86 eV and 0.48 eV, respectively. Such differences came from the formation of strong halogen bonds between the methoxy groups of the substrate and the proximal iodine of the linker. As expected for UiO-66-*m*-I, where the iodine atom is placed further away from the catalytic site, the activation free energy (0.85 eV) was essentially the same as that for UiO-66 (0.86 eV).

Node functionalisation is also an appealing approach to enhance catalytic activity. In that sense, DFT computations reported the hydrolysis of GB *via* NU-1000 nodes functionalised with single-metal atoms, M_1-NU-1000, where M stands for metals with oxidation states (II), (III), and (IV).[114] Adsorption energies and activation free energies for the nucleophilic attack and the product desorption steps narrowed down the candidates from 36 to 19 systems. Turnover frequency (TOF) estimations pointed towards Cu(III) as the most promising catalyst (1000 times faster than the parent NU-1000), followed by Fe(III), W(IV), and Ni(II). Additional studies were carried out on bimetallic M_2-oxo clusters (M = Mg, Al, V, Cr, Mn, Fe, Co, Ni, Cu, Zn) for the gas-phase hydrolysis of DMNP.[115] Using the product desorption as the initial criterion (*i.e.*, the determining step for the NU-1000 system), the $Al_2O_2(OH)_2$ cluster was selected for further gas-phase mechanistic studies. Nevertheless, the large energy barrier found for the water nucleophilic attack rendered a material with similar activity to the parent NU-1000. The inclusion of one spectator water molecule did decrease the transition state energy by 0.18 eV, but such an effect was unlikely to affect the overall kinetics of NU-1000, which was limited by product desorption.

5.3.3 Dimerisation of Ethylene

The controlled oligomerisation of ethylene, directly linked to co-monomer synthesis in polymer production, is one of the few industrial processes that still relies on homogeneous catalysts. For that reason, this process is an appealing target to explore in heterogeneous MOF catalysis. The most common reaction pathway is the so-called Cossee–Arlman mechanism (see Scheme 5.5). After catalyst activation, the mechanism entails ethylene coordination and its subsequent insertion to produce a butyl intermediate, β-hydride elimination to form a metal–hydride, 1-butene displacement by ethylene, and second

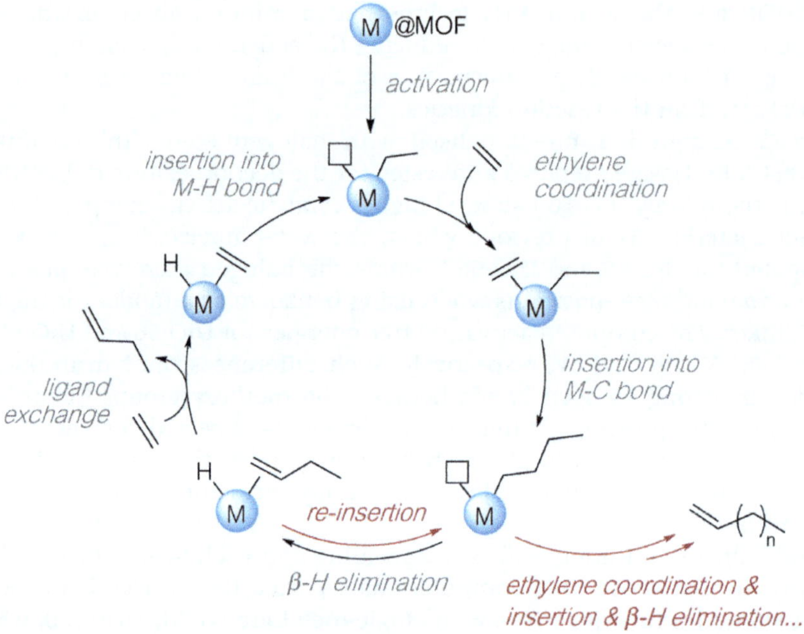

Scheme 5.5

insertion to regenerate the catalyst. Oligomers can be obtained if re-insertion of 1-butene and ethylene coordination compete with β-hydride elimination. The alternative metallacycle mechanism *via* concerted reductive coupling is usually operative for early transition metal catalysts.

As representative examples, well-defined molecular Ni complexes anchored to MOFs exhibited excellent catalytic activity towards ethylene dimerisation.[116,117] Likewise, the Ni-containing MFU-4*l* MOF showed superior activity and selectivity compared to homogeneous counterparts.[118,119]

A variety of transition metal species deposited on the nodes of NU-1000 were active towards ethylene oligomerisation, such as Ir,[120] Rh,[121] and Ni.[122] DFT calculations were carried out on cluster models of M_1-NU-1000 based on Co(II) and Ni(II).[123] Three different reaction mechanisms were computed for Ni-NU-1000: non-activated Philips-type dimerisation, metallacycle pathway, and Et_2AlCl-activated Cossee–Arlman route; but only the latter was energetically feasible. Similar trends were found for the enthalpy profiles of Co and Ni-based catalysts, where ethylene insertion into the M–C bond was the RDS with activation enthalpies of 1.04 eV and 0.68 eV, respectively (see Figure 5.16).

Subsequent studies on Ni-NU-1000 indicated that periodic models were in agreement with molecular clusters and Ni_1 and Ni_4-hydroxo species (see Figure 5.17) presented similar catalytic activity.[124] Under that premise, a thorough catalyst screening was then carried out on M_1-NU-1000 models (17 candidates).[125] The RDS depended on the nature of the metal and its oxidation state, thus the evaluation of full catalytic cycles was recommended

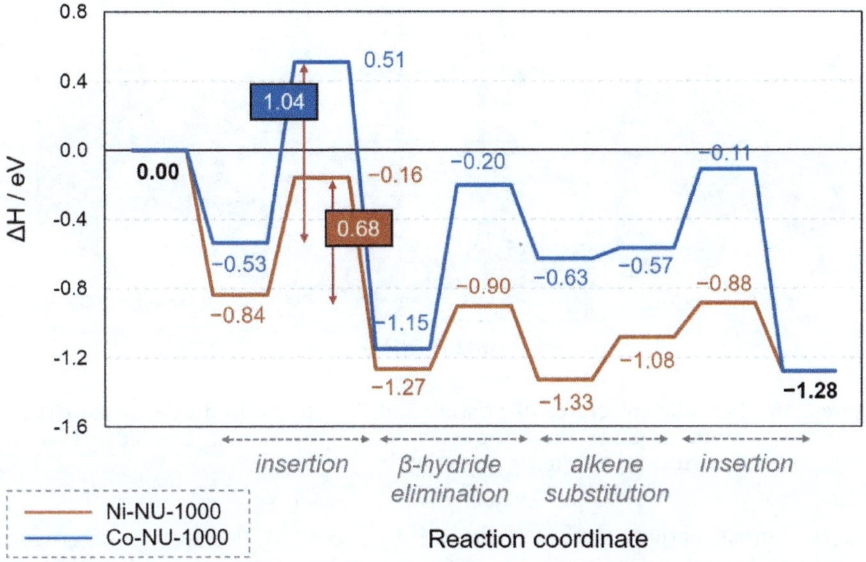

Figure 5.16 Enthalpy profiles for ethylene dimerisation catalysed by Ni- and Co-NU-1000 nodes.[123]

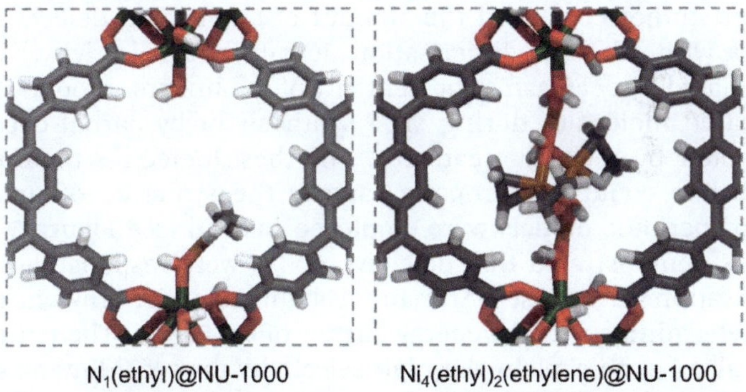

Figure 5.17 Periodic structures of $[Ni_1]$ and $[Ni_4]$ species deposited on NU-1000 nodes for ethylene dimerisation.[124] Atom legend: Zr = green, Ni = brown, O = red, C = grey, H = white.

when exploring a variety of potential catalytic species. In line with experiments, the best catalyst was Ni(II)-NU-1000, with an activation free energy of 0.60 eV (with ethylene insertion as the RDS). Other good candidates were those based on Cr(II), Co(III), and Pd(II) with activation free energies of 0.76 eV (ethylene insertion), 0.65 eV (β-hydride elimination), and 0.71 eV (1-butene desorption), respectively. Related M_2–hydroxide and M_1-NU-1000 models were employed to build scaling relationships using the free energies of intermediates M–hydride–ethylene and M–ethyl as main descriptors.[126] In line with the above-mentioned reports, microkinetic modelling indicated that Ni-NU-1000

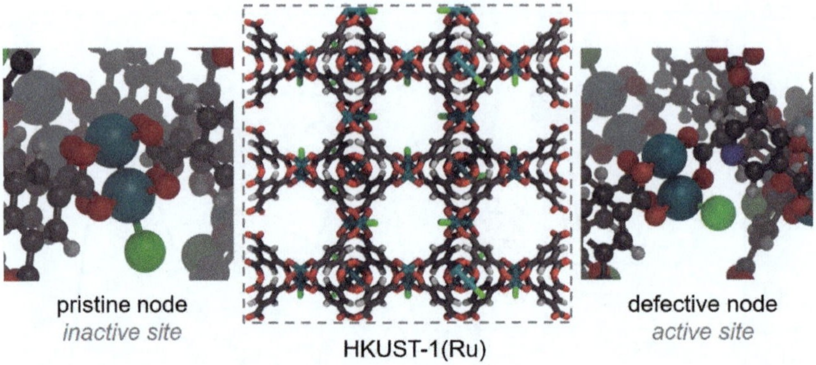

Figure 5.18 Periodic structures of pristine and defective mixed valence Ru-BTC.[124] Atom legend: Ru = turquoise, Cl = light green, O = red, N = blue, C = grey, H = white.

was the most active catalyst followed by Co-NU-1000, and a degree of rate control analysis pointed out different determining steps for different metal catalysts.

Initial tests on mixed valence Ru-BTC MOFs indicated that the controlled inclusion of defects had a positive impact on catalytic performance.[127] Subsequent studies correlated the amount of engineered defects *via* missing linkers with ethylene dimerisation activity and selectivity.[128] Defects were modulated by decreasing the denticity of the linkers through the use of pyridine-based molecules during MOF synthesis or by partial decarboxylation promoted by thermal treatment. For these defective materials, molecular clusters could not properly capture the structure of unsaturated nodes, thus periodic models were employed instead (see Figure 5.18). DFT simulations demonstrated that defective nodes were responsible for ethylene dimerisation *via* a Cossee–Arlman mechanism, where ethylene insertion was rate determining with an energy barrier of *ca.* 0.9 eV. The nature of the node was also key to explain the high selectivity towards dimers, since the 1-butene desorption from the bimetallic site was 0.65 eV more favourable than that from a monometallic model.

5.3.4 Oxidation of Alkanes

The selective functionalisation of light alkanes, for either transport or bulk chemical production, is a cornerstone process in industry. Pioneering experiments demonstrated the catalytic activity of Fe-diluted Mg-MOF-74 towards ethane to ethanol oxidation by N_2O.[129] Detailed electronic structure calculations suggested the participation of high-spin Fe(IV)=O intermediates in the catalytic cycle. A complete mechanistic study was later reported using cluster models at M06-L level.[130] The accuracy of the DF to properly describe the spin states of Fe was benchmarked against MR methods. The proposed

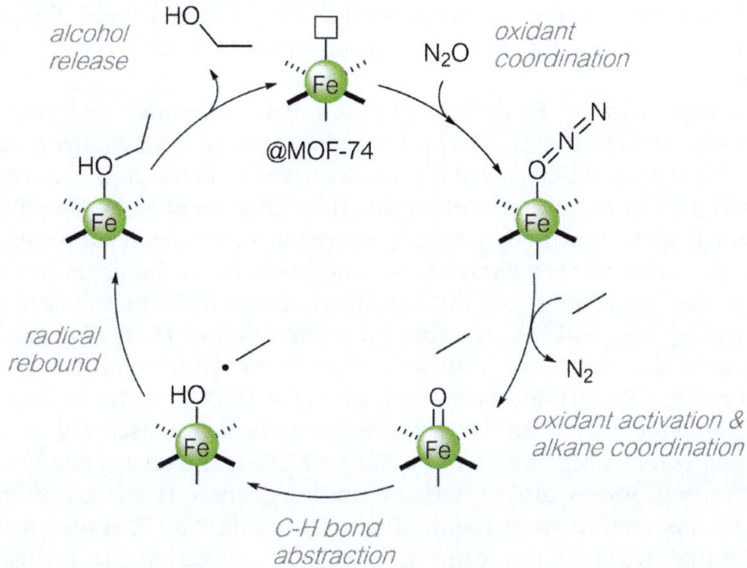

Scheme 5.6

mechanism at the Fe(II) centre of Fe-diluted Mg-MOF-74 (see Scheme 5.6) involved N_2O coordination and activation to form an oxo intermediate, C–H bond abstraction, radical rebound to form ethanol, and product release. Formation of Fe(IV)=O was the RDS with an activation enthalpy of 0.85 eV. Hydrogen abstraction from ethane by the oxo moiety required 0.60 eV and yielded ethyl and Fe(III)–OH radicals, which rapidly rebound *via* hydroxylation to form ethanol. Alternative pathways *via* desaturation (second H abstraction to form ethylene) and dissociation (release of the ethyl radical) presented higher activation enthalpies. Lastly, decoordination of ethanol to regenerate the Fe(II) catalytic site required 0.84 eV, but it was unlikely to compete with N_2O activation (0.85 eV) when entropic effects were included. Independently and simultaneously, a similar mechanism was reported using a quantum mechanics/molecular mechanics approach (B3LYP/UFF).[131]

The reactivity of the oxo-centred $FeZn_3O(pyrazole)_6$ cluster was computationally tested for ethane oxidation.[132] After initial N_2O activation to form the active Fe–oxo site (0.92 eV), ethane C–H bond abstraction followed a σ pathway with a collinear Fe–O–H angle (*via* 0.98 eV) over the alternative π pathway with a bent Fe–O–H angle (*via* 1.17 eV). Similarly, Fe-based oxo-centred nodes from the MIL family were explored due to their structural similarities with the Fe centres of MOF-74, *i.e.*, a square-pyramidal environment with one open metal site.[133] Cluster models capped with formate linkers were employed and spin-state energetics at M06-L level were compared to MR methods. Overall, a similar reaction pathway was found (see Scheme 5.6), where N_2O activation was the RDS with an activation enthalpy of 1.45 eV, and C–H bond abstraction of ethane and methane required 0.41 eV and 0.62 eV, respectively. The predicted feasibility of this reaction was later confirmed by experiments for the conversion of propane.[134]

The incorporation of Fe sites in Al-based MIL-53 (formed by [AlO_6] octahedra connected by BDC linkers) provided a material with catalytic activity towards the H_2O_2-mediated oxidation of methane to methanol.[135] According to periodic DFT simulations at PBE level, the exchange of Fe by Al within the MOF structure was thermodynamically more favoured than its deposition on extra-framework sites (see Figure 5.19). The reaction mechanisms were then computed for the mixed-metal MIL-53(Al,Fe) (mono- and dimeric Fe species) as well as Fe-based MIL-53. The initial coordination of H_2O_2 displaced one carboxylate linker from Fe. Homolytic O–O dissociation followed by water removal created Fe–OH and Fe–O–Fe species (or Al–O–Fe). The bridging oxo group abstracted a H from CH_4, yielding a radical CH_3 (see Figure 5.19), which rapidly recombined with OH to form CH_3OH. The energy profiles were similar for both mono- and dimeric Fe catalysts, where H_2O_2 activation and C–H bond abstraction steps required *ca.* 1.1 eV and 0.8 eV, respectively. As for the Fe-based MIL-53, the resulting oxo specie was very stable, resulting in a higher C–H bond abstraction barrier of 1.33 eV. Complete and comprehensive studies of the full catalytic network were reported shortly afterwards.[136] These calculations revealed the origin of the high selectivity, as the over-oxidation of methane presented higher barriers than methanol release. However, it was found that the catalyst also promoted the undesired H_2O_2 oxidation side-reaction, suggesting the need for alternative oxidants.

With the great variety of MOFs, including both synthesised and hypothetical structures, screening studies soon populated the literature. Building on the above-mentioned results of Fe-based MOF-74,[130] a screening algorithm was designed to detect feasible open metal sites able to accommodate an Fe=O moiety for the N_2O-mediated oxidation of ethane.[137] After an

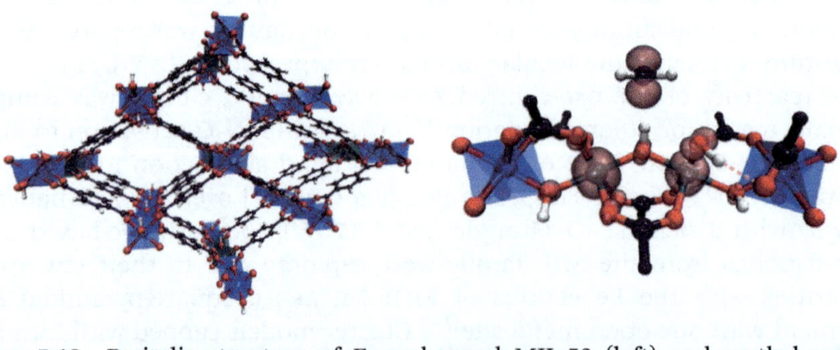

Figure 5.19 Periodic structure of Fe-exchanged MIL-53 (left) and methyl radical intermediate (and its corresponding spin density) formed after C–H bond abstraction (right).
Adapted from ref. 136 with permission from the Royal Society of Chemistry.

automated screening of >5000 MOFs followed by a supervised refinement of few structures (including geometry and transition state optimisations), seven Fe-based MOFs were identified as promising candidates. Further catalytic tests using Fe-BTT (where BTT stands for 1,3,5-benzenetristetrazolate) confirmed the potential of these predictions. On a similar note, a computational screening of Fe-based MOFs was performed for the same reaction using a bottom-up approach, where different linkers were connected to Fe centres to assemble a variety of potential candidates.[138] A thorough study indicated that the α LUMO energy of the Fe=O intermediate could be used as single descriptor to analyse and predict reactivity. The activation enthalpies of Fe oxidation *via* N_2O decreased as Fe 3d levels decreased, while the activation enthalpies of C–H bond abstraction behaved otherwise. These trends created a volcano-type diagram, where an optimal range of Fe 3d states was found to balance the activation energies of the catalytic cycle.

A fully automated workflow was designed to perform high-throughput periodic DFT simulations, including volume relaxation, ionic relaxation, spin states, and adsorption energies.[139] The protocol screened >800 MOFs to identify open metal sites for the oxidative methane activation. Among all the MOFs tested, the largest energy barrier usually involved the oxidation of the metal centre *via* N_2O activation. Rather than focusing on the C–H bond abstraction step, this work pointed towards the search of active sites that can be readily oxidised. Soon after, this workflow was employed to screen a subset of 60 MOFs for the same reaction.[140] The formation of metal–oxo active sites was successfully employed to build scaling relationships between their reactivity towards C–H bonds as well as their thermodynamic stability. Such formation energy should be neither too endothermic (low stability but high reactivity) nor too exothermic (high stability but low reactivity), as expected from the Sabatier principle. Finally, the coupling of the active site formation with bond-forming or bond-breaking events was proposed to circumvent the above-mentioned scaling relationship. Indeed, for Cr- and Mo-BTC MOFs the metal–oxo formation encompassed the metal–metal bond breaking at the node, and the predicted reactivity did not follow the previous linear correlation. This situation might also extrapolate to materials with reversible linker mobility.

The functionalisation of NU-1000 nodes with CuO species proved very effective for the aerobic oxidation of methane to methanol, which involved catalyst activation with O_2, methane activation, and steam treatment to release the product.[141] Computational studies addressed both the speciation and structure of Cu-based compounds as well as the complete reaction mechanism.[142] Metal-decorated clusters bearing benzoate linkers near the deposition sites were modelled at M06-L level. The structures of the Cu-decorated materials had $[Cu(OH)_2]$ and $[Cu_2(OH)_4]$ fragments anchored to the nodes. Starting from them, a range of different species were evaluated after activation with O_2. The computed phase diagram showed that the most stable species contained hydroxo groups, where only minor contributions

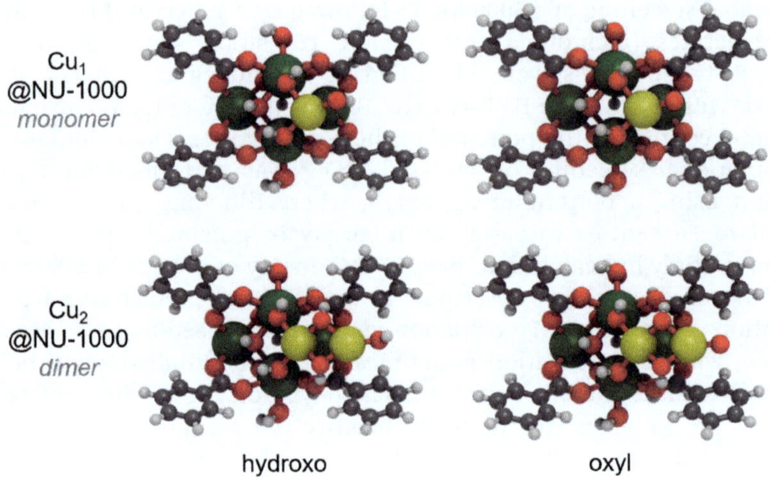

Figure 5.20 Speciation of [Cu] species deposited on NU-1000 nodes for methane to methanol oxidation.[142] Atom legend: Zr = green, Cu = dark yellow, O = red, C = grey, H = white.

came from those bearing oxyl (see Figure 5.20). According to the activation free energies for C–H bond abstraction and rebound steps, the dominant Cu–hydroxo species exhibited low activity. However, the participation of mono- and dinuclear Cu–oxyl species with activation free energies of 0.75 eV and 0.87 eV, respectively, did explain the experimental results. Similar [Cu_2O_2] units were optimised at the B3LYP level in cluster models within the MOF-808 framework.[143] As a final remark, the methane C–H bond abstraction step was also studied on bimetallic CuM species (M = Fe, Co, Ni, Cu, Zn) deposited on NU-1000 nodes.[144] It was found that activation free energies correlated well with reaction free energies as well as with key bond lengths in transition state structures.

So far, this section has focused on the selective conversion of alkanes to alcohols. However, the oxidation of alkane towards alkenes through an oxidative dehydrogenation process is also crucial for industry, particularly the propane-to-propylene transformation. In this line, NU-1000 nodes were decorated with Co atoms which exhibited high activity and selectivity towards propylene under mild conditions.[145] Cluster-based calculations at M06-L level were carried out to gain insight into the reaction mechanism (see Scheme 5.7). Catalyst activation was expected to provide a mononuclear Co(III) centre and one oxyl group attached to the node. The C–H bond abstraction of the secondary hydrogen of propane had an activation enthalpy of 0.55 eV and resulted in an intermediate bearing Co–isopropyl and oxyl moieties. The hydrogen abstraction by the latter (0.69 eV) was more favourable than the isopropyl migration to form the C–O bond of isopropanol (1.00 eV), which agreed with the kinetic control observed in the experiments. A dinuclear Co cluster supported on NU-1000 was also found to be competent for this reaction and the

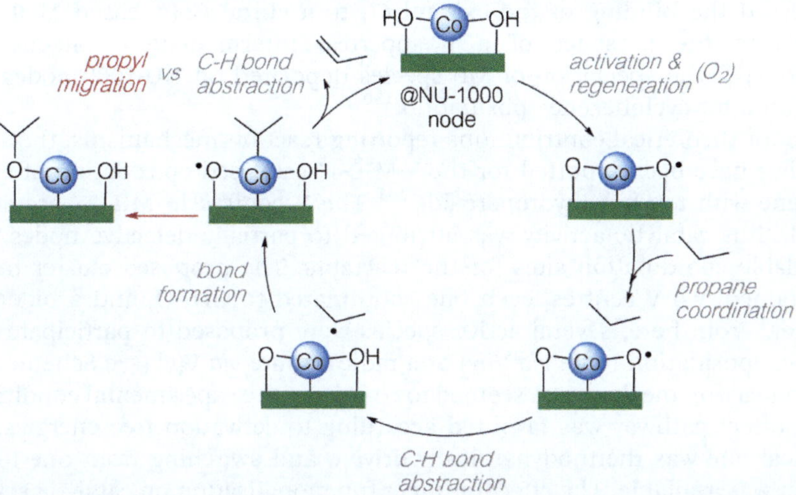

Scheme 5.7

computed activation enthalpies for CoM derivatives (M = Mo, Ti, Mn, Fe, Ni, Cu, Zn, Ag) correlated well with simple descriptors, such as spin densities and CM5 atomic charges of oxyl groups.[146]

Fe_2M oxo-centred nodes were computationally evaluated for propane dehydrogenation promoted by N_2O.[147] The reaction mechanism is similar to that for alkane oxidation (see Scheme 5.6) where the radical rebound is substituted by a second C–H bond abstraction. For the Fe_3 cluster, activation of N_2O to form the Fe=O intermediate required 1.52 eV and the subsequent relative C–H bond abstraction barrier was only 0.21 eV. These two steps were screened using Fe_2M species (M = Ti, V, Cr, Mn, Fe, Co, Ni, Cu, Zn), where M was considered both the active site and a spectator metal. The resulting energy barriers were correlated to the oxo formation energy; as this descriptor became more endergonic, the N_2O activation step increased and the C–H bond abstraction step decreased. Overall, the most plausible materials for synthesis were those based on Mn, Fe, Co, and Ni. The predicted activity trends were further confirmed by experiments using cyclohexane as substrate.

5.3.5 Oxidation of Alkenes

The oxidation of alkenes is an important process that leads to a variety of products for industrial use. To narrow down this general view, the present section focuses on the epoxidation of alkenes. This process has a rich reaction network and a typical product distribution contains epoxides, carbonyl groups, and diols. Several computational studies have shed light on different aspects of this reaction. An organometallic [Mn] compound, confined within the pores of MIL-101, was able to oxidise dihydronaphthalene in a similar way as its homogeneous analogue;[148] periodic DFT studies analysed the binding of the oxidant O_2 to a chiral Co(II)-based MOF and indicated the presence of a η^1-superoxo intermediate in alkene oxidations;[149] the speciation of Mo species deposited on NU-1000 nodes was explored for cyclohexene epoxidation.[150]

As for theoretical contributions reporting reaction mechanisms, thorough studies have been reported for the V-MIL-47-catalysed epoxidation of cyclohexene with *tert*-butylhydroperoxide.[151] The V centres in MIL-47 are saturated, thus catalytic activity was attributed to partially defective nodes with available coordination sites for the reactants. The proposed cluster model contained two V centres, each one coordinated to OH, O, and a bidentate linker. From here, several active species were proposed to participate in a direct epoxidation route *via* V(IV) or a radical route *via* V(V) (see Scheme 5.8). Both reaction mechanisms seemed to co-exist at the experimental conditions. The direct pathway was favoured according to activation free energies, the radical one was thermodynamically driven, and switching from one to the other was available. The effect of linker functionalisation on catalytic activity was later evaluated for this system.[152] Periodic structures were essential to properly define the active site, but exploring full reaction mechanisms was only feasible with extended cluster models. In line with experimental trends, the computed reactivity of linker-functionalised V-MIL-47 MOFs were higher than those of the parent material.

Cluster models of Fe-BTC were employed to appraise the oxidation of ethylene using N_2O as oxidant (see Scheme 5.9).[153] The N_2O decomposition was the RDS with a barrier of 1.03 eV. Ethylene then reacted with Fe=O forming an ethyleneoxy intermediate. The subsequent C–O coupling to yield ethylene oxide occurred rapidly with a barrier of 0.22 eV, while the hydride shift to generate

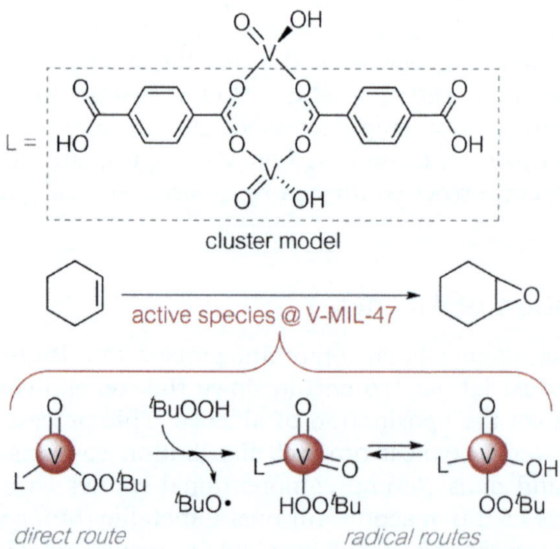

Scheme 5.8

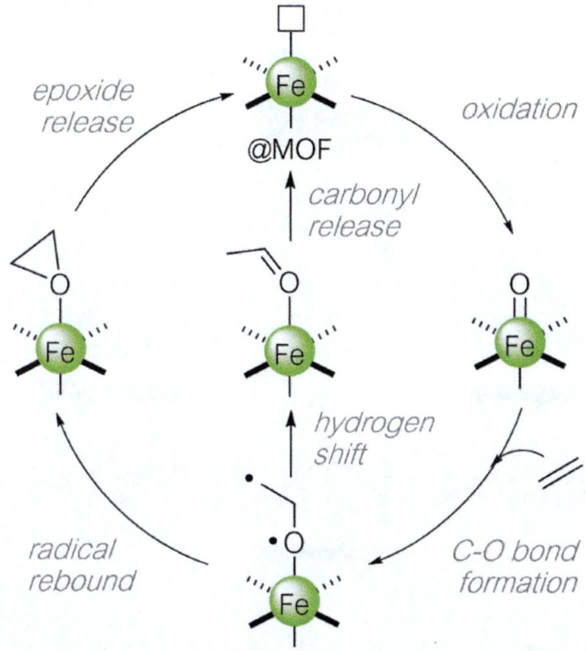

Scheme 5.9

acetaldehyde had a larger barrier of 0.62 eV. The desorption of these product species required around 0.96 eV. Overall, Fe-BTC was proposed as feasible catalyst for the selective oxidation of ethylene to ethylene oxide. A similar approach was carried out to address the aerobic oxidation of propylene.[154]

5.3.6 Others

The following paragraphs will cover a variety of chemical reactions that do not fall under the umbrella of the previous sections, but still provide a comprehensive picture of the wide scope of computational modelling in MOF and COF catalysis.

Ethylene hydrogenation (usually coupled with ethylene dimerisation) was computationally studied for NU-1000-based materials decorated with different transition metals, *e.g.*, Ir,[120] Rh,[121] and Ni.[122] Regarding Pt, a typical heterogeneous hydrogenation catalyst, few-atom clusters were deposited on NU-1000 nodes and showed high activity and resistance towards sintering.[155] Several configurations and spin states were considered for Pt_n clusters ($n = 1$–5), and the hydrogenation mechanism was finally evaluated using a Pt_4 model (see Scheme 5.10). After initial ethylene coordination, H_2 activation occurred across a Pt–Pt bond, forming one terminal H and one bridging H. Next, ethylene inserted into the terminal Pt–H bond with an activation free energy of 1.18 eV. The bridging H migrated to a terminal position to promote ethyl insertion *via* 0.95 eV and yielded ethane as final product.

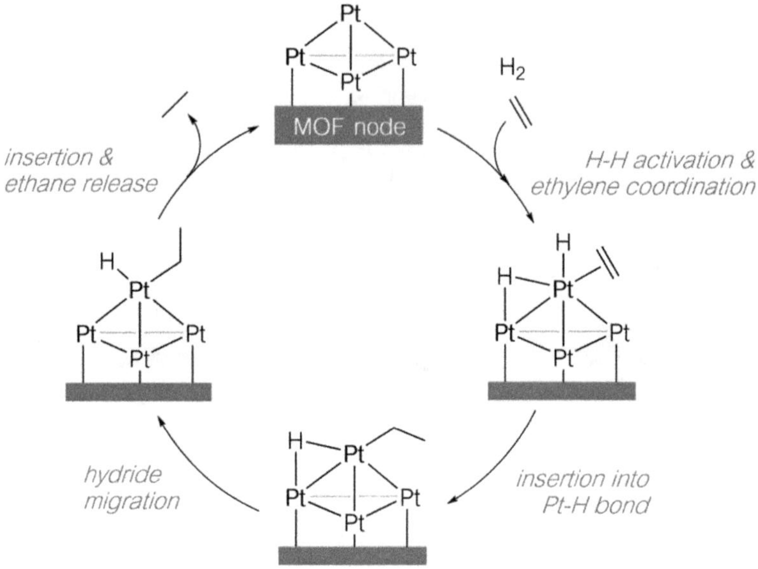

Scheme 5.10

The cyclisation of citronellal to isopulegol was reported for UiO-66 and UiO-66-NO_2 materials.[156] The computed activation free energies were 0.75 eV and 0.63 eV, respectively, and agreed with the experimental trend. The higher catalytic activity observed for UiO-66-NO_2 was attributed to enthalpic contributions leading to a stronger adsorption of the reactant on the node as well as electrostatic interactions between the reactant and the nitro groups of the linkers.

NU-1000 nodes decorated with first-row transition metals (M = Fe, Co, Ni, Cu, Zn) were tested for the acceptor-less dehydrogenation of cyclohexanol to cyclohexanone.[157] The reaction mechanism for Fe-NU-1000 went through proton transfer, β-hydride elimination, and H–H bond formation (RDS), where the μ_3-O group of the node was proposed to participate in hydrogen accepting and donating processes (see Figure 5.21). Among the materials considered, the Co- and Ni-based catalysts provided the lowest activation free energies for the RDS (1.20 eV and 1.15 eV, respectively).

Defective UiO-66 materials were examined for the Oppenauer oxidation of prenol with furfural by means of periodic DFT.[158] The catalysts exhibited an amphoteric nature over the course of the reaction; the Zr atoms behaved as Lewis acid sites by coordinating the substrates while the μ_3-O group acted as a Brønsted base site by abstracting an acidic H. The hydrated nodes were more active than the dehydrated ones due to the greater basicity of its μ_3-O group. Such behaviour was clearly reflected on the initial alcohol deprotonation, which presented activation free energies of 0.08 eV and 0.52 eV, respectively. The theoretical predictions were later confirmed by experiments.

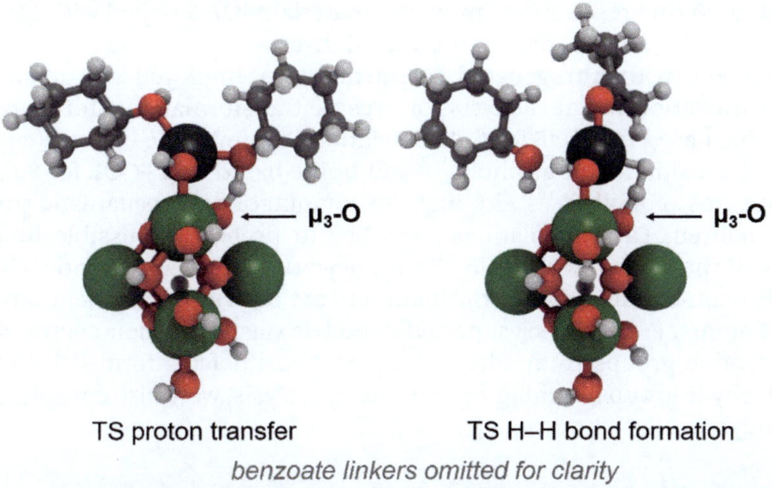

Figure 5.21 Transition state structures of proton transfer and H–H bond formation steps promoted by the μ_3-O group in Fe-NU-1000 node.[157] Atom legend: Zr = green, Fe = dark green, O = red, C = grey, H = white.

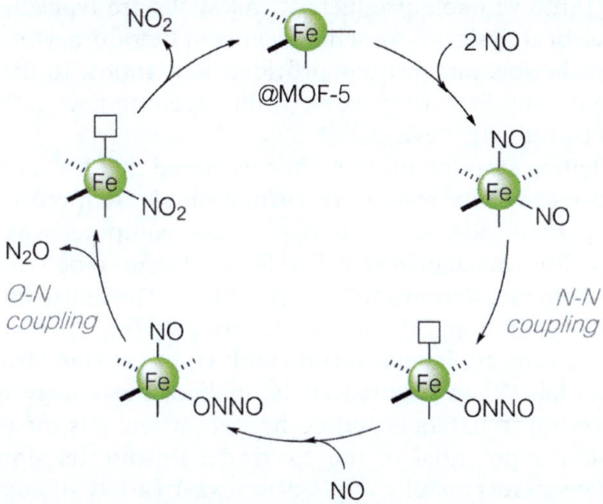

Scheme 5.11

The NO disproportionation towards N_2O and NO_2 *via* Fe-MOF-5[159] was computationally addressed using periodic structures at PBE+U level (see Scheme 5.11).[160] The Fe centre first coordinated up to two NO molecules which underwent N–N coupling to form a η^1-*O*-hyponitrite intermediate. The freshly-formed vacant site bound a third NO molecule which subsequently reacted with hyponitrite *via* O–N coupling to form N_2O (easily released) and NO_2. Final removal of NO_2 from the Fe centre regenerated the catalytic site.

According to this reaction pathway, the material MOF-5 doped with Mn was predicted to be more active than the Fe derivative.

COFs built with nitrogenated fragments were employed as supports to anchor transition metal catalysts for organic transformations, for instance, bipyridine-based Ir(I)-COF for C–H borylation,[161] bipyridine-based Ir(III)-COF for aerobic oxidation of alcohols,[162] and imine-based Pd(II)-COF for Suzuki–Miyaura cross coupling.[163] Although no computational mechanistic studies were reported, DFT simulations were key to probe the possible binding modes of the complexes within the framework. For Ir compounds, cluster models demonstrated their coordination to the bipyridine fragment through both N atoms. For Pd species, periodic models suggested their coordination to two imine groups from adjacent layers. Local defects formed by amine and aldehyde groups, coming from imine hydrolysis, were also considered in the study.

5.3.7 Electrocatalysis

So far, only thermal-based reactions have been considered. However, in the search for renewable energy sources, many efforts are focused on the development of electrochemical processes to convert simple molecules (H_2O, CO_2, N_2) into valuable products.[164] As MOFs are typically insulators, reports on electrocatalysis are somehow scarce in literature. However, recent progress in synthetic strategies has provided new routes to design porous materials with conductivity properties,[61] thus opening new paths for applications in electrocatalytic devices.[165]

From a modelling standpoint, it is often assumed that the kinetic barriers of proton and electron transfers are surmountable at room temperature; therefore, only thermodynamic energetics are computed. As such, discussion on reaction mechanisms will refer to potential-determining steps (PDS) rather than rate-determining steps (RDS). The most employed approach is the Computational Hydrogen Electrode (CHE),[166] which states that protons and electrons are in equilibrium with H_2 gas at standard conditions and zero potential. When a potential is applied, the energy of all states involved in electron transfers is shifted by $-eU$, where e is the charge of an electron and U the potential of the electrode. Despite its simplicity, this approach has been successfully applied to a vast variety of electrocatalytic systems. More complete (and thus costly) methodologies would require the inclusion of kinetics[167] and solvation.[168] This section will survey representative computational studies of MOF-catalysed electrochemical reactions under the CHE approximation.

One of the most important electrochemical reactions is water splitting. This process generates clean H_2 from H_2O by coupling the hydrogen evolution reaction (HER) with oxygen evolution reaction (OER). The HER involves two proton-coupled electron transfers (PCETs). The overpotential depends on the adsorption free energy of hydrogen on the electrocatalyst, which should be neither too weak nor too strong according to the Sabatier principle. Thus, most computational studies use this parameter as a descriptor to evaluate and predict reactivity. Compared to state-of-the-art Pt electrocatalysts, conductive MOFs containing Earth-abundant metals are promising alternatives.[165] As a case example, 1D Co–dithiolene was reported as an efficient electrocatalyst for HER.[169] Periodic PBE+U calculations identified the S atom as the catalytic site and evaluated the impact of alkali cations from the electrolyte *via* adsorption energy, band structure, and DOS simulations. Further studies were undertaken to predict reactivity trends for different first-row transition metals.[170] After benchmarking PBE+U with HSE06, hydrogen adsorption free energies and electronic structure calculations pointed towards Cr- and Ni-dithiolene materials as the best candidates for HER.

OER is a more complex process that requires four PCETs and it is the bottleneck of the water oxidation reaction.[171] The water nucleophilic attack mechanism (see Scheme 5.12) proceeds *via* four steps, each one involving one PCET: metal–hydroxo formation, metal–oxo formation, water nucleophilic attack to the oxo moiety yielding a hydroperoxo intermediate, and dioxygen release. The localised and isolated nature of the catalytic sites within MOFs makes the dinuclear pathway (*i.e.*, coupling of two metal–oxo moieties to form a peroxo species) less feasible. Compared to typical noble metal oxides, first-row transition-metal-based MOFs hold promise for OER, particularly those containing a mixture of metals at the nodes.

Ultrathin 2D MOFs bearing both Ni(II) and Co(II) atoms showed outstanding catalytic activity towards OER compared to the single-metal analogues.[172] PBE+U simulations on a periodic cell were performed on Ni-, Co-, and mixed NiCo models of a MOF surface, which was designed by minimising the number of truncated metal–O bonds. Computations predicted the formation of M–OOH as PDS with low overpotentials for the mixed NiCo material when Co was the catalytic site. According to DOS

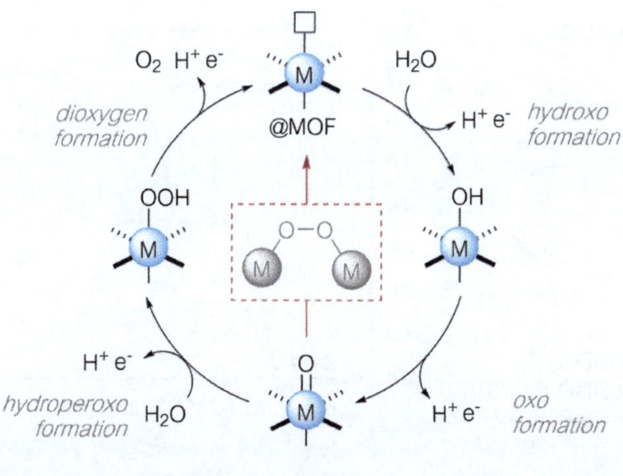

Scheme 5.12

calculations, this behaviour was attributed to charge transfer from Ni to Co centres. The benefit of mixing metals within the MOF nodes has also been observed in later OER reports using finite models.[173] PBE + U simulations on Fe_2M carboxylate clusters pointed towards the conversion of Fe–OH to Fe–O as PDS. Computed overpotentials for different metal clusters, where the reactivity still took place on the Fe sites, followed the order: Fe > Zn > Ni > Co.

Similar to previous studies in catalysis, the MOF electrochemical activity can be tuned by defect engineering, for instance, by 'introducing' missing linkers.[174] The layered-pillared CoBDC MOF is formed by Co(ii) octahedra with hydroxide and BDC linkers. According to periodic PBE + U calculations, the substitution of one ditopic linker by the monotopic carboxyferrocene (CoBDC-Fc) populated new electronic states near the Fermi level, thus increasing the conductivity. The OER mechanism was then computed at Co in pristine CoBDC, Co in CoBDC-Fc, and unsaturated Co in CoBDC-Fc. The activity of Co in bulk phase for both pristine and modified CoBDC gave free energies greater than 3.5 eV for the formation of Co–OH. However, the coordinatively unsaturated Co centre in CoBDC-Fc stabilised all OER intermediates and yielded a significantly lower PDS of 1.85 eV related to the formation of Co–O (see Figure 5.22).

The inverse process of OER, namely oxygen reduction reaction (ORR), is key to control the efficiency of fuel cells. Conductive 2D MOFs containing M–N linkages (M = first-row transition metal) have shown properties comparable to state-of-the-art Pt electrocatalysts, such as the Ni_3(hexaiminotriphenylene)$_2$. A detailed report on this MOF supported the participation of the linker, rather than the metal ion, in the ORR mechanism.[175] According to B3LYP calculations on a cluster model, the O_2 binding on the β-C of the triphenylene unit was favoured over the Ni centre by 0.3 eV (0.5 eV when including electron transfer). Although there is no direct Ni–O_2 bonding, the

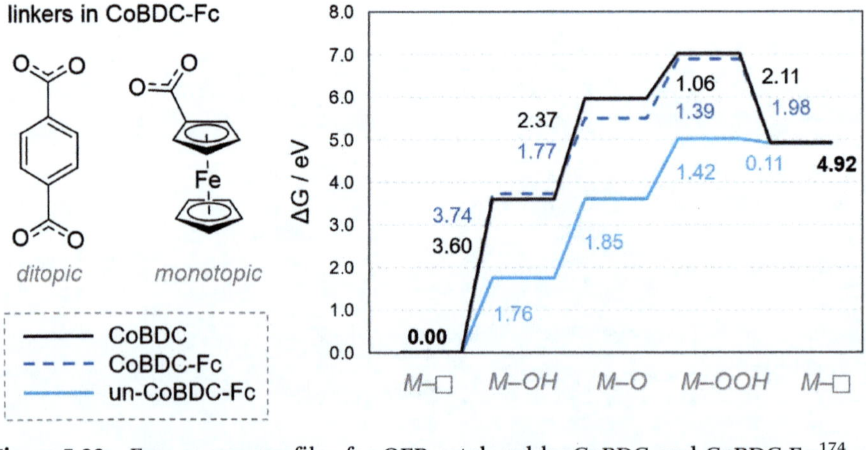

Figure 5.22 Free energy profiles for OER catalysed by CoBDC and CoBDC-Fc.[174]

metal centre did help in the process by accepting spin density. Later on, a periodic DFT study at PBE level suggested a direct participation of metal centres in ORR intermediates, where the Co-based MOF followed a $4e^-$ pathway (*via* M–O and M–OH intermediates) and the Ni-based analogue followed a $2e^-$ route (*via* OOH).[176] Similar DFT studies were performed for thiol-derived materials (M = Fe–Ni, Ru–Pd, Os–Pt), where Co was predicted to be the most active site for ORR with an overpotential of 0.30 V.[177]

Metal-porphyrin-based 2D COFs were also explored for OER and ORR activity using periodic models at the PBE+U level.[178] Several TM-COF systems were evaluated (TM = Sc, Ti, V, Cr, Mn, Fe, Co, Ni, Cu, Zn) using a model that contained one layer of the material. The adsorption energies of M–O and M–OH species correlated with the overpotential following a volcano-shaped diagram, showcasing Fe as the most balanced catalyst. Interestingly, the crystal field stabilisation energy, a magnitude related to the d orbital splitting of transition metals, worked well as a descriptor to predict the electrocatalytic activity of these materials. Using similar monolayer models, a variety of transition metals were placed in different coordination environments within COFs and tested for ORR.[179] Among them, Cu and Ni derivatives appeared as the best performing materials.

This last paragraph covers the nitrogen reduction reaction (NRR). The development of active and selective electrochemical devices that could substitute the current energy-consuming Haber–Bosch process is a major challenge in current catalysis.[180] Most reports focus on metal surfaces,[181,182] but computational works are starting to appear in recent MOF literature. For example, DFT studies of the NRR were reported for a 2D MOF based on transition metal centres and hexaaminobenzene.[183] An initial screening of MOFs bearing different metals (Co, Ni, Cu, and Mo) was carried out, where the adsorption of N_2/H_2O and the conductivity were used as criteria to predict potential electrocatalytic activity. Among the transition metals evaluated, the Mo-based material was the most promising. The PDS for both distal and alternating mechanisms (see Scheme 5.13) was the first hydrogenation of Mo–N_2 to form Mo–N_2H with a low overpotential of 0.18 V. The competing HER showed a larger potential, thus the desired NRR would dominate.

5.4 Conclusions and Perspectives

The present chapter has covered the main concepts to simulate porous materials at the quantum mechanical level. The impact of both the method and the model has been addressed through the discussion of case studies, and the importance of computational tools in spectroscopy and catalysis has been showcased with representative and recent literature.

Quantum mechanics has proven to be crucial to characterise materials and elucidate reaction pathways in catalysis and electrocatalysis, yet this is only the beginning. Photocatalysis with porous materials promises a bright future[184] and computational techniques have the potential to boost its progress. The binding energy of an electron in MOFs allows one to evaluate

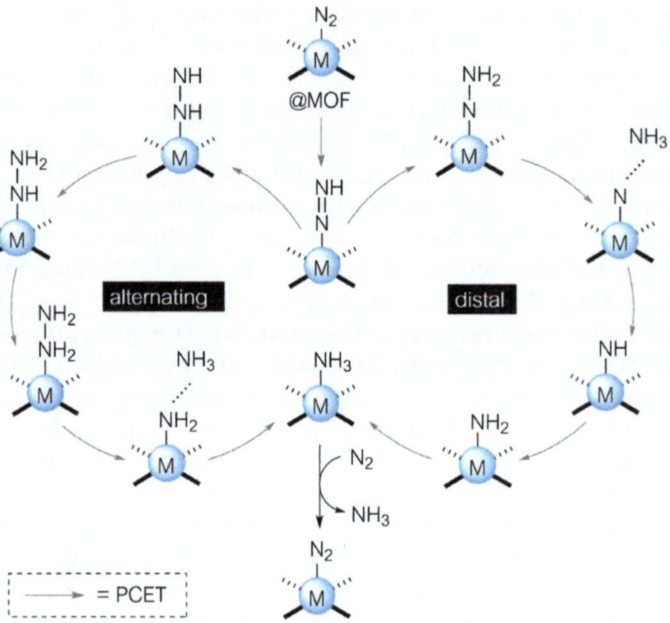

Scheme 5.13

their thermodynamic feasibility in photochemical processes,[185] and the calculation of band gaps, excited states, and charge transfer processes have already proved useful to guide the rational design of photocatalysts.[186] Nevertheless, computation of the photocatalytic reaction mechanisms remains a challenge for the community.

Finally, by combining supercomputing power with data processing and mining, the current computational frontier is leaning towards the development of statistical learning. Algorithms for the design of porous materials with focus on gas adsorption,[187] gas storage,[188] band gaps,[189] structure-mechanical stability,[190,191] and organic cages[192] have been already reported. Despite the complex picture of catalytic networks, where a huge number of variables should be considered (activity, selectivity, stability, off-cycle species, defects, transport limitations, *etc.*), statistical learning in catalysis[193] is a promising strategy to accelerate the discovery of new and unforeseen porous materials.

Abbreviations

BDC	benzene-1,4-dicarboxylate
BPDC	4,4′-biphenyl-dicarboxylate
BTC	1,3,5-benzene-tricarboxylate
CHE	computational hydrogen electrode
COF	covalent organic framework
DF	density functional

DFT	density functional theory
DOBDC	2,5-dihydroxybenzene-1,4-dicarboxylate
DOS	density of states
GGA	generalised gradient approximation
HER	hydrogen evolution reaction
LDA	local density approximation
MOF	metal–organic framework
MR	multireference
NRR	nitrogen reduction reaction
OER	oxygen evolution reaction
ORR	oxygen reduction reaction
PCET	proton-coupled electron transfer
PDS	potential-determining step
RDS	rate-determining step
SIE	self-interaction error
TD-DFT	time-dependent density functional theory
TOF	turnover frequency

References

1. O. M. Yaghi, *ACS Cent. Sci.*, 2019, **5**, 1295.
2. H. Furukawa, K. E. Cordova, M. O'Keeffe and O. M. Yaghi, *Science*, 2013, **341**, 1230444.
3. C. S. Diercks and O. M. Yaghi, *Science*, 2017, **355**, 923.
4. C. E. Wilmer, M. Leaf, C. Y. Lee, O. K. Farha, B. G. Hauser, J. T. Hupp and R. Q. Snurr, *Nat. Chem.*, 2012, **4**, 83.
5. H.-C. Zhou, J. R. Long and O. M. Yaghi, *Chem. Rev.*, 2012, **112**, 673.
6. C. H. Hendon, A. J. Rieth, M. D. Korzyński and M. Dincă, *ACS Cent. Sci.*, 2017, **3**, 554.
7. K. Geng, T. He, R. Liu, S. Dalapati, K. T. Tan, Z. Li, S. Tao, Y. Gong, Q. Jiang and D. Jiang, *Chem. Rev.*, 2020, **120**, 8814.
8. P. G. Boyd, Y. Lee and B. Smit, *Nat. Rev. Mater.*, 2017, **2**, 17037.
9. P. Z. Moghadam, A. Li, S. B. Wiggin, A. Tao, A. G. P. Maloney, P. A. Wood, S. C. Ward and D. Fairen-Jimenez, *Chem. Mater.*, 2017, **29**, 2618.
10. Y. G. Chung, J. Camp, M. Haranczyk, B. J. Sikora, W. Bury, V. Krungleviciute, T. Yildirim, O. K. Farha, D. S. Sholl and R. Q. Snurr, *Chem. Mater.*, 2014, **26**, 6185.
11. D. Nazarian, J. S. Camp, Y. G. Chung, R. Q. Snurr and D. S. Sholl, *Chem. Mater.*, 2017, **29**, 2521.
12. Y. G. Chung, E. Haldoupis, B. J. Bucior, M. Haranczyk, S. Lee, H. Zhang, K. D. Vogiatzis, M. Milisavljevic, S. Ling, J. S. Camp, B. Slater, J. I. Siepmann, D. S. Sholl and R. Q. Snurr, *J. Chem. Eng. Data*, 2019, **64**, 5985.
13. D. Ongari, A. V. Yakutovich, L. Talirz and B. Smit, *ACS Cent. Sci.*, 2019, **5**, 1663.

14. P. Hohenberg and W. Kohn, *Phys. Rev.*, 1964, **136**, B864.
15. W. Kohn and L. J. Sham, *Phys. Rev.*, 1965, **140**, A1133.
16. J. P. Perdew and K. Schmidt, *AIP Conf. Proc.*, 2001, **577**, 1.
17. A. M. Reilly and A. Tkatchenko, *Chem. Sci.*, 2015, **6**, 3289.
18. S. Grimme, A. Hansen, J. G. Brandenburg and C. Bannwarth, *Chem. Rev.*, 2016, **116**, 5105.
19. K. D. Vogiatzis, M. V. Polynski, J. K. Kirkland, J. Townsend, A. Hashemi, C. Liu and E. A. Pidko, *Chem. Rev.*, 2019, **119**, 2453.
20. C. J. Cramer and D. G. Truhlar, *Phys. Chem. Chem. Phys.*, 2009, **11**, 10757.
21. C. A. Gaggioli, S. J. Stoneburner, C. J. Cramer and L. Gagliardi, *ACS Catal.*, 2019, **9**, 8481.
22. J. Sauer, *Acc. Chem. Res.*, 2019, **52**, 3502.
23. D. Nazarian, P. Ganesh and D. S. Sholl, *J. Mater. Chem. A*, 2015, **3**, 22432.
24. F. Formalik, M. Fischer, J. Rogacka, L. Firlej and B. Kuchta, *J. Chem. Phys.*, 2018, **149**, 064110.
25. I. Choudhuri and D. G. Truhlar, *Molecules*, 2020, **25**, 1552.
26. E. A. Pidko, *ACS Catal.*, 2017, **7**, 4230.
27. R. B. Getman, Y.-S. Bae, C. E. Wilmer and R. Q. Snurr, *Chem. Rev.*, 2012, **112**, 703.
28. S. Odoh, C. J. Cramer, D. G. Truhlar and L. Gagliardi, *Chem. Rev.*, 2015, **115**, 6051.
29. F. X. Coudert and A. H. Fuchs, *Coord. Chem. Rev.*, 2016, 307, 211.
30. G. Fraux and F. X. Coudert, *Chem. Commun.*, 2017, **53**, 7211.
31. S. M. J. Rogge, A. Bavykina, J. Hajek, H. Garcia, A. I. Olivos-Suarez, A. Sepúlveda-Escribano, A. Vimont, G. Clet, P. Bazin, F. Kapteijn, M. Daturi, E. V. Ramos-Fernandez, F. X. Llabrés i Xamena, V. Van Speybroeck and J. Gascon, *Chem. Soc. Rev.*, 2017, **46**, 3134.
32. V. Bernales, M. A. Ortuño, D. G. Truhlar, C. J. Cramer and L. Gagliardi, *ACS Cent. Sci.*, 2018, **4**, 5.
33. B. Lukose, A. Kuc, J. Frenzel and T. Heine, *Beilstein J. Nanotechnol.*, 2010, **1**, 60.
34. V. Van Speybroeck, K. Hemelsoet, L. Joos, M. Waroquier, R. G. Bell and C. R. A. Catlow, *Chem. Soc. Rev.*, 2015, **44**, 7044.
35. G. Li and E. A. Pidko, *ChemCatChem*, 2019, **11**, 134.
36. C. Chizallet, *ACS Catal.*, 2020, **10**, 5579.
37. J. D. Evans, K. E. Jelfs, G. M. Day and C. J. Doonan, *Chem. Soc. Rev.*, 2017, **46**, 3286.
38. A. J. Howarth, A. W. Peters, N. A. Vermeulen, T. C. Wang, J. T. Hupp and O. K. Farha, *Chem. Mater.*, 2017, **29**, 26.
39. L. Hanna and J. V. Lockard, *J. Phys.: Condens. Matter*, 2019, **31**, 483001.
40. I. M. Alecu, J. Zheng, Y. Zhao and D. G. Truhlar, *J. Chem. Theory Comput.*, 2010, **6**, 2872.
41. B. Civalleri, F. Napoli, Y. Noël, C. Roetti and R. Dovesi, *CrystEngComm*, 2006, **8**, 364.

42. S. Bordiga, C. Lamberti, G. Ricchiardi, L. Regli, F. Bonino, A. Damin, K.-P. Lillerud, M. Bjorgen and A. Zecchina, *Chem. Commun.*, 2004, **20**, 2300.
43. L. Valenzano, B. Civalleri, S. Chavan, G. T. Palomino, C. O. Areán and S. Bordiga, *J. Phys. Chem. C*, 2010, **114**, 11185.
44. N. Nijem, P. Canepa, L. Kong, H. Wu, J. Li, T. Thonhauser and Y. J. Chabal, *J. Phys.: Condens. Matter*, 2012, **24**, 424203.
45. H. Wu, J. M. Simmons, G. Srinivas, W. Zhou and T. Yildirim, *J. Phys. Chem. Lett.*, 2010, **1**, 1946.
46. B. Supronowicz, A. Mavrandonakis and T. Heine, *J. Phys. Chem. C*, 2013, **117**, 14570.
47. L. Valenzano, B. Civalleri, S. Chavan, S. Bordiga, M. H. Nilsen, S. Jakobsen, K. P. Lillerud and C. Lamberti, *Chem. Mater.*, 2011, **23**, 1700.
48. S. Chavan, J. G. Vitillo, D. Gianolio, O. Zavorotynska, B. Civalleri, S. Jakobsen, M. H. Nilsen, L. Valenzano, C. Lamberti, K. P. Lillerud and S. Bordiga, *Phys. Chem. Chem. Phys.*, 2012, **14**, 1614.
49. M. Vandichel, J. Hajek, F. Vermoortele, M. Waroquier, D. De Vos and V. Van Speybroeck, *CrysEngComm*, 2015, **17**, 395.
50. M. Vandichel, J. Hajek, A. Ghysels, A. De Vos, M. Waroquier and V. Van Speybroeck, *CrysEngComm*, 2016, **18**, 7056.
51. N. Planas, J. E. Mondloch, S. Tussupbayev, J. Borycz, L. Gagliardi, J. T. Hupp, O. K. Farha and C. J. Cramer, *J. Phys. Chem. Lett.*, 2014, **5**, 3716.
52. A. E. Platero-Prats, A. Mavrandonakis, L. C. Gallington, Y. Liu, J. T. Hupp, O. K. Farha, C. J. Cramer and K. W. Chapman, *J. Am. Chem. Soc.*, 2016, **138**, 4178.
53. M. D. Allendorf, C. A. Bauer, R. K. Bhakta and R. J. T. Houk, *Chem. Soc. Rev.*, 2009, **38**, 1330.
54. A. D. Laurent and D. Jacquemin, *Int. J. Quantum Chem.*, 2013, **113**, 2019.
55. C. Latouche, D. Skouteris, F. Palazzeti and V. Barone, *J. Chem. Theory Comput.*, 2015, **11**, 3281.
56. M. Ji, X. Lan, Z. Han, C. Hao and J. Qiu, *Inorg. Chem.*, 2012, **51**, 12389.
57. Y.-Y. Liu, R. Decadt, T. Bogaerts, K. Hemelsoet, A. M. Kaczmarek, D. Poelman, M. Waroquier, V. Van Speybroeck, R. Van Deun and P. Van Der Voort, *J. Phys. Chem. C*, 2013, **117**, 11302.
58. L. Wilbraham, F.-X. Coudert and I. Ciofini, *Phys. Chem. Chem. Phys.*, 2016, **18**, 25176.
59. K. Hendrickx, D. E. P. Vanpoucke, K. Leus, K. Lejaeghere, A. Van Yperen-De Deyne, V. Van Speybroeck, P. Van Der Voort and K. Hemelsoet, *Inorg. Chem.*, 2015, **54**, 10701.
60. A. Van Yperen-De Deyne, K. Hendrickx, L. Vanduyfhuys, G. Sastre, P. Van Der Voort, V. Van Speybroeck and K. Hemelsoet, *Theor. Chem. Acc.*, 2016, **135**, 102.
61. L. Sun, M. G. Campbell and M. Dincă, *Angew. Chem., Int. Ed.*, 2016, **55**, 3566.

62. F. S. Hegner, J. R. Galán-Macarós and N. López, *Inorg. Chem.*, 2016, **55**, 12851.
63. I. Choudhuri and D. G. Truhlar, *J. Phys. Chem. C*, 2019, **123**, 17416.
64. J. L. Bao, L. Gagliardi and D. G. Truhlar, *J. Phys. Chem. Lett.*, 2018, **9**, 2353.
65. M. Fuentes-Cabrera, D. M. Nicholson, B. G. Sumpter and M. Widom, *J. Chem. Phys.*, 2005, **123**, 124713.
66. E. Flage-Larsen, A. Røyset, J. H. Cavka and K. Thorshaug, *J. Phys. Chem. C*, 2013, **117**, 20610.
67. K. Hendrickx, J. J. Joos, A. De Vos, D. Poelman, P. F. Smet, V. Van Speybroeck, P. Van Der Voort and K. Lejaeghere, *Inorg. Chem.*, 2018, **57**, 5463.
68. K. L. Svane, J. K. Bristow, J. D. Gale and A. Walsh, *J. Mater. Chem. A*, 2018, **6**, 8507.
69. L. Sun, C. H. Hendon, S. S. Park, Y. Tulchinsky, R. Wan, F. Wang, A. Walsh and M. Dincă, *Chem. Sci.*, 2017, **8**, 4450.
70. B. Himmetoglu, A. Floris, S. de Gironcoli and M. Cococcioni, *Int. J. Quantum Chem.*, 2014, **114**, 14.
71. H. J. Kulik, *J. Chem. Phys.*, 2015, **142**, 240901.
72. G. W. Mann, K. Lee, M. Cococcioni, B. Smit and J. B. Neaton, *J. Chem. Phys.*, 2016, **144**, 174104.
73. A. Oliveira, G. F. Lima and H. A. Abreu, *Chem. Phys. Lett.*, 2018, **691**, 283.
74. M. Kurmoo, *Chem. Soc. Rev.*, 2009, **38**, 1353.
75. E. D. Bloch, W. L. Queen, R. Krishna, J. M. Zadrozny, C. M. Brown and J. R. Long, *Science*, 2012, **335**, 1606.
76. P. Canepa, Y. J. Chabal and T. Thonhauser, *Phys. Rev. B*, 2013, **87**, 094407.
77. J. Borycz, J. Paier, P. Verma, L. E. Darago, D. J. Xiao, D. G. Truhlar, J. R. Long and L. Gagliardi, *Inorg. Chem.*, 2016, **55**, 4924.
78. P. Verma, R. Maurice and D. G. Truhlar, *J. Phys. Chem. C*, 2016, **120**, 9933.
79. R. Maurice, P. Verma, J. M. Zadrozny, S. Luo, J. Borycz, J. R. Long, D. G. Truhlar and L. Gagliardi, *Inorg. Chem.*, 2013, **52**, 9379.
80. P. Verma, R. Maurice and D. G. Truhlar, *J. Phys. Chem. C*, 2015, **119**, 28499.
81. J. Lee, O. K. Farha, J. Roberts, K. A. Scheidt, S. T. Nguyen and J. T. Hupp, *Chem. Soc. Rev.*, 2009, **38**, 1450.
82. A. Corma, H. García and F. X. Llabrés i Xamena, *Chem. Rev.*, 2010, **110**, 4606.
83. J. Gascon, A. Corma, F. Kaptejin and F. X. Llabrés i Xamena, *ACS Catal.*, 2014, **4**, 361.
84. D. Yang and B. C. Gates, *ACS Catal.*, 2019, **9**, 1779.
85. A. Bavykina, N. Kolobov, I. S. Khan, J. A. Bau, A. Ramirez and J. Gascon, *Chem. Rev.*, 2020, **120**, 8468.
86. P. Valvekens, F. Vermoortele and D. De Vos, *Catal. Sci. Technol.*, 2013, **3**, 1435.

87. V. Van Speybroeck, K. De Wispelaere, J. Van Der Mynsbrugge, M. Vandichel, K. Hemelsoet and M. Waroquier, *Chem. Soc. Rev.*, 2014, **43**, 7326.
88. M. H. Mahyuddin, Y. Shiota, A. Staykov and K. Yoshizawa, *Acc. Chem. Res.*, 2018, **51**, 2382.
89. J. Gascon, U. Aktay, M. D. Hernandez-Alonso, G. P. M. van Klink and F. Kapteijn, *J. Catal.*, 2009, **261**, 75.
90. R. Cortese and D. Duca, *Phys. Chem. Chem. Phys.*, 2011, **13**, 15995.
91. M. Opanasenko, A. Dhakshinamoorthy, M. Shamzhy, P. Nachtigall, M. Horáček, H. Garcia and J. Čejka, *Catal. Sci. Technol.*, 2013, **3**, 500.
92. M. Položij, M. Rubeš, J. Čejka and P. Nachtigall, *ChemCatChem*, 2014, **6**, 2821.
93. P. Valvekens, M. Vandichel, M. Waroquier, V. Van Speybroeck and D. De Vos, *J. Catal.*, 2014, **317**, 1.
94. J. Hajek, M. Vandichel, B. Van de Voorde, B. Bueken, D. De Vos, M. Waroquier and V. Van Speybroeck, *J. Catal.*, 2015, **331**, 1.
95. C. Caratelli, J. Hajek, F. G. Cirujano, M. Waroquier, F. X. Llabrés i Xamena and V. Van Speybroeck, *J. Catal.*, 2017, **352**, 401.
96. J. Luis-Barrerra, R. Cano, G. Imani-Shakibaei, J. Heras-Domingo, J. Pérez-Carvajal, I. Imaz, D. Maspoch, X. Solans-Monfort, J. Alemán and R. Mas-Ballesté, *Catal. Sci. Technol.*, 2019, **9**, 6007.
97. J. B. DeCoste and G. W. Peterson, *Chem. Rev.*, 2014, **114**, 5695.
98. N. S. Bobbitt, M. L. Mendonca, A. J. Howarth, T. Islamoglu, J. T. Hupp, O. K. Farha and R. Q. Snurr, *Chem. Soc. Rev.*, 2017, **46**, 3357.
99. T. Islamoglu, Z. Chen, M. C. Wasson, C. T. Buru, K. O. Kirlikovali, U. Afrin, M. R. Mian and O. K. Farha, *Chem. Rev.*, 2020, **120**, 8130.
100. Y. Bai, Y. Dou, L.-H. Xie, W. Rutledge, J.-R. Li and H.-C. Zhou, *Chem. Soc. Rev.*, 2016, **45**, 2327.
101. M. Rimoldi, A. J. Howarth, M. R. DeStefano, L. Lin, S. Goswami, P. Li, J. T. Hupp and O. K. Farha, *ACS Catal.*, 2017, **7**, 997.
102. S. Royuela, R. Gil-San Millán, M. J. Mancheño, M. M. Ramos, J. L. Segura, J. A. R. Navarro and F. Zamora, *Materials*, 2019, **12**, 1974.
103. J. E. Mondloch, M. J. Katz, W. C. Isley III, P. Ghosh, P. Liao, W. Bury, G. W. Wagner, M. G. Hall, J. B. DeCoste, G. W. Peterson, R. Q. Snurr, C. J. Cramer, J. T. Hupp and O. K. Farha, *Nat. Mater.*, 2015, **14**, 512.
104. H. Chen, P. Liao, M. L. Mendonca and R. Q. Snurr, *J. Phys. Chem. C*, 2018, **122**, 12362.
105. D. Troya, *J. Phys. Chem. C*, 2016, **120**, 29312.
106. G. Wang, C. Sharp, A. M. Plonka, Q. Wang, A. I. Frenkel, W. Guo, C. Hill, C. Smith, J. Kollar, D. Troya and J. R. Morris, *J. Phys. Chem. C*, 2017, **121**, 11261.
107. I. Stassen, B. Bueken, H. Reinsch, J. F. M. Oudenhoven, D. Wouters, J. Hajek, V. Van Speybroeck, N. Stock, P. M. Vereecken, R. Van Schaijk, D. De Vos and R. Ameloot, *Chem. Sci.*, 2016, **7**, 5827.
108. J. A. Harvey, J. A. Greathouse and D. F. Sava Gallis, *J. Phys. Chem. C*, 2018, **122**, 26889.

109. J. A. Harvey, C. J. Pearce, M. G. Hall, E. J. Bruni, J. B. DeCoste and D. F. Sava Gallis, *Dalton Trans.*, 2019, **48**, 16153.
110. M. R. Momeni and C. J. Cramer, *ACS Appl. Mater. Interfaces*, 2018, **10**, 18435.
111. M. R. Momeni and C. J. Cramer, *Chem. Mater.*, 2018, **30**, 4432.
112. T. Islamoglu, M. A. Ortuño, E. Proussaloglou, A. J. Howarth, N. A. Vermeulen, A. Atilgan, A. M. Asiri, C. J. Cramer and O. K. Farha, *Angew. Chem., Int. Ed.*, 2018, **57**, 1949.
113. M. Kalaj, M. R. Momeni, K. C. Bentz, K. S. Barcus, J. M. Palomba, F. Paesani and S. M. Cohen, *Chem. Commun.*, 2019, **55**, 3481.
114. M. L. Mendonca and R. Q. Snurr, *ACS Catal.*, 2020, **10**, 1310.
115. H. Chen and R. Q. Snurr, *ACS Appl. Mater. Interfaces*, 2020, **12**, 14631.
116. J. Canivet, S. Aguado, Y. Schuurman and D. Farrusseng, *J. Am. Chem. Soc.*, 2013, **135**, 4195.
117. S. T. Madrahimov, J. R. Gallagher, G. Zhang, Z. Meinhart, S. J. Garibay, M. Delferro, J. T. Miller, O. K. Farha, J. T. Hupp and S. T. Nguyen, *ACS Catal.*, 2015, **5**, 6713.
118. E. D. Metzger, C. K. Brozek, R. J. Comito and M. Dincă, *ACS Cent. Sci.*, 2016, **2**, 148.
119. E. D. Metzger, R. J. Comito, C. H. Hendon and M. Dincă, *J. Am. Chem. Soc.*, 2017, **139**, 757.
120. D. Yang, C. A. Gaggioli, E. Conley, M. Babucci, L. Gagliardi and B. C. Gates, *J. Catal.*, 2020, **382**, 165.
121. V. Bernales, D. Yang, J. Yu, G. Gümüşlü, C. J. Cramer, B. C. Gates and L. Gagliardi, *ACS Appl. Mater. Interfaces*, 2017, **9**, 33511.
122. Z. Li, N. M. Schweitzer, A. B. League, V. Bernales, A. W. Peters, A. B. Getsoian, T. C. Wang, J. T. Miller, A. Vjunov, J. L. Fulton, J. A. Lercher, C. J. Cramer, L. Gagliardi, J. T. Hupp and O. K. Farha, *J. Am. Chem. Soc.*, 2016, **138**, 1977.
123. V. Bernales, A. B. League, Z. Li, N. M. Schweitzer, A. W. Peters, R. K. Carlson, J. T. Hupp, C. J. Cramer, O. K. Farha and L. Gagliardi, *J. Phys. Chem. C*, 2016, **120**, 23576.
124. J. Ye, L. Gagliardi, C. J. Cramer and D. G. Truhlar, *J. Catal.*, 2017, **354**, 278.
125. J. Ye, L. Gagliardi, C. J. Cramer and D. G. Truhlar, *J. Catal.*, 2018, **360**, 160.
126. S. Pellizzeri, M. Barona, V. Bernales, P. Miró, P. Liao, L. Gagliardi, R. Q. Snurr and R. B. Getman, *Catal. Today*, 2018, **312**, 49.
127. W. Zhang, M. Kauer, O. Halbherr, K. Epp, P. Guo, M. I. Gonzalez, D. J. Xiao, C. Wiktor, F. X. LIabrés i Xamena, C. Wöll, Y. Wang, M. Muhler and R. A. Fischer, *Chem. Eur. J.*, 2016, **22**, 14297.
128. I. Agirrezabal-Telleria, I. Luz, M. A. Ortuño, M. Oregui-Bengoechea, I. Gandarias, N. López, M. A. Lail and M. Soukri, *Nat. Commun.*, 2019, **10**, 2076.
129. D. J. Xiao, E. D. Bloch, J. A. Mason, W. L. Queen, M. R. Hudson, N. Planas, J. Borycz, A. L. Dzubak, P. Verma, K. Lee, F. Bonino,

V. Crocellà, J. Yano, S. Bordiga, D. G. Truhlar, L. Gagliardi, C. M. Brown and J. R. Long, *Nat. Chem.*, 2014, **6**, 590.
130. P. Verma, K. D. Vogiatzis, N. Planas, J. Borycz, D. J. Xiao, J. R. Long, L. Gagliardi and D. G. Truhlar, *J. Am. Chem. Soc.*, 2015, **137**, 5770.
131. H. Hirao, W. K. H. Ng, A. M. P. Moeljadi and S. Bureekaew, *ACS Catal.*, 2015, **5**, 3287.
132. S. Impeng, S. Siwaipram, S. Bureekaew and M. Probst, *Phys. Chem. Chem. Phys.*, 2017, **19**, 3782.
133. J. G. Vitillo, A. Bhan, C. J. Cramer, C. C. Lu and L. Gagliardi, *ACS Catal.*, 2019, **9**, 2970.
134. M. C. Simons, J. G. Vitillo, M. Babucci, A. S. Hoffman, A. Boubnov, M. L. Beauvais, Z. Chen, C. J. Cramer, K. W. Chapman, S. R. Bare, B. C. Gates, C. C. Lu, L. Gagliardi and A. Bhan, *J. Am. Chem. Soc.*, 2019, **141**, 18142.
135. D. Y. Osadchii, A. I. Olivos-Suarez, Á. Szécsényi, G. Li, M. A. Nasalevich, I. A. Dugulan, P. S. Crespo, E. J. M. Hensen, S. L. Veber, M. V. Fedin, G. Sankar, E. A. Pidko and J. Gascon, *ACS Catal.*, 2018, **8**, 5542.
136. Á. Szécsényi, G. Li, J. Gascon and E. A. Pidko, *Chem. Sci.*, 2018, **9**, 6765.
137. K. D. Vogiatzis, E. Haldoupis, D. J. Xiao, J. R. Long, J. I. Siepmann and L. Gagliardi, *J. Phys. Chem. C*, 2016, **120**, 18707.
138. P. Liao, R. G. Getman and R. Q. Snurr, *ACS Appl. Mater. Interfaces*, 2017, **9**, 33484.
139. A. S. Rosen, J. M. Notestein and R. Q. Snurr, *J. Comput. Chem.*, 2019, **40**, 1305.
140. A. S. Rosen, J. M. Notestein and R. Q. Snurr, *ACS Catal.*, 2019, **9**, 3576.
141. T. Ikuno, J. Zheng, A. Vjunov, M. Sanchez-Sanchez, M. A. Ortuño, D. R. Pahls, J. L. Fulton, D. M. Camaioni, Z. Li, D. Ray, B. L. Mehdi, N. D. Browning, O. K. Farha, J. T. Hupp, C. J. Cramer, L. Gagliardi and J. A. Lercher, *J. Am. Chem. Soc.*, 2017, **139**, 10294.
142. J. Zheng, J. Ye, M. A. Ortuño, J. L. Fulton, O. Y. Gutiérrez, D. M. Camaioni, R. K. Motkuri, Z. Li, T. E. Webber, B. L. Mehdi, N. D. Browning, R. L. Penn, O. K. Farha, J. T. Hupp, D. G. Truhlar, C. J. Cramer and J. A. Lercher, *J. Am. Chem. Soc.*, 2019, **141**, 9292.
143. J. Baek, B. Rungtaweevoranit, X. Pei, M. Park, S. C. Pakra, Y.-S. Liu, R. Matheu, S. A. Alshmimri, S. Alshehri, C. A. Trickett, G. A. Somorjai and O. M. Yaghi, *J. Am. Chem. Soc.*, 2018, **140**, 18208.
144. D. R. Pahls, M. A. Ortuño, P. H. Winegar, C. J. Cramer and L. Gagliardi, *Inorg. Chem.*, 2017, **56**, 8739.
145. Z. Li, A. W. Peters, V. Bernales, M. A. Ortuño, N. M. Schweitzer, M. R. DeStefano, L. C. Gallington, A. E. Platero-Prats, K. W. Chapman, C. J. Cramer, L. Gagliardi, J. T. Hupp and O. K. Farha, *ACS Cent. Sci.*, 2017, **3**, 31.
146. M. C. Simons, M. A. Ortuño, V. Bernales, C. A. Gaggioli, C. J. Cramer, A. Bhan and L. Gagliardi, *ACS Catal.*, 2018, **8**, 2864.
147. M. Barona, S. Ahn, W. Morris, W. Hoover, J. M. Notestein, O. K. Farha and R. Q. Snurr, *ACS Catal.*, 2020, **10**, 1460.

148. T. Bogaerts, A. Van Yperen-De Deyne, Y.-Y. Liu, F. Lynen, V. Van Speybroeck and P. Van Der Voort, *Chem. Commun.*, 2013, **49**, 8021.
149. G. Tuci, G. Giambastiani, S. Kwon, P. C. Stair, R. Q. Snurr and A. Rossin, *ACS Catal.*, 2014, **4**, 1032.
150. H. Noh, Y. Cui, A. W. Peters, D. R. Pahls, M. A. Ortuño, N. A. Vermeulen, C. J. Cramer, L. Gagliardi, J. T. Hupp and O. K. Farha, *J. Am. Chem. Soc.*, 2016, **138**, 14720.
151. K. Leus, M. Vandichel, Y.-Y. Liu, I. Muylaert, J. Musschoot, S. Pyl, H. Vrielinck, F. Callens, G. B. Marin, C. Detavernier, P. V. Wiper, Y. Z. Khimyak, M. Waroquier, V. Van Speybroeck and P. Van Der Voort, *J. Catal.*, 2012, **285**, 196.
152. M. Vandichel, S. Biwas, K. Leus, J. Paier, J. Sauer, T. Verstraelen, P. Van Der Voort, M. Waroquier and V. Van Speybroeck, *ChemPlusChem*, 2014, **79**, 1183.
153. T. Maihom, S. Choomwattana, S. Wannakao, M. Probst and J. Limtrakul, *ChemPhysChem*, 2016, **17**, 3416.
154. T. Maihom, M. Sawangphruk, M. Probst and J. Limtrakul, *Phys. Chem. Chem. Phys.*, 2018, **20**, 6726.
155. I. S. Kim, Z. Li, J. Zheng, A. E. Platero-Prats, A. Mavrandonakis, S. Pellizzeri, M. Ferrandon, A. Vjunov, L. C. Gallington, T. E. Webber, N. A. Vermeulen, R. L. Penn, R. B. Getman, C. J. Cramer, K. W. Chapman, D. M. Camaioni, J. L. Fulton, J. A. Lercher, O. K. Farha, J. T. Hupp and A. B. F. Martinson, *Angew. Chem., Int. Ed.*, 2018, **57**, 909.
156. F. Vermoortele, M. Vandichel, B. Van de Voorde, R. Ameloot, M. Waroquier, V. Van Speybroeck and D. De Vos, *Angew. Chem., Int. Ed.*, 2012, **51**, 4887.
157. M. A. Ortuño, V. Bernales, L. Gagliardi and C. J. Cramer, *J. Phys. Chem. C*, 2016, **120**, 24697.
158. J. Hajek, B. Bueken, M. Waroquier, D. De Vos and V. Van Speybroeck, *ChemCatChem*, 2017, **9**, 2203.
159. C. K. Brozek, J. T. Miller, S. A. Stoian and M. Dincă, *J. Am. Chem. Soc.*, 2015, **137**, 7495.
160. J. Jover, C. K. Brozek, M. Dincă and N. López, *Chem. Mater.*, 2019, **31**, 8875.
161. N. Tahir, F. Muniz-Miranda, J. Everaert, P. Tack, T. Heugebaert, K. Leus, L. Vincze, C. V. Stevens, V. Van Speybroeck and P. Van Der Voort, *J. Catal.*, 2019, **371**, 135.
162. S. Abednatanzi, P. G. Derakhshandeh, P. Tack, F. Muniz-Miranda, Y.-Y. Liu, J. Everaert, M. Meledina, F. V. Bussche, L. Vincze, C. V. Stevens, V. Van Speybroeck, H. Vrielinck, F. Callens, K. Leus and P. Van Der Voort, *Appl. Catal., B*, 2020, **269**, 118769.
163. I. Romero-Muñiz, A. Mavrandonakis, P. Albacete, A. Vega, V. Briois, F. Zamora and A. E. Platero-Prats, *Angew. Chem., Int. Ed.*, 2020, **59**, 13013.

164. Z. W. Seh, J. Kibsgaard, C. F. Dickens, I. Chorkendorff, J. K. Nørskov and T. F. Jaramillo, *Science*, 2017, **355**, eaad4998.
165. C. A. Downes and S. C. Marinescu, *ChemSusChem*, 2017, **10**, 4374.
166. J. K. Nørskov, J. Rossmeisl, A. Logadottir, L. Lindqvist, J. R. Kitchin, T. Bligaard and H. Jónsson, *J. Phys. Chem. B*, 2004, **108**, 17886.
167. K. Chan and J. K. Nørskov, *J. Phys. Chem. Lett.*, 2015, **6**, 2663.
168. J. A. Gauthier, S. Ringer, C. F. Dickens, A. J. Garza, A. T. Bell, M. Head-Gordon, J. K. Nørskov and K. Chan, *ACS Catal.*, 2019, **9**, 920.
169. Y. Wang, X. Liu, J. Liu, M. Al-Mamum, A. W.-C. Liew, H. Yin, W. Wen, Y. L. Zhong, P. Liu and H. Zhao, *ACS Appl. Energy Mater.*, 2018, **1**, 1688.
170. J. Liu, H. Yin, P. Liu, S. Chen, S. Yin, W. Wang, H. Zhao and Y. Wang, *J. Phys. Chem. Lett.*, 2019, **10**, 6955.
171. J. Soriano-López, W. Schmitt and M. García-Melchor, *Curr. Opin. Electrochem.*, 2018, **7**, 22.
172. S. Zhao, Y. Wang, J. Dong, C.-T. He, H. Yin, P. An, K. Zhao, X. Zhang, C. Gao, L. Zhang, J. Lv, J. Wang, J. Zhang, A. M. Khattak, N. A. Khan, Z. Wei, J. Zhang, S. Liu, H. Zhao and Z. Tang, *Nat. Energy*, 2016, **1**, 16184.
173. X.-L. Wang, L.-Z. Dong, M. Qiao, Y.-J. Tang, J. Liu, Y. Li, S.-L. Li, J.-X. Su and Y.-Q. Lan, *Angew. Chem., Int. Ed.*, 2018, **57**, 9660.
174. Z. Xue, K. Liu, Q. Liu, Y. Li, M. Li, C.-Y. Su, N. Ogiwara, H. Kobayashi, H. Kitagawa, M. Liu and G. Li, *Nat. Commun.*, 2019, **10**, 5048.
175. E. M. Miner, S. Gul, N. D. Ricke, E. Pastor, J. Yano, V. K. Yachandra, T. Van Voorhis and M. Dincă, *ACS Catal.*, 2017, **7**, 7726.
176. Y. Lian, W. Yang, C. Zhang, H. Sun, Z. Deng, W. Xu, L. Song, Z. Ouyang, Z. Wang, J. Guo and Y. Peng, *Angew. Chem., Int. Ed.*, 2020, **59**, 286.
177. P. Zhang, X. Yang, W. Gao, X. Hou, J. Mi, L. Liu, J. Huang, M. Dong and C. Stampfl, *Catal. Sci. Technol.*, 2018, **8**, 3666.
178. C.-Y. Lin, L. Zhang, Z. Zhao and Z. Xia, *Adv. Mater.*, 2017, **29**, 1606635.
179. K. Iwase, S. Nakanishi, M. Miyayama and K. Kamiya, *ACS Appl. Energy Mater.*, 2020, **3**, 1644.
180. A. J. Martín, T. Shinagawa and J. Pérez-Ramírez, *Chem*, 2018, **5**, 263.
181. E. Skúlason, T. Bligaard, S. Gudmundsdóttir, F. Studt, J. Rossmeisl, F. Abid-Pedersen, T. Vegge, H. Jónsson and J. K. Nørskov, *Phys. Chem. Chem. Phys.*, 2012, **14**, 1235.
182. J. H. Montoya, C. Tsai, A. Vojvodic and J. K. Nørskov, *ChemSusChem*, 2015, **8**, 2180.
183. Q. Cui, G. Qin, W. Wang, K. R. Geethalakshmi, A. Du and Q. Sun, *J. Mater. Chem. A*, 2019, **7**, 14510.
184. H. Wang, H. Wang, Z. Wang, L. Tang, G. Zeng, P. Xu, M. Chen, T. Xiong, C. Zhou, X. Li, D. Huang, Y. Zhu, Z. Wanga and J. Tang, *Chem. Soc. Rev.*, 2020, **49**, 4135.
185. K. T. Butler, C. H. Hendon and A. Walsh, *J. Am. Chem. Soc.*, 2014, **136**, 2703.

186. X.-P. Wu, I. Choudhuri and D. G. Truhlar, *Energy Environ. Mater.*, 2019, **2**, 251.
187. G. Borboudakis, T. Stergiannakos, M. Frysali, E. Klontzas, I. Tsamardinos and G. E. Froudakis, *npj Comput. Mater.*, 2017, **3**, 40.
188. N. S. Bobbitt and R. Q. Snurr, *Mol. Simul.*, 2019, **45**, 1069.
189. Y. He, E. D. Cubuk, M. D. Allendorf and E. J. Reed, *J. Phys. Chem. Lett.*, 2018, **9**, 4562.
190. P. Z. Moghadam, S. M. J. Rogge, A. Li, C.-M. Chow, J. Wieme, N. Moharrami, M. Aragones-Anglada, G. Conduit, D. A. Gomez-Gualdron, V. Van Speybroeck and D. Fairen-Jimenez, *Matter*, 2019, **1**, 219.
191. J. D. Evans and F. X. Coudert, *Chem. Mater.*, 2017, **29**, 7833.
192. L. Turcani, R. L. Greenaway and K. E. Jelfs, *Chem. Mater.*, 2019, **31**, 714.
193. J. R. Kitchin, *Nat. Catal.*, 2018, **1**, 230.

CHAPTER 6

Machine Learning in Porous Materials

Z. JENSEN AND E. OLIVETTI*

Department of Materials Science & Engineering, Massachusetts Institute of Technology, 02139, Cambridge, MA, USA
*Email: elsao@mit.edu

6.1 Introduction

As described in previous chapters, porous materials cover a wide range of materials chemistry and structure and offer a landscape of opportunities in applications across a broad range of domains. The porous material space includes zeolites, metal–organic frameworks (MOFs), covalent organic frameworks (COFs), porous organic cages and porous carbon materials. Historically, porous, high surface area materials have been divided by inorganic (zeolites and ordered mesoporous oxides) and organic–inorganic materials (MOFs), purely organic materials, including COFs, conjugated microporous polymers (CMPs), and even aerogels or hydrogels. Porous materials from organic building blocks containing high surface areas have been assembled with hydrogen bonding (HOFs), supramolecular interactions (SOFs) or hypercoordinated main group elements (SiCOFs).[1] Porous materials can also be defined by the continuum of variation in their short-range crystallinity *versus* long-range disorder. For example, porous molecular crystals have weak intermolecular interactions and offer ease of synthesis and applicability where amorphous porous phases are desired.[2] These materials are unlike MOFs and COFs in that they do not have fixed framework-types and are not formed by strong coordination or covalent bonds. And there are even opportunities in the

Inorganic Materials Series No. 8
Computer Simulation of Porous Materials: Current Approaches and Future Opportunities
Edited by Kim Jelfs

Published by the Royal Society of Chemistry, www.rsc.org

structure of metal–organic nanocapsules that further self-assemble into hierarchical structures. As this list has just begun to describe, the design space for porous materials is unquestionably vast. The application space is also astounding, including roles in separation, adsorption, ion exchange, catalysis and spectroscopic properties. Understanding and prediction around many of these applications have been discussed in previous chapters. Here our point is to emphasise that the opportunities are seemingly endless, and the computational capabilities described throughout this book have pointed to simulation methods to inform design coupled to resulting properties and performance. This multitude of opportunities means that the space of materials discovery is essentially infinite, and robust methods to filter the space by interesting application or synthesisability are invaluable. For MOFs alone, upwards of 88 000 structures are present within the Cambridge Structural Database (CSD). Figure 6.1 shows the spread of the MOF space covered by known databases, just for MOFs. The time-intensity of experimental enquiry limits the pace of materials discovery across this combinatorial space and in some cases simulation-based approaches may prove too computationally intense to pursue in a brute-force manner. However, pursuing scientific and engineering enquiry around porous materials by linking experiment with simulation and data-driven methods holds significant promise to address a host of material needs.

This chapter will cover the opportunity in design and synthesis of porous materials made possible through data-driven machine learning (ML) methods.

Turning to the tools of ML in porous materials development has been of interest for many decades, beginning first with design of experiment efforts to focus enquiry on the most promising experiments.[8–10] There is tremendous promise in this approach to accelerate the discovery process. But why this moment and why this class of materials? Machine learning has gained traction across every dimension of science and materials development and synthesis is no exception. The availability of tools in machine learning has democratised the space, improved algorithms imply that a broader space of problems can be investigated, workflow management tools invite links to autonomous experimentation, and a broad range of data have become available in materials databases.[11]

The use of ML in porous materials development is motivated by the potential limit of current methods given that the infinite tunability of porous materials and a wide range of applications are accessible, but only a subset has been achieved. In zeolites this is referred to as the zeolite conundrum; in MOFs researchers have commented on the potential stemming from the combinatorial opportunity including the choice of metal cluster, the organic linker as well as specifying topology. Although less well explored through ML methods, COFs have well-defined pores, excellent stability and fine-tuned physicochemical properties that make them appealing candidates for data-driven methods.[12] The fact that the space of porous materials can be comprehensively and combinatorically described, and that there are patterns to leverage inherent to the molecules, frameworks and crystal packing, presents a

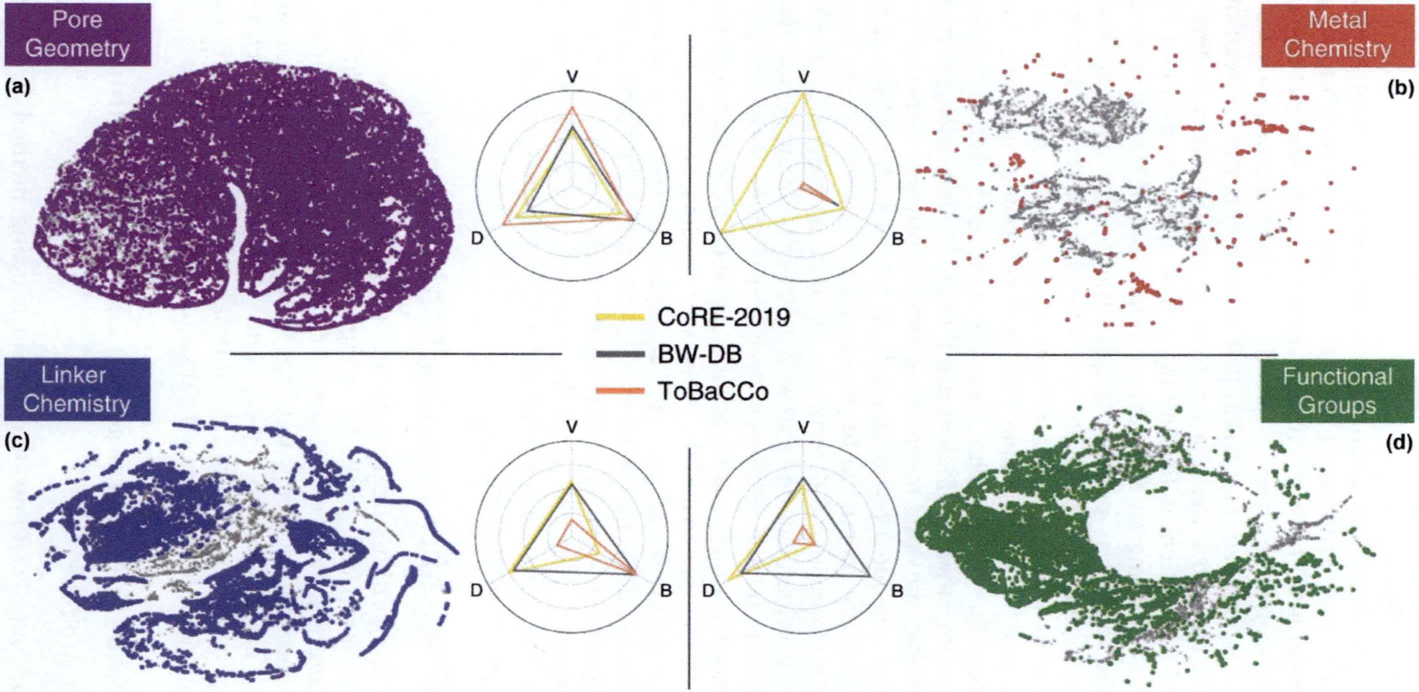

Figure 6.1 This figure demonstrates the breadth of MOF attributes from three MOF databases, CoRE-2019,[3,4] BW-DB (Boyd and Woo)[5] and ToBaCCo[6,7] plotted using a t-SNE, which is a dimensionality reduction visualisation approach. The (a) pore geometry, (b) metal chemistry, (c) linker chemistry, and (d) functional groups are mapped to 2D spaces. The radar charts show diversity metrics according to variety (V), which measures the number of bins that are sampled when MOF space is binned, balance (B), which reflects the evenness of the distribution among the bins and disparity is how spread out the bins are.

structure that is ideal for data-driven methods to increase the efficiency and effectiveness of search and discovery.

However, we also must remain vigilant to stay ahead of the hype curve and be sure that we are not overpromising what the domain has to offer using ML for porous materials. The very same reasons that ML has become so popular also underscores the need to consider carefully the methods that are used, validations pursued, and controlled experiments run. Instead of declaring that ML can unlock the treasure trove to porous materials development, we should invest effort in building robust datasets governed by FAIR principles and building an algorithmic foundation of how to achieve the most scientific insight as possible around the design, synthesis and application of porous materials.[13] FAIR-data principles emphasise development of metadata to describe datasets that are findable, meaning the metadata are indexed in a searchable resource; accessible, meaning the metadata are accessible even when data are not open; interoperable, meaning the metadata provide sufficient detail to enable referencing across datasets, and reusable; meaning metadata are sufficiently descriptive.

This chapter will describe a few basics behind ML as an introduction to the method, the goals for its pursuit in porous materials development, and then cover the development of descriptors for porous materials, briefly mention database development, and end with a forward-looking perspective on the opportunities to pursue in the next decade. The scope of this chapter will not pursue basic ML methods in detail or detailed descriptions of the examples of ML in porous materials. Interested readers are directed to general method descriptions and the many examples covered in reviews on the topic by Jablonka *et al.*[14] and Chong *et al.*[15] In terms of the scope of ML methods used in porous materials, we will briefly touch on the use of ML in structure and property prediction including sorption and catalysis, while offering a deeper look into the potential for ML to inform synthesis and crystallisation of porous materials.

6.2 General Machine Learning Methods

Supervised learning algorithms are the tools primarily used when applying ML in the porous material space.[14] In the following, we discuss several different types of these algorithms along with guidelines and practical considerations when using this method, including training, optimisation and evaluation. Interested readers are directed to several reviews on the use of machine learning and the development of materials informatics for further background on applications of relevant methods in the domain of materials science.[16,17]

6.2.1 Overview of Common Machine Learning Algorithms

6.2.1.1 Linear Models

The simplest and most well-known machine learning method is linear regression. The dependent variable (y) is modelled as a combination of linear coefficients (β) assigned to each of the independent variables (X) with error (ε).

$$y = X\beta + \varepsilon \tag{6.1}$$

Linear regression typically minimises the sum of the mean squared loss between the predicted and true value of y to determine the best coefficients.

While simple, linear regression can be a powerful tool for modelling a linearly dependent physical phenomenon. It is important to note that features can be engineered to become linearly related to the dependent variable, *e.g.* taking the logarithm of a feature. However, it is still a relatively simple method that struggles with modelling data that contain complex, nonlinear relationships between the variables.

Two additional issues with linear regression occur when the number of features is large or the data are sparse and when there is multicollinearity in the data. Lasso regression addresses the first issue by adding an L1 regularisation term to the least squares minimisation. This regularisation forces many of the coefficients to zero, effectively performing variable selection and decreasing the number of features considered by the model. Ridge regression addresses the second issue by adding an L2 regularisation term that adds bias into the estimates while decreasing the variance. This often results in more accurate predictions overall by decreasing the overfitting. These two regularisation techniques can be combined in a single model, referred to as elastic-net, which often works best when modelling data in practice.

While the above methods refer to regression problems, linear models can also be used in classification. The simplest linear classification model is logistic regression. This model can also be combined with L1 and L2 regularisation terms in a similar fashion to regression models. The cost function that it is optimised for each of the described linear models is shown in Table 6.1.

Instance-based models are another simple ML algorithm. These models take the weighted average of the nearest training data to make the next prediction. The most common of these is k-nearest neighbours, which predicts the dependent variable given the weighted average of the k-closest values within the feature space. These models do not have to learn parameters, only the value of k needs to be selected, making them quick to train and predict as well as simple to interpret.

6.2.1.2 Kernel Models

Kernel models map the data into a higher dimensional space where the relationships between the features and dependent variable becomes linear, allowing for easier modelling as described in the linear model section. This mapping is done through a kernel function, K, that transforms the data into a higher dimensional space. This prediction equation is shown in eqn (6.2). The most common kernel is the radial basis function shown in eqn (6.3).

Table 6.1 The cost function to be optimised by model form for linear models.

Model	Cost function
Linear regression	$\sum_{i=1}^{n}\left(y_i - \sum_{j=1}^{p} x_{ij}\beta_j\right)^2$
Ridge	$\sum_{i=1}^{n}\left(y_i - \sum_{j=1}^{p} x_{ij}\beta_j\right)^2 + \alpha\sum_{j=1}^{p}\beta_j^2$
Lasso	$\sum_{i=1}^{n}\left(y_i - \sum_{j=1}^{p} x_{ij}\beta_j\right)^2 + \alpha\sum_{j=1}^{p}\lvert\beta_j\rvert$
Elastic net	$\sum_{i=1}^{n}\left(y_i - \sum_{j=1}^{p} x_{ij}\beta_j\right)^2 + \alpha_1\sum_{j=1}^{p}\beta_j^2 + \alpha_2\sum_{j=1}^{p}\lvert\beta_j\rvert$
Logistic	$\alpha\sum_{i=1}^{n}\log(\exp(-y_i(X_i^T\beta + c)) + 1)$

These methods can be very powerful when dealing with nonlinear data or when the number of features is high relative to the number of data points.

$$y = \sum_{i}^{n} \alpha_i K(x_i', x) \tag{6.2}$$

$$K(x, x') = \exp\left(-\gamma\|x - x'\|^2\right) \tag{6.3}$$

The two most commonly used kernel methods are support vector machines and kernel ridge regression. These methods are very similar, both learning a linear function relating the features to the dependent variable in the kernel-transformed space. They differ only in their loss functions. Support vector machines typically use epsilon-insensitive loss while kernel ridge regression uses a ridge loss function. Epsilon-insensitive loss is an L1 loss function indicating that support vector machines may perform better when dealing with sparse data.

6.2.1.3 Bayesian Models

Bayesian models employ the Bayesian probability view to model probabilities building upon Bayes' rule:

$$P(\theta|D) = \frac{P(D|\theta)^* P(\theta)}{P(D)} \tag{6.4}$$

Eqn (6.4) describes the posterior distribution (probability of the model parameters given the data, $P(\theta|D)$) as a function of the likelihood (probability of the data given the model parameters, $P(D|\theta)$), the prior (probability of the model parameters $P(\theta)$), and the evidence (probability of the data, $P(D)$).

These models are often used in active learning (Bayesian optimisation) because of their inherent ability to describe the uncertainty associated with a prediction. Active learning is a special ML workflow where the model participates in an iterative search problem to find the optimal point in a feature space. The most commonly used Bayesian models are naïve Bayes and Gaussian processes.

6.2.1.4 *Tree Models and Ensemble Methods*

Tree models, namely decision trees, are another commonly used method especially for nonlinear data. These models create a flowchart-like, tree structure splitting each node on feature values that give the best prediction accuracy using metrics such as Gini impurity or information gain.

A clever way of increasing model performance is to use a collection of models, or ensemble, rather than a single model. In practice, ensemble models often outperform single estimator models and are some of the most commonly used algorithms in chemistry and materials science. One common ensemble creation technique is bagging (bootstrap aggregation), which generates a new, unique training set for each model in the ensemble by drawing from the original dataset with replacement until reaching the original number of data points. Each model is then trained on its unique training set and the average prediction (voting) becomes the overall ensemble model prediction. The random forest algorithm is a widely used example of bagging using an ensemble of decision tree models. In addition to bagging, each decision tree in a random forest is trained using a random subsection of the total number of features to increase the differences between trees.

Another ensemble technique is to iteratively train the models composing the ensemble, referred to as boosting. Two common boosting algorithms are AdaBoost and Gradient Boosting Decision Trees, both of which use decision trees as the models in the ensemble.

6.2.1.5 *Neural Networks*

Neural networks are some of the most popular ML algorithms. They act as function approximators that combine nonlinear transformations of the data by linking the input and output layers to a number of inner transformation nodes called hidden layers. Neural networks are trained iteratively through gradient descent where the error is calculated through the backpropagation algorithm. Neural networks can be many layers deep (deep learning), which typically requires large amounts of data in order to develop accurate models. However, they are useful for certain types of problems including modelling sequence, image and text data as well as generating new data.

Several deep learning neural network variants exist for dealing with specific types of data and problems. Deep learning models are very good at providing value to unstructured (non/less-featured) data because the first layers of the network essentially learn a feature representation. This makes them good choices for working with data such as images or text. For images, convolutional neural networks are typically used while for sequences and text-based data, recurrent neural networks such as gated recurrent units (GRU) and long-short-term-memory (LSTM) models are more appropriate. Another class of advanced neural network models are autoencoders where the data are compressed into a small dimensional space, called a latent space, and then recreated. By learning a distribution on top of this latent space, new data can be generated that are consistent with the underlying distribution of the data. These types of generative models are called variational autoencoders (VAE). Another class of generative models are adversarial networks (GAN). These models consist of two neural networks, one to generate new data from a latent space, and one to try and determine whether the generated data are real. The goal of the first model to learn to create data that can trick the second model into thinking they are true data. A final class of neural network discussed here are graph neural networks.[18] These networks apply neural network frameworks on graphs to solve problems such as labelling graph nodes or learning a feature representation over a graph. These models can be especially useful for dealing with physical concepts that map well to graphs such as crystal structures[19] and synthesis procedures.[20]

6.2.2 Practical Considerations

6.2.2.1 How to Choose an ML Algorithm

Choosing which algorithm to deploy can often seem arbitrary and challenging. It is usually best to start with a simple model and build up to more complicated solutions when the model is not performing adequately. However, there are some basic guidelines that can help in model selection based on the problem you need to solve and the structure of the data. The most important part of any ML workflow is to define the problem. Certain ML algorithms are much better suited to specific types of problems. One important consideration is the level of interpretability required from the model. Often in scientific applications of ML, the 'why' can be as important as the 'what' being predicted by the model. Linear, tree and instance-based models are very interpretable while neural networks and kernel models are considered to be 'black-box' models in which the 'why' for a prediction is generally hidden from the user. Often the user has to balance the model performance against the interpretability, which is one reason for the popularity of tree ensemble methods as they combine strong performance and some interpretability although not as much as a single decision tree. Another example is active learning. Active learning requires uncertainty predictions along with each model prediction making Bayesian models and ensembles appropriate. A final example is generative models where new data can be generated from a latent space that represents a compressed feature space. Typically, neural networks are used for these types of models.

6.2.2.2 Data Considerations

Many qualities of the data can help in model selection including the size, type and relationships contained within the data.[21] Model size is the easiest way to determine appropriate models. Neural networks typically require large amounts of data, typically greater than 10 000 data points, to provide accurate predictions, especially for deep neural networks, while other models typically provide more value on smaller datasets. The type of data is also very important. Image, sequence and text data are typically best modelled using complex recurrent and/or convolutional neural networks. Other types of numerical and categorical data are much more flexible in terms of which model is appropriate. A final consideration is the nature of the data. Often this point is not overly applicable if not much is known about the data. However, especially in science, there is often physical intuition behind the data. For example, if the data (or transformations of the data) are known to linearly correlate, linear models will often perform well. Another example is if instead of being linearly dependent, the data are more related through threshold values (often the case with synthesis data where excess reactant no longer has any effect), then decision tree or ensemble methods may give the best performance. Determining these types of relationships often comes from domain knowledge in the field, for example, for ideal gases, pressure is linearly related to the temperature of the gas.

6.2.2.3 Model Considerations

Splitting the data is an important aspect of model development. Data used for testing the model should never be used when training the model because it will usually overestimate the model's ability to generalise to new, unseen data. In practice, this means the data should typically be split into three sets: train, validation and test. The train set is used to train the model, the validation set is used to optimise the hyperparameters of the model, and the test set is used to evaluate the model's performance.

How to assign data to each set is also important and relies heavily on the task your model is designed to accomplish. The most common approach is to split the data randomly, resulting in all three sets having the same underlying distribution. While this split often typically results in the highest performance measurement, it often does not reflect a true test of the purpose of the model especially in science. For example, if we wanted to maximise the property of a material, a much more accurate evaluation of our model's performance would be to leave out the highest values of the property to see if the model can accurately predict them using only the data associated with lower values. This split would give a much better representation of our model's ability to extrapolate to materials with properties higher than seen in the data set. Another common option is setting an entire cluster as the test set ('Leave one cluster out').[22] This option tests if the model finds new regions of the feature space that have an interesting property rather than only interpolating to nearby data points. A final example is splitting the data by time. This split is often popular with generative models used to predict new compounds. Training a model on only older data and trying to predict newer points tests a model's ability to generate accurate predictions moving forward in time.[23] Performing these types of problem-informed and domain-informed splits helps develop an accurate representation of a model's usefulness and has led to significant changes in the materials field such as adapting active learning methodologies for property optimisation instead of single model approaches.[24]

When training a model, the main goal is to minimise the bias/variance trade-off.[25] Bias is an error resulting from incorrect assumptions by the learning algorithm while variance is an error resulting from noise or fluctuations in the data. Underfitting the model is when the model does not capture the patterns in the data leading to high bias and low variance. Overfitting occurs when the model does too well at learning the underlying data, capturing noise in the data along with the actual pattern, leading to low bias but high variance.

Underfitting occurs when the model is not able to capture the patterns in the data. The main ways to counter underfitting are to choose a more complicated model or engineer more descriptive features. Often, simple algorithms, such as linear regression, cannot model complex data accurately if there is significant nonlinearity in the data. Another option is to transform the data into more information-rich features. This gives the model more information to learn the underlying pattern.

Overfitting occurs when the model learns the data 'too well' (*i.e.* performs well on the training set but poorly on the testing set), which causes the model to generalise to new data poorly. Different techniques exist for different types of ML models to help minimise overfitting, referred to as regularisation. In linear models, Lasso and ridge terms are both forms of regularisation while decision trees use pruning instead of a regularisation term. Neural networks have several common techniques including batch normalisation and dropout layers. In addition, models that are trained iteratively like neural networks can exhibit overfitting if trained too long. Early stopping is a technique that attempts to stop training at the optimal point by examining the performance on a held-out validation set.

ML models typically have hyperparameters, values that need to be set by the user. These values can have a huge effect on the overall performance of a model. These parameters are almost always set empirically, through testing of different models trained with different hyperparameter values with the held-out validation set. Different search algorithms can be used depending on the number and ranges of the hyperparameters. For simpler models such as linear or decision tree based models, a brute-force grid search is often used, while for more complicated neural networks that require significant time to train, Bayesian optimisation is often used.

Optimisation algorithms often employ cross validation in which the training and validation data are split into *n*-folds. Then, for each chosen value of the hyperparameters, *n*-models are trained using each of the *n*-folds as the validation set for one of the models. The performance of these models is averaged and then compared against the other hyperparameter combinations. Typically, the best performing hyperparameters are selected for use on the model.

6.2.3 Model Evaluation

Many different metrics are available to evaluate model performance, several of which are described below. These metrics are split into classification and regression categories. While typically these values are reported on test sets, they are also used on validation sets when optimising models and can be used on train sets to determine under or overfitting.

6.2.3.1 Classification Model Evaluation

Accuracy is the simplest of the classification metrics. It is defined as the number of correct classifications over the number of classifications. While commonly used, this metric can give a false indication of model performance, especially if dealing with unbalanced data. For example, if a researcher tries to identify successful from failed synthesis routes, they might have 99 failed routes for every one successful route. If the model learns to only predict failed for every synthesis route that would lead to 99% accuracy even though the model will never find a successful synthesis route.

Instead of accuracy, a better family of metrics are precision, recall and F1. Precision is the predicted true positives over all predicted positives (*i.e.* how many predicted positives are true positives), recall is predicted true positives over all positives (*i.e.* how many of the true positives are predicted as positive), and F1 is the weighted average of the two. These metrics provide much more insight into model performance than accuracy and provide useful metrics even when the classes are unbalanced.

Receiver operating characteristic (ROC) curves are a common way to visualise classification performance (Figure 6.2). They are plotted as the false-positive rate *versus* the true-positive rate. A well-performing model has its curve in the upper left portion of this graph, a random guess for each classification is a straight diagonal across, and models worse than random are in the lower right. The area under this curve gives a good measure of model performance with higher area indicating a better model.

6.2.3.2 Regression Model Evaluation

The most commonly used evaluation for regression models is mean squared error (MSE). MSE is the normalised sum of the squares of the true minus the predicted value. Often the square root is taken giving root mean squared error (RMSE) to put the evaluation in the same units as the prediction. Mean absolute error (MAE) is the normalised sum of the absolute value of the true

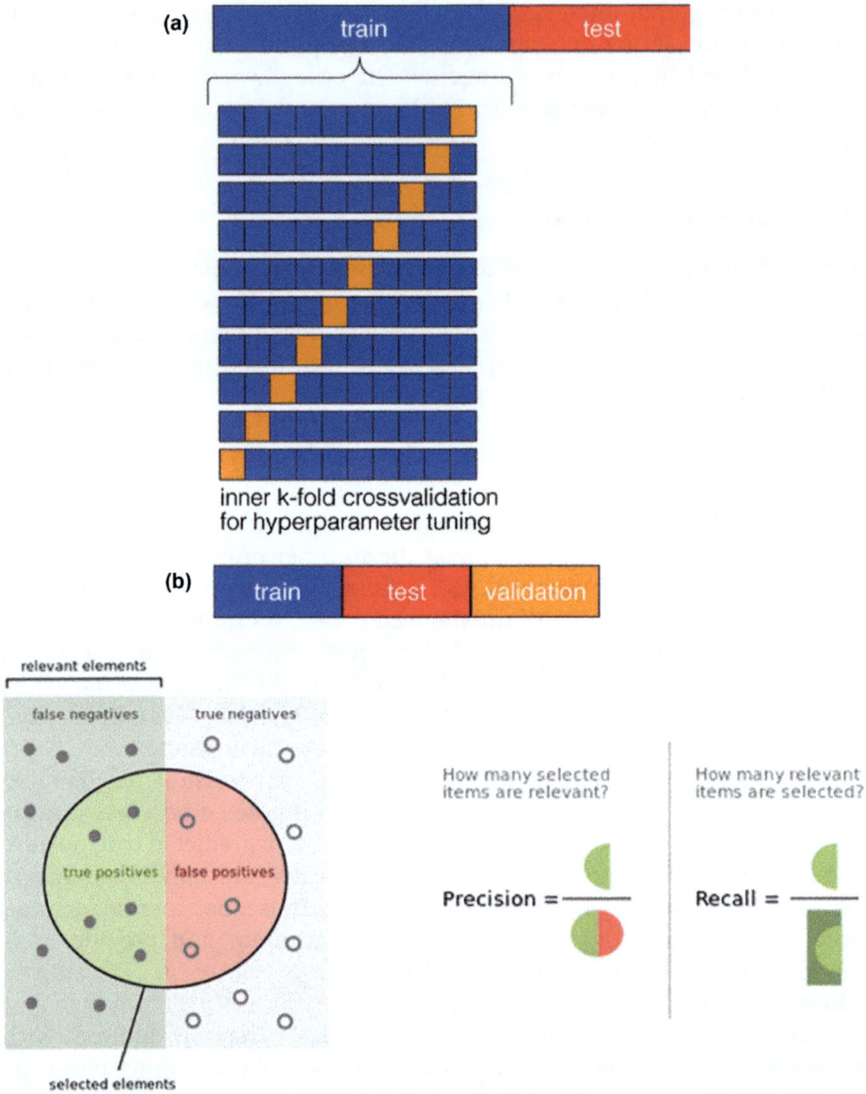

Figure 6.2 (a) Illustration of cross validation used to prevent under and overfitting of a model. (b) Model evaluation metrics including accuracy, precision, recall, F1 scores and receiver operating characteristic (ROC) curve.

minus predicted value. The main practical difference between MSE and MAE is the treatment of outliers. Due to the square term, MSE punishes outliers more than MAE. If error severity does not increase with distance from the true value, MAE can be a better metric to use than MSE.

$$\mathrm{MSE} = \frac{1}{n}\sum_{i=1}^{n}(y_{\mathrm{true}} - y_{\mathrm{predicted}})^2 \quad (6.5)$$

$$\mathrm{MAE} = \frac{1}{n}\sum_{i=1}^{n}\left|y_{\mathrm{true}} - y_{\mathrm{predicted}}\right| \quad (6.6)$$

There are ongoing needs for the use of machine learning in materials science around data sharing, algorithm development, and, perhaps of greatest interest, linking physics and physics-based simulations with machine learning.[26] The more we are able to integrate theory into the design of these models, we may be less constrained by the paucity of data as we may derive more value from the data we do have and are better able to link across physics-based models that are used at varying length scales.[27] Another large goal for the field is to shift from materials informatic-type approaches, which include multiple linear models that are trained on small datasets, to larger complex nonlinear models trained on heterogeneous, large datasets.

6.3 Rationale for Machine Learning in Porous Materials

6.3.1 Defining Characteristics of Porous Materials

Pore geometry and chemistry are both crucial in the application space of porous materials. Pore size correlates with surface area, which governs the surface accessible for an application. Pore volumes influence end functions such as size-based permeability and selectivity and the interconnectivity of the pores and degree of structural tortuosity. Geometric control for porous materials is obtainable by hard templating or tunable using soft templates; the size and order of mesopores at the molecular level is dictated by molecular self-assembly or directed assembly.[28,29]

From a chemical perspective, surface functional groups govern wettability to liquids, binding affinity to molecules, and selective interaction with certain gases. This chemical tunability can be achieved through the starting materials themselves or though post-processing where moieties are used to modify surface chemistries.

There are significant challenges for the investigation of machine learning approaches in porous materials relative to other fields, implying that off-the-shelf tools are not always immediately applicable, but where instead domain experts need to work closely with computer scientists to further advance algorithm development. These challenges include complex topologies relative to other crystalline materials, potential for intricate chemistries including both organic and inorganic species, constraints related to periodic boundary conditions, and non-unique representations of a unit cell leading to ambiguous assignment of lattice parameters.[30] Furthermore, the chemical environment within the pore is important, including the relationship between the porous molecule itself. Given the charged nature of some species, careful attention must be paid to counter ions. Brute-force screening is computationally expensive, so very often ML approaches use combinations of molecular simulations and machine learning.

6.3.2 Applications of ML in Porous Materials

In their most basic form, data-driven methods enable collection and analysis of enormous combinations of material metrics including compositions, morphologies, pore sizes/volumes and surface chemistries. Machine learning leveraging large datasets can provide a tool to investigate and predict phenomena that are difficult to explicitly model mathematically. If models are effectively designed they can then inform correlation and patterns found in new data; for example, adsorption capacity for a given gas at specific thermodynamic conditions for a new molecule. To first order, much like any other scientific enquiry, there are relationships between inputs (for example, properties, composition, synthesis and structure) and outputs (*i.e.* activity, selectivity and stability). The use of ML in this way provides the most benefit in the absence of theoretical models or in the case of particularly complex relationships. However, even when phenomenological models exist, ML might be used to provide more rapid results relative to simulation approaches. Generally, supervised learning can provide an inferred relationship given examples without explicitly knowing the underlying principles, but one can also pursue some hybrid analysis by further constraining the underlying relationship with domain knowledge. Models have been used to predict properties including gas absorption, storage and separation, catalysis, drug delivery, chemiresistive sensors and supercapacitors.[14,15] ML in porous systems has been used to inform synthesis of the materials and the 'holy grail', to design a material for a specific property.

6.4 Applications of ML to Porous Materials

6.4.1 Model Framing

The goals in developing a robust ML algorithm are to develop an appropriate dataset for model training and testing, to establish high-quality predictors and then to apply the proper ML method. Expressed in another way, the general approach to learning includes developing an engineered set of features or representation (choice of an ML model function that is tunable), and combining it with the correct evaluation approaches or choice of goodness of fit given a particular optimisation (choice of tuning algorithm). Section 6.2 discussed the general approaches to machine learning in materials science providing insight into performance evaluation measures such as F1, ROC and MAE, as well as data and model considerations. Within this section, we will focus briefly on data and then develop the feature engineering or descriptor development for porous materials, an area of significant specificity to this class of materials. One overall framing that provides insight into data, representation and model performance is whether we consider a bottom-up self-assembly based building block approach, a top-down topology-based construction or perhaps most interestingly an approach that provides insightful combinations of both.

6.4.1.1 Data for Porous Materials Development

Most broadly, data may include experimental data, computational information and data extracted from the published literature. Recent examples of the latter include work by Park *et al.*[31] In each case we should aim to confirm that data are collected using specified, basic protocols including consistent model choice or experimental methods corresponding to FAIR data principles. The community is still young in its ability to build up, and make use of, broad, diverse datasets. Ensuring that the data we are delivering to a model provides a comprehensive search space and that we can operate across disparate datasets remains a significant challenge. Few large open databases of porous materials properties exist (exceptions may include Materials Project,[32] and, more recent examples,[33] although the former is sparse for porous materials) and training data are expensive to generate; therefore quite often we have to confront bias in datasets and leverage approaches that help facilitate use of small datasets.[34] Recent work has explored the limitations in researchers' ability to capture diversity in the chemical design space.[35] There have also been pursuits in the porous materials space to augmenting datasets by creating new diffraction patterns, for example, by merging the main peaks of the predicted patterns with shuffled noise from patterns measured in the laboratory.[36]

For the most part in ML, models for porous materials are developed using training sets derived from computational or hypothetical databases of COFs, MOFs and zeolites. An example of a primary database includes hMOFs from Wilmer,[37] which is a database of over 130k structures that were created by combining together building blocks of MOFs. Another critical dataset is that by Chung *et al.*, also known as the computation-ready experimental MOF database (CoRE-MOF).[3] This set includes a wide range of topological features in pore shapes and sizes and consists of 5109 experimentally synthesised MOFs. Other MOF databases include the ToBaCCo generated database,[6] the MOFs extracted from the Cambridge Structure Database (CSD),[38] and more recently CORE-COF.[39] For zeolites, there are databases developed by Deem as well as the standardised sets available through the International Zeolite Association (IZA).[40]

However, because of the way these databases are constructed, they can only cover a limited part of the design space. There are other limits including defects present in 'real world' materials and kinetic phenomena that work against thermodynamics stabilities. There are specific challenges with leveraging computational datasets for applications to the prediction of synthesisable porous materials, including the removal of solvents, addition of hydrogen, resolution of partial occupancy and other structural distortions (particularly relevant in mechanical property prediction).

6.4.1.2 Descriptors

As stated above, one critical aspect for an ML model is to define the specifications of the inputs. This may be called a descriptor or representation. The most significant obstacle for the prediction of stable and chemically valid three-dimensional inorganic crystal structures is the choice of computational representation, as this governs the relevance of the model that is output. The requirements for a viable descriptor are as follows: their specification must be invariant with respect to transformations (in other words, it must conform to requirements around periodicity, translational, rotational and permutation symmetry). This means that the number of atoms should not influence the prediction. The descriptor must also be equivariant, conserved and non-degenerate, in other words, different structures should not share the same descriptor. On the more pragmatic side, a good descriptor allows for generalisation, is efficient to be computed and is continuous.[14] Our aim in this section is to summarise the categories of descriptors; we refer interested readers to relevant mathematical considerations of descriptors found in the work by Jablonka *et al.*[14] We consider geometric, chemical and energy-based descriptors along with considerations of local *versus* global descriptors.

Categories of descriptors in the porous materials space have been broken down in a few different ways. First, they have considered **geometric** effects including density, pore diameter, and gravimetric or volumetric surface area, as well as pore geometry (in MOFs for example).[41] Other geometric descriptors of interest have included void fraction, pore limiting diameter, largest cavity diameter and pore size distribution. For example, the structural diversity of zeolitic imidazolate frameworks induces a wide variety of their properties, including geometric characteristics (*e.g.* channel dimensionality, connectivity and pore space), which are directly relevant for adsorption-related applications, but also in physical properties such as mechanical behaviour. Beyond these geometric properties that may be directly related to physisorption, geometric considerations must also be informed by the interactive effects that can lead to multicollinearity. Neglecting these interactive effects can lead to problems with overfitting or defining tautological, degenerate approaches.

Beyond these geometric considerations, it is of great import to include information about **chemical** descriptors. The diversity in chemical environment within the pores of MOFs relative to zeolites, for example, motivate attention to chemical descriptors for that class of porous materials.

The simplest form for a chemical representation of a MOF is the choice of metal cluster, organic linker and topology in a vectorised format for computational consideration.[42] For each of the chemical species one might capture atom/building block count, which might also include atomic properties, such as atomic number, electronegativity, electron affinity, ionisation energy, *etc.* Chemical descriptors can also include atomic-property weighted radial distribution frequencies (RDF, which also includes an element of geometric information), degree of unsaturation, metal/halogen to carbon ratio, nitrogen to oxygen ratios and van der Waals volume. One problem found in fingerprinting approaches for MOFs is that they require the assignment of bonds and bond orders, which is not trivial for solid structures, especially for experimental structures that might contain disorder or incorrect protonation. An example for chemical descriptors may be one that includes information about the metal, to be able to describe metal-centred transitions, information about the ligands, and the functional groups, to be able to describe ligand-centred transitions.[43] A typical approach is to encode the nodes, the linkers, and the functional groups separately by using correlations on a structure graph scaled by chemically-aligned heuristics such as electronegativity. One might also consider augmenting this set of features with additional descriptors for the linkers, such as the number of aromatic rings or double bonds.

Another chemical representation involves graph-based approaches. Recent work used reticular framework representations, which are tuples that consist of edges, vertices and topology for a given structure to define an organic vertex.[15] In other graph-based approaches, molecular graphs enabled researchers to describe building blocks and then kernel-based techniques were used to investigate similarities.[44]

Considerations between geometric and chemistry-based considerations are critical as the behaviour of a catalyst will include a pore-shape effect as well as a functional group effect. Examples of combined descriptors related to zeolites have included space group, Si–O bond length and Si–O–Si angle distribution Other informative features indicate effective channel width including diameter of the smallest aperture, largest included sphere, largest free sphere, available or accessible pores and channel dimensionality. Then, when we substitute with other chemical species this can become generalisable. For MOFs, the representation must include information about crystalline materials containing complex organic molecules. Complementing this with chemical information has been used to inform the hypothesis that electrostatic polarisation from a local zeolite environment can be adjusted by the Si/Al ratio and thus enhance binding strength.[45]

Energy-based descriptors are also valuable. These directly consider the interaction between the chemical species used in synthesis, adsorption or catalysis, and the porous framework often translated to a potential energy surface. Interaction energy-based descriptors such as heat adsorption (for example, for an energy histogram the interaction energy between gas and framework is binned and used as input) help account for interactions that would take place between adsorbates and MOFs.[46] In these histogram-based approaches the probability of a single particle (probe atom) being adsorbed by the material is considered as a descriptor. The descriptor then depends on the interaction energy at a particular point that can be computed using the Boltzmann factor and then a total probability can be calculated by averaging the Boltzmann factor over a large number of different sampling points.[47] These works have noted that the community will need more advanced descriptors as the guest–host interactions become strong (CO_2, H_2S and H_2). In this case, the interaction depends less on chemical composition but on the underlying potential energy surface instead – there is therefore a key opportunity in reducing the cost of molecular simulations for materials screening. Fanourgakis *et al.* also used average Boltzmann factors as descriptors but expanded this work to include a range of polar and nonpolar probe atoms. This approach accounted for different ways charge may be distributed in an adsorbate including dispersion, position and electrostatics.[47] This approach showed improvement in predictive performance relative to previous approaches through better energy-based accounting of void fraction, density and surface area.

Beyond this consideration of geometry with and without considerations of chemistry or energy, descriptors can be thought of on a **local** level, which asserts that the total property of a compound can be broken down or decomposed into a sum of contributions of local environments. Local descriptors could capture the composition, for example, and be centred on an atom. This is challenging for porous materials because both pore-centred descriptors as well as atom-centred descriptors are important. The Smooth Overlap of Atomic Positions (SOAP) approach provides some quasi-local information as it is a projection of the atomic density onto a four-dimensional sphere showing a smooth similarity measure of local environments.

Juxtaposed with local descriptor development would be those that move beyond the local scale to more **global** considerations around pore size and shape, interpenetration and ligand structure,[48] pore limiting diameters, accessible surface areas or persistent homology fingerprints. As mentioned above, these more geometric properties have local consideration, but must also capture long-range effects. Here we begin to consider development in topological descriptors (more detail provided in the next paragraph), such as persistent homology (an algebraic tool which describes topological features with a set of persistent barcodes[49]), which is used to describe the pore environment of nanoporous materials including pore connectivity, morphology and cavity size thereby providing a quantitative description of the pore context. Persistent homology has been used to compute topological signatures of channels and voids constituting the porosity of a material where standard global metrics are overlaid on topological signatures to cluster crystal structures or topological signatures for supervised learning of properties.

In the case of nanoporous materials, a fundamental question in developing a descriptor is how to systematically characterise the similarity of pore structures. As mentioned above, topological data analysis (TDA) studies the shape of big and high-dimensional data to discover structure and to identify subgroups. Often the methods used in TDA develop coarse grid-computed interactions between the adsorbate and the resulting framework. These grid-computed interactions can be summarised in histograms, which are then used to generate models. However, work thus far has used only simple silica-based chemistries thereby exploring only variations in pore topology as opposed to chemical variation.[50] In this case, the approach was used to predict performance and then TDA was used to detect high-performing structures with topological similarity for a particular application and thereby estimate crude approximations of target properties. The authors assert that this may be extensible to other applications by adjusting the size of the target gas molecule and to other porous materials by taking into account information about energy or charge.

This large set of options shows that the different approaches and descriptors are encoding different kinds of structural similarities and can hence be seen as complementary, and suitable for different applications. For example, SOAP, while elegant to inform the potential energy surface of the molecules (done through molecular simulations), may be computationally intense when linked to other descriptor development. The persistent homology method is not sensitive enough to the subtle changes in atomic configurations to be able to map them to a potential energy surface. On the other hand, for the cases where long-range distances are involved – for example, where the aim is to classify pore shapes and molecular packing – then we need higher-level descriptors that are invariant to exact lattice arrangement. For such purposes, persistent homology may help encode geometric similarity. Other applications have begun to use graph neural networks to build stronger representations that accelerate property prediction.[51]

6.4.2 ML-based Prediction in Porous Materials

We briefly listed above some of the major applications of ML methods in porous materials, but here we develop a few more specific examples towards property prediction, the inverse design problem and synthesis.

6.4.2.1 *Examples of Property Prediction*

ML methods have been used to screen potential porous materials for their use in gas storage, most commonly for methane and hydrogen using adsorption isotherms. Mixture isotherms have also been used to obtain data for gas separations leveraging both thermodynamic properties and transport such as diffusion coefficients. For the most part, these studies have predicted one point on a gas adsorption isotherm, but recent work has also predicted full isotherms for zeolites. One of the most common pursuits in property prediction is around mechanical stability, most specifically predicting bulk and shear moduli, but chemical reactivity has also been pursued.[52,53] There have been preliminary efforts to predict broader property sets such as electrical conductivity for batteries, optical properties (*e.g.*, photocatalysis, sensing, luminescence, optoelectronics), or catalyst supports for extraordinary reactant diffusion and product selectivity.[28,54]

To more significantly leverage the capability of ML, one approach would be to develop models trained with consideration for their generality, for example, to make adsorption predictions of an arbitrary gas molecule for a variety of different conditions. However, there are challenges in filtering model results if the proper physical constraints are not considered, as modelling adsorption for an arbitrary molecule in an arbitrary structure may be impractical. Across these property prediction approaches, one area of interest is to learn whether a material discovered for one application might be applied to another. One potential limitation of taking a 'building block' approach to ML model development (*i.e.* considering structural building units in zeolites, for example) is that transferability becomes difficult if the building blocks of one material class are not relevant to the class of interest.

6.4.2.2 *Inverse Design*

Beyond predicting performance of porous materials in a specific application, another desire is the pursuit of inverse design, where the property would be specified, and the resulting combination of chemistry and structure would be proposed. This is of great interest towards finely tuned, user-desired final properties during materials generation and is often described as the 'holy grail' of ML in any scientific domain. There are just a few examples of this application of ML in porous materials, but the few examples deserve attention because of their potential value to the community.

Work outlining the promise for inverse design approaches to avoid conventional methods that require brute-force generation of porous materials was performed by Kim *et al.*[30] In this work, the authors focused on generating pure silica zeolite structures and used a generative adversarial neural network in which a neural network generates model results optimised for tricking a discriminator network. Their contribution leveraged material and energy representations to generate many possible candidates in both the energy and materials space, building on previous work by the same group that focused just on the energy dimension to capture the interaction energy between the adsorbate and zeolite in methane storage applications.[55] They then confirmed that the adsorption properties of the energy shape GAN-generated structure had a similar distribution to those of zeolites used in the dataset, but the generative nature of the model means that one obtains a more diverse dataset than the existing structures. However, the early method was not a generative approach, as they could not generate actual material structures, so the more recent work in ZeoGAN was a significant advance. In addition to considering inputs in both energy and materials dimensions, the materials grid was also further divided into silicon and oxygen grids, whose positions are represented by Gaussian functions, potentially pointing to approaches for more complex chemistries.

Work by Zhang *et al.* has structured an approach where user-desired properties for a given metal node and topology net can be used to design new MOFs. They were successful in increasing the performance of existing MOFs for methane storage and carbon capture applications.[56] This was done through a recurrent neural network that was used to predict the value of the child node derived from parent node's values. There is a valuable link across many of these contributions where simulations are used to calculate a baseline value to reward generative models, therefore the types of advances discussed in previous chapters around simulation form a strong foundation for the continued use of ML approaches in tandem with computational, first-principle approaches.

Another example is found in the development of an ML-based supramolecular variational autoencoder (SmVAE) by Yao *et al.*[57] RFcode (a representation for reticular framework, as described above) enabled MOF deconstruction and reconstruction (the characterising feature of an autoencoder) with individual treatment of edges, vertices and topologies. Validation of these methods (outside of the time-consuming step of synthesising them) can prove challenging and in this case, the authors looked at the latent-space neighbours of a known MOF to explore structural similarity and the phase transition from two additional known MOFs. The relevance of this method in the inverse design problem is to have the model trained to predict material performance as well as deconstruct and reconstruct materials. In order to enable this, a loss function for the prediction of target properties should also be defined. These results hold promise as some of the generated structures showed the highest known performance to date (Figure 6.3 – SmVAE).

6.4.2.3 Predicting the Synthesis of Porous Materials

There is tremendous opportunity to limit the combinatorial space of porous materials to direct development of materials with predictable, efficient and effective performance. Another framing of opportunity can be found in then understanding how to make the proposed materials. Broadly, materials synthesis focuses on determination of the relative thermodynamic stability (or minima) of a proposed compound through various steps of the processing and a fully predictive model will select precursor materials and solvents as well as predict the temperature and reaction time to generate the material of interest. If thermodynamics entirely determined the situation, simulation and *ab initio* methods would likely prove sufficient if we are able to define the right metrics for thermodynamic stability. However, synthesisability is a complex phenomenon involving kinetic considerations as well, and thermodynamics is just one contributor.[58] The consideration of whether the thermodynamically predicted product is kinetically relevant is true for most of materials synthesis,[59] but may pose particular challenges for porous materials because of the complexity offered by structure directing agents or modifiers.

There will also be myriad practical considerations including the availability of precursors and the nature of experimental conditions.[60] The complex nature of the synthesis space also invites the use of data-driven methods to focus

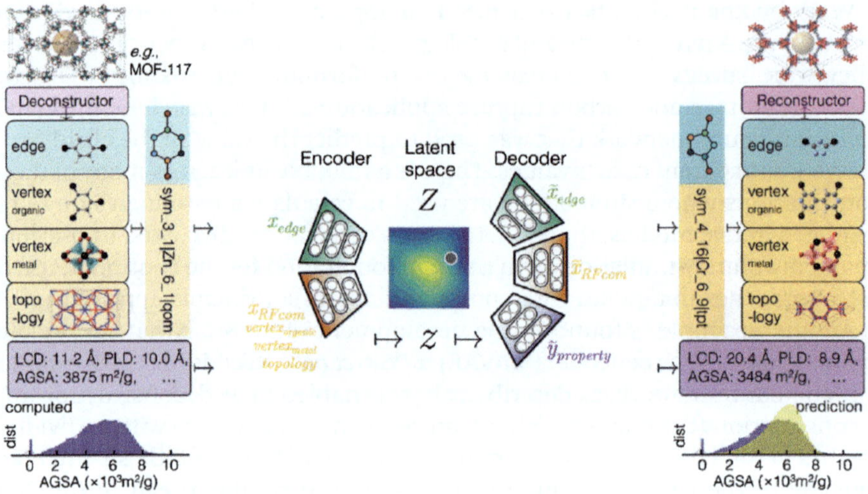

Figure 6.3 Schematic diagram of the automated reticular framework discovery platform empowered by the supramolecular variational autoencoder (SmVAE). Individual modules encode and decode each part of the RFcode. The model leverages an additional component that learns to predict properties from the latent space.
Reproduced from ref. 57 with permission from Springer Nature, Copyright 2021.

on the prediction of key synthesis variables in an efficient way. These approaches may lead to systematic approaches to capture chemical intuition. However, again one must take a measured approach to what one expects. Even known synthesis conditions for one porous material are not typically transferable to new materials, which makes it difficult to draw a general synthetic route for these materials. This may indicate limits for traditional ML approaches, as learning outside a representative dataset poses challenges. Furthermore, in experimental work individual experts obtain chemical intuition and knowledge from both successful and failed experiments.[60] Often these approaches require incorporation of unreported data and as numerous researchers have pointed out, gaining broad access to failed experiments or negative data would prove invaluable to this line of scientific enquiry. Quite pointedly, the current understanding of the self-assembly procedure, *i.e.* kinetics and energetics of framework bond formation, nucleation and crystal growth, has remained too limited to guide synthesis and broad access to interoperable data would be a tremendous advance for the community. There may be opportunities to link learning in the synthesis of new porous materials. For example, prediction of catalytic activity may inform binding energy for molecules, and tools developed to predict catalytic activity may also be interesting to inform the role of templating molecules in synthesis.

Significant advances in synthesis theory warrant discussion as they may lead to increased understanding of the kinetics that would be transferable to the porous materials domain. Work by Jang *et al.* used semi-supervised approaches, linked with first-principles modelling to learn structures with a higher probability of synthesisability through a crystal-likeness score.[58] Their model learns frequent crystal structural motifs that appear in already-synthesised compounds using graph-based structural encoding pointing to a structure-synthesisability relationship. For this work, ML has been used to model the probability of synthesis based on partial supervised learning of a material database. This contributes to a structural pattern of synthesisable materials that would be generalisable for all crystal types, not just one class of materials. Work by Aykol *et al.* developed metrics based on thermodynamic limits of energy to inform the likelihood of polymorph synthesis, whose baseline is set by the amorphous state of that chemistry.[61] Work by a subset of these authors looked at crystal structure prediction with structural information using a deep neural net.[62] Here a network analysis of stable crystals was developed from phase diagrams to predict the synthesisability of a not-yet-synthesised crystal. These methods may prove valuable within the porous materials context but have yet to be directly applied to this domain. There has been significant work in synthetic accessibility score development in organic molecules, which informs a statistical rule-based scoring method based on molecular structure. The specific metrics yield a lower score for frequently appearing fragments in already-synthesised drug datasets and a higher score for the presence of nonstandard structural features implying complexity in synthesis. There are still algorithm needs for these approaches in molecular design,[63] and this domain is still lacking for crystal structure synthesisability.

The challenges in applying these more general approaches to porous materials synthesis include our ability to decompose porous structure in terms of different functional groups or fragments as is done in molecules to structural motifs or local environment of atoms.

6.4.2.3.1 Porous Material Synthesis Examples. The requirements, in terms of synthetic knowledge, vary across porous materials, so we frame our examples individually across the porous materials classes with examples drawn from MOFs and zeolites where work has been focused.[64] Szczypinski *et al.* have delineated porous synthesis routes into the direct synthesis of a framework material or a solid-phase formulation of molecular materials, where for the latter the authors acknowledge that the properties of the resulting material depend on the properties of the discrete units as well as the collective.[60]

MOF synthesis involves self-assembly of the structural building blocks. The building blocks are typically commercially available, so the self-assembly process itself is where insight can be furthered. In other words, for MOFs the ligands and metal nodes are not the limiting factor for successful synthesis, but rather the optimal conditions for the ligands and metals to self-assemble into crystals.

ML has been used to inform synthesis conditions for MOFs towards optimised crystallinity and phase purity (BET surface area was also captured) of HKUST-1, focused on nine input variables outlining the synthesis conditions, including differences in solvent composition, temperature and synthesis methods (conventional heating, microwave, electrochemistry, mechanochemistry, ultrasonic approaches).[65] In this case, researchers used a robotic synthesiser to capture parameters for microwave heating through the use of a genetic algorithm. Solvent and water concentration were identified as important features through random forest, regression-based approaches and the model learnings were applied to systems with different metal ions, although these enquiries are still early. This work laid out a framework where genetic algorithm approaches were used to optimise the synthesis conditions while, in parallel, an ML algorithm learnt the relative importance of the experimental variables, which could rank the next generations of the genetic algorithm *in silico* and return to experiment when convergence is reached. Others have pursued solvothermal crystallisation of metal–organic nanocapsules *via* a similar approach.[66]

On the other hand, for zeolites there is interest both in the design of the templating organic structure directing agent (OSDA) molecules in addition to the conditions for crystallisation. Crystallisation is an interfacial phenomenon where the nucleation and crystallisation of solute molecules is mediated by interactions through organic and inorganic cations with structure directing agents (SDAs). These SDAs, which typically are amines and ammonium cations of sizes and shapes mapped to the geometry of porous channels/cages, will direct pore formation. Zeolite discovery remains based on expert intuition and on trial-and-error approaches with minimal control over the resulting structures; however, significant advances have been made through links with high-throughput experimental analysis requiring concomitant topological, textural and chemical analysis at commensurate time scales. Work has incorporated local interatomic distances that reflect geometries and local linkages but this is without consideration of host–guest effects.[67]

The main challenge in zeolite synthesis is the incomplete understanding of the molecular-level interactions and the kinetic and thermodynamic driving forces that govern the adsorption and binding specificity of OSDAs to precursors leading to specific nucleation/crystallisation events. A fundamental understanding of these processes and the ability to *a priori* control crystallisation requires synergistic research efforts to probe atomic to macroscopic length scales. The complex and multidimensional space of zeolite synthesis requires drawing inferences from incomplete and imperfect information, for which ML methods are very well-suited to replace the intuition-based, trial-and-error approaches traditionally used to guide experimentation.[68] For OSDA design, Deem and co-workers calculated the stabilisation energy of various OSDAs inside zeolite beta to facilitate the search for novel OSDAs.[69] Schwalbe-Koda has built a novel way to inform the interzeolite transition opportunities facilitating synthesis from one zeolite to another as a more facile mode of synthesis.[70]

As was mentioned above, one of the limitations of the use of ML to direct porous materials synthesis is the lack of data. Literature-based approaches have proven valuable, including work by the authors to inform the synthesis of germanium zeolites[71] and work by Muraok *et al.* to predict phase based on synthesis conditions; both contributions used decision tree framing (Figure 6.4).[72] In the latter contribution, the authors explored the chemical space of OSDA-free synthesis of aluminosilicate zeolites through a similarity network. The contribution was to extract synthesis descriptors from ML models to identify structure descriptors with appropriate importance to establish synthesis–structure relationships. In this way, the authors demonstrated that pattern recognition may provide the most effective approach for materials that are synthesised through kinetically controlled pathways (true for much of the porous materials space), because it is difficult to formulate first-principles embodiments for the complex energy landscape. Furthermore, the building unit concept as leveraged in MOFs is difficult in zeolites as these units may not be present in the intermediate mixtures.[73]

6.5 Outlook

The use of ML approaches in materials development comes with significant opportunities, but the community is still early on in this exploration and we must be careful to avoid certain pitfalls that have been described throughout this chapter. Here we describe several directions for advancement in the use of ML in porous materials, but also look more broadly at ML in materials development.

We must be able to take a critical outlook at results to understand if learning stems from sensible chemical and physical relationships or improvements in model architecture *versus* that based on improved hyperparameters' optimisation. Being comprehensive in reporting around model performance and setup will help build confidence in these approaches.

There can always be improved and continued development of descriptors that captures the complexity and ambiguity of porous materials and becomes specific to each material class. For example, modelling individual building blocks as superatoms, which can then be used in conjunction with currently employed atomic-level descriptors for a hierarchical, in-depth consideration of a given porous material.[15]

Continued efforts to link ML models with simulation approaches may serve to increase the efficiency and accuracy of those simulations; for example, the use of neural network potentials, which offer accuracy improvement over generalised force fields.[74] These have recently been applied in MOFs.[75,76]

Generalisation also plays a role in generative models where the purely atomic approach previously seen in zeolites would be inefficient for MOFs. Instead, a generative ML architecture specialised for MOFs would leverage their hierarchical structure. Structure prediction activities need to improve along dimensions that capture long-range effects.[77] As mentioned above, there are still improvement opportunities in the domain of synthesisability and chemical stability where, compared with chemical reactivity, the

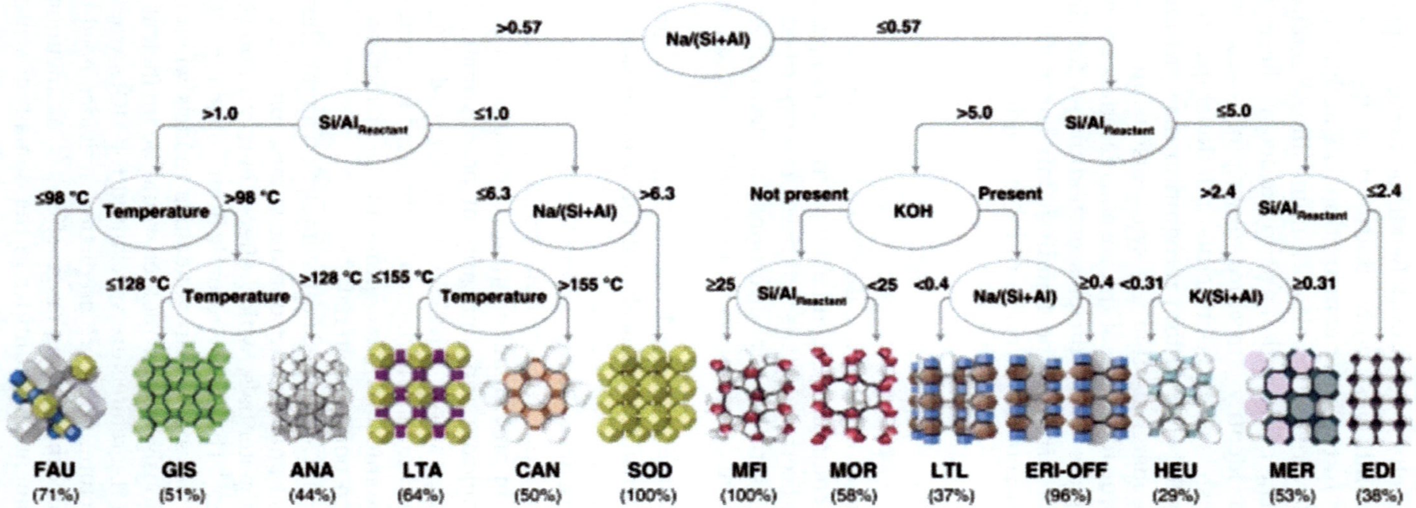

Figure 6.4 Decision trees used in informing synthesis for zeolites. The percentages represent the fraction that the dominant phases appear in the completed tree.
Reproduced from ref. 73, https://doi.org/10.1038/s41467-019-12394-0, under the terms of the CC BY 4.0 license, http://creativecommons.org/licenses/by/4.0/.

solid-state community lags, likely because of the lack of large open databases of properties and the expense associated with the development of training data. In particular, efforts to include external thermodynamic variables such as conditions of synthesis are of interest.

Within synthesisability, there will be continued advances in autonomous exploration that should be leveraged by the community.[78,79] When combined with robotic laboratory systems, ML systems can enable autonomous research. In this case, the desired properties of the materials system are inputs, and the ML-controlled setup can autonomously search processing and composition space in a quest to discover the next material with the desired properties. Use of ML in autonomous high-throughput analyses began with the use of an ANN to analyse important synthesis parameters based on experimental data by Corma *et al.*[8]

Chemometrics combined with automation may enable process efficiency and improved outcome quality,[80] enabling the chemist to focus on research. The needs in this domain are on the hardware side, but also on the information flow side, as robust workflows need to be established for these methods to prove valuable. This will continue to provide live feedback to the robotic system that might capture chemical intuition in the synthesis of metal–organic frameworks.

Finally, the overall needs for ML development for materials design more generally, but also applicable to porous materials, include informing uncertainties in prediction so they can be captured and appropriately propagated throughout the models. This will assist in not only improving the models themselves, but also increasing the communities' trust in the models. Given the expense of manually labelled data in the scientific domain due to the relative expertise needed, further development of unsupervised methods would prove quite valuable. Even shifting towards semi-supervised approaches wherein the ML model itself would predict and determine the labels for a large set of unlabelled data (by learning from a small subset of labelled data) could lead to more efficient understanding of the intrinsic properties of these materials.

Likely the most critical need is to move effectively as a modelling community from delivering just a final set of results to including the infrastructure needed to obtain and reproduce the results. This is a mode accepted by the computer science community that should pervade throughout the scientific community.

References

1. A. Thomas, *Nat. Commun.*, 2020, **11**, 11–13.
2. S. M. Moosavi, H. Xu, L. Chen, A. I. Cooper and B. Smit, *Chem. Sci.*, 2020, **11**, 5423–5433.
3. Y. G. Chung, J. Camp, M. Haranczyk, B. J. Sikora, W. Bury, V. Krungleviciute, T. Yildirim, O. K. Farha, D. S. Sholl and R. Q. Snurr, *Chem. Mater.*, 2014, **26**, 6185–6192.

4. Y. G. Chung, E. Haldoupis, B. J. Bucior, M. Haranczyk, S. Lee, H. Zhang, K. D. Vogiatzis, M. Milisavljevic, S. Ling and J. S. Camp, *J. Chem. Eng. Data*, 2019, **64**, 5985–5998.
5. P. G. Boyd, A. Chidambaram, E. García-Díez, C. P. Ireland, T. D. Daff, R. Bounds, A. Gładysiak, P. Schouwink, S. M. Moosavi and M. M. Maroto-Valer, *Nature*, 2019, **576**, 253–256.
6. Y. J. Colón, D. A. Gómez-Gualdrón and R. Q. Snurr, *Cryst. Growth Des.*, 2017, **17**, 5801–5810.
7. D. A. Gómez-Gualdrón, Y. J. Colón, X. Zhang, T. C. Wang, Y.-S. Chen, J. T. Hupp, T. Yildirim, O. K. Farha, J. Zhang and R. Q. Snurr, *Energy Environ. Sci.*, 2016, **9**, 3279–3289.
8. A. Corma, M. Moliner, J. M. Serra, P. Serna, M. J. Díaz-Cabañas and L. A. Baumes, *Chem. Mater.*, 2006, **18**, 3287–3296.
9. K. Choi, D. Gardner, N. Hilbrandt and T. Bein, *Angew. Chem., Int. Ed.*, 1999, **38**, 2891–2894.
10. D. E. Akporiaye, I. M. Dahl, A. Karlsson and R. Wendelbo, *Angew. Chem., Int. Ed.*, 1998, **37**, 609–611.
11. D. Ongari, L. Talirz and B. Smit, *ACS Cent. Sci.*, 2019, **6**, 1890–1900.
12. J. Warnan and E. Reisner, *Angew. Chem., Int. Ed.*, 2020, **59**, 17344–17354.
13. M. D. Wilkinson, M. Dumontier, I. J. Aalbersberg, G. Appleton, M. Axton, A. Baak, N. Blomberg, J.-W. Boiten, L. B. da Silva Santos and P. E. Bourne, *Sci. Data*, 2016, **3**, 1–9.
14. K. M. Jablonka, D. Ongari, S. M. Moosavi and B. Smit, *Chem. Rev.*, 2020, **120**, 8066–8129.
15. S. Chong, S. Lee, B. Kim and J. Kim, *Coord. Chem. Rev.*, 2020, **423**, 213487.
16. R. Ramprasad, R. Batra, G. Pilania, A. Mannodi-Kanakkithodi and C. Kim, *npj Comput. Mater.*, 2017, **3**, 54.
17. D. Morgan and R. Jacobs, *Annu. Rev. Mater. Res.*, 2020, **50**, 71–103.
18. F. Scarselli, M. Gori, A. C. Tsoi, M. Hagenbuchner and G. Monfardini, *IEEE Trans. Neural Networks*, 2008, **20**, 61–80.
19. T. Xie and J. C. Grossman, *Phys. Rev. Lett.*, 2018, **120**, 145301.
20. S. A. Malik, R. E. A. Goodall and A. A. Lee, *Chem. Mater.*, 2020, **33**, 616–624.
21. A. Y.-T. Wang, R. J. Murdock, S. K. Kauwe, A. O. Oliynyk, A. Gurlo, J. Brgoch, K. A. Persson and T. D. Sparks, *Chem. Mater.*, 2020, **32**, 4954–4965.
22. B. Meredig, E. Antono, C. Church, M. Hutchinson, J. Ling, S. Paradiso, B. Blaiszik, I. Foster, B. Gibbons and J. Hattrick-Simpers, *Mol. Syst. Des. Eng.*, 2018, **3**, 819–825.
23. E. Kim, Z. Jensen, A. van Grootel, K. Huang, M. Staib, S. Mysore, H.-S. Chang, E. Strubell, A. McCallum, S. Jegelka and E. Olivetti, *J. Chem. Inf. Model.*, 2020, **60**, 1194–1201.
24. L. Ward, M. Aykol, B. Blaiszik, I. Foster, B. Meredig, J. Saal and S. Suram, *MRS Bull.*, 2018, **43**, 683–689.
25. P. Mehta, M. Bukov, C.-H. Wang, A. G. R. Day, C. Richardson, C. K. Fisher and D. J. Schwab, *Phys. Rep.*, 2019, **810**, 1–124.

26. N. Wagner and J. M. Rondinelli, *Front. Mater.*, 2016, **3**, 28.
27. B. Meredig, *Chem. Mater.*, 2019, **31**, 9579–9581.
28. T. Liu and G. Liu, *Nat. Commun.*, 2020, **11**, 10–12.
29. D. Ongari, P. G. Boyd, S. Barthel, M. Witman, M. Haranczyk and B. Smit, *Langmuir*, 2017, **33**, 14529–14538.
30. B. Kim, S. Lee and J. Kim, *Sci. Adv.*, 2020, **6**, 1–8.
31. S. Park, B. Kim, S. Choi, P. G. Boyd, B. Smit and J. Kim, *J. Chem. Inf. Model.*, 2018, **58**, 244–251.
32. A. Jain, S. P. Ong, G. Hautier, W. Chen, W. D. Richards, S. Dacek, S. Cholia, D. Gunter, D. Skinner and G. Ceder, *APL Mater.*, 2013, **1**, 11002.
33. A. S. Rosen, S. M. Iyer, D. Ray, Z. Yao, A. Aspuru-Guzik, L. Gagliardi, J. Notestein and R. Q. Snurr, *Matter*, 2021, **4**, 1578–1597.
34. Y. Zhang and C. Ling, *npj Comput. Mater.*, 2018, **4**, 25.
35. S. M. Moosavi, A. Nandy, K. M. Jablonka, D. Ongari, J. P. Janet, P. G. Boyd, Y. Lee, B. Smit and H. J. Kulik, *Nat. Commun.*, 2020, **11**, 1–10.
36. H. Wang, Y. Xie, D. Li, H. Deng, Y. Zhao, M. Xin and J. Lin, *J. Chem. Inf. Model.*, 2020, **60**, 2004–2011.
37. C. E. Wilmer, M. Leaf, C. Y. Lee, O. K. Farha, B. G. Hauser, J. T. Hupp and R. Q. Snurr, *Nat. Chem.*, 2012, **4**, 83–89.
38. C. R. Groom, I. J. Bruno, M. P. Lightfoot and S. C. Ward, *Acta Crystallogr., Sect. B: Struct. Sci., Cryst. Eng. Mater.*, 2016, **72**, 171–179.
39. D. Ongari, A. V. Yakutovich, L. Talirz and B. Smit, *ACS Cent. Sci.*, 2019, **5**, 1663–1675.
40. C. Baerlocher, D. Olson and W. Meier, 2001.
41. R. L. Martin, B. Smit and M. Haranczyk, *J. Chem. Inf. Model.*, 2012, **52**, 308–318.
42. G. Borboudakis, T. Stergiannakos, M. Frysali, E. Klontzas, I. Tsamardinos and G. E. Froudakis, *npj Comput. Mater.*, 2017, 1–6.
43. K. M. Jablonka, S. M. Moosavi, M. Asgari, C. Ireland, L. Patiny and B. Smit, *ChemRxiv*, 2020, 1–25.
44. H. Ohno and Y. Mukae, *J. Phys. Chem. C*, 2016, **120**, 23963–23968.
45. Y. Gu, Z. Liu, C. Yu, X. Gu, L. Xu, Y. Gao and J. Ma, *J. Phys. Chem. C*, 2020, **124**, 9314–9328.
46. B. J. Bucior, N. S. Bobbitt, T. Islamoglu, S. Goswami, A. Gopalan, T. Yildirim, O. K. Farha, N. Bagheri and R. Q. Snurr, *Mol. Syst. Des. Eng.*, 2019, **4**, 162–174.
47. G. S. Fanourgakis, K. Gkagkas, E. Tylianakis and G. Froudakis, *J. Phys. Chem. C*, 2020, **124**, 7117–7126.
48. Y. G. Chung, D. A. Gómez-Gualdrón, P. Li, K. T. Leperi, P. Deria, H. Zhang, N. A. Vermeulen, J. F. Stoddart, F. You and J. T. Hupp, *Sci. Adv.*, 2016, **2**, e1600909.
49. Y. Lee, S. D. Barthel, P. Dłotko, S. M. Moosavi, K. Hess and B. Smit, *Nat. Commun.*, 2017, **8**, 15396.
50. Y. Lee, S. D. Barthel, S. M. Moosavi and K. Hess, *J. Chem. Theory Comput.*, 2018, **14**, 4427–4437.

51. A. Raza, A. Sturluson, C. M. Simon and X. Fern, *J. Phys. Chem. C*, 2020, **124**, 19070–19082.
52. J. D. Evans and F. X. Coudert, *Chem. Mater.*, 2017, **29**, 7833–7839.
53. P. Z. Moghadam, S. M. J. Rogge, A. Li, C. M. Chow, J. Wieme, N. Moharrami, M. Aragones-Anglada, G. Conduit, D. A. Gomez-Gualdron, V. Van Speybroeck and D. Fairen-Jimenez, *Matter*, 2019, **1**, 219–234.
54. Y. He, E. D. Cubuk, M. D. Allendorf and E. J. Reed, *J. Phys. Chem. Lett.*, 2018, **9**, 4562–4569.
55. S. Lee, B. Kim and J. Kim, *J. Mater. Chem. A*, 2019, **7**, 2709–2716.
56. X. Zhang, K. Zhang and Y. Lee, *ACS Appl. Mater. Interfaces*, 2019, **12**, 734–743.
57. Z. Yao, N. S. Bobbitt, B. J. Bucior, G. H. Kumar, S. P. Collins, T. Burns, T. K. Woo, O. K. Farha and Q. Snurr, *Nat. Mach. Intell.*, 2021, **3**, 76–86.
58. J. Jang, G. H. Gu, J. Noh, J. Kim and Y. Jung, *J. Am. Chem. Soc.*, 2020, **142**, 18836–18843.
59. S. Lee, A. Yang, T. P. Moneypenny and J. S. Moore, *J. Am. Chem. Soc.*, 2016, **138**, 2182–2185.
60. F. T. Szczypiński, S. Bennett and K. E. Jelfs, *Chem. Sci.*, 2021, **12**, 830–840.
61. M. Aykol, S. S. Dwaraknath, W. Sun and K. A. Persson, *Sci. Adv.*, 2018, **4**, 1–8.
62. M. Aykol, V. I. Hegde, L. Hung, S. Suram, P. Herring, C. Wolverton and J. S. Hummelshøj, *Nat. Commun.*, 2019, **10**, 1–7.
63. W. Gao and C. W. Coley, *J. Chem. Inf. Model.*, 2020, **60**, 5714–5723.
64. I. G. Clayson, D. Hewitt, M. Hutereau, T. Pope and B. Slater, *Adv. Mater.*, 2020, **2002780**, 1–47.
65. S. M. Moosavi, A. Chidambaram, L. Talirz, M. Haranczyk, K. C. Stylianou and B. Smit, *Nat. Commun.*, 2019, **10**, 539.
66. Y. Xie, C. Zhang, X. Hu, C. Zhang, S. P. Kelley, J. L. Atwood and J. Lin, *J. Am. Chem. Soc.*, 2020, **142**, 1475–1481.
67. Y. Li, J. Yu and R. Xu, *Angew. Chem., Int. Ed.*, 2013, **52**, 1673–1677.
68. M. Moliner, Y. Román-Leshkov and A. Corma, *Acc. Chem. Res.*, 2019, **52**, 2971–2980.
69. F. Daeyaert, F. Ye and M. W. Deem, *Proc. Natl. Acad. Sci. U. S. A.*, 2019, **116**, 3413–3418.
70. D. Schwalbe-Koda, Z. Jensen, E. Olivetti and R. Gómez-Bombarelli, *Nat. Mater.*, 2019, **18**, 1177–1181.
71. Z. Jensen, E. Kim, S. Kwon, T. Z. H. Gani, Y. Román-Leshkov, M. Moliner, A. Corma and E. Olivetti, *ACS Cent. Sci.*, 2019, **5**, 892–899.
72. K. Muraoka, Y. Sada, D. Miyazaki, W. Chaikittisilp and T. Okubo, *Nat. Commun.*, 2019, **10**, 1–11.
73. B. A. Helfrecht, R. Semino, G. Pireddu, S. M. Auerbach and M. Ceriotti, *J. Chem. Phys.*, 2019, **151**, 154112.
74. O. T. Unke and M. Meuwly, *J. Chem. Theory Comput.*, 2019, **15**, 3678–3693.

75. M. Eckhoff and J. Behler, *J. Chem. Theory Comput.*, 2019, **15**, 3793–3809.
76. A. Raza, F. Waqar, A. Sturluson, C. Simon and X. Fern, 2020, arXiv:2012.03723.
77. A. Grisafi and M. Ceriotti, *J. Chem. Phys.* 2019, **151**, 204105.
78. B. Burger, P. M. Maffettone, V. V Gusev, C. M. Aitchison, Y. Bai, X. Wang, X. Li, B. M. Alston, B. Li, R. Clowes, N. Rankin, B. Harris, R. S. Sprick and A. I. Cooper, *Nature*, 2020, **583**, 237–241.
79. P. S. Gromski, A. B. Henson, J. M. Granda and L. Cronin, *Nat. Rev. Chem.*, 2019, **3**, 119–128.
80. P. S. Gromski, J. M. Granda and L. Cronin, *Trends Chem.*, 2020, **2**, 4–12.

Subject Index